Understanding Information and Its Role as a Tool

In Memory of Mark Burgin

World Scientific Series in Information Studies
(ISSN: 1793-7876)

Founding Series Editor: Mark Burgin

Series Editors:
Gordana Dodig-Crnkovic *(Mälardalen University, Sweden and Chalmers University of Technology, Sweden)*
Marcin J Schroeder *(Akita International University, Japan)*

International Advisory Board:
Søren Brier *(Copenhagen Business School, Copenhagen, Denmark)*
Tony Bryant *(Leeds Metropolitan University, Leeds, United Kingdom)*
Wolfgang Hofkirchner *(Vienna University of Technology, Austria)*
William R King *(University of Pittsburgh, Pittsburgh, USA)*
Teresa Guarda *(Universidad Estatal de la Península de Santa Elena, Ecuador)*
José María Díaz Nafría *(Universidad a Distancia de Madrid, Spain)*

Published:

Vol. 17 *Understanding Information and Its Role as a Tool:*
In Memory of Mark Burgin
edited by Marcin J Schroeder & Wolfgang Hofkirchner

Vol. 16 *Probability, Information, and Physics: Problems with Quantum Mechanics in the Context of a Novel Probability Theory*
by Paolo Rocchi

Vol. 15 *Chaos, Information, and the Future of Physics: The Seaman-Rössler Dialogue with Information Perspectives by Burgin and Seaman*
by William Seaman, Otto E Rössler & Mark Burgin

Vol. 14 *The Logic of the Third: A Paradigm Shift to a Shared Future for Humanity*
by Wolfgang Hofkirchner

Vol. 13 *Ontological Information: Information in the Physical World*
by Roman Krzanowski

Vol. 12 *Trilogy of Numbers and Arithmetic*
Book 1: History of Numbers and Arithmetic: An Information Perspective
by Mark Burgin

More information on this series can also be found at https://www.worldscientific.com/series/wssis

World Scientific Series in Information Studies — **Vol. 17**

UNDERSTANDING INFORMATION AND ITS ROLE AS A TOOL

In Memory of Mark Burgin

Editors

Marcin J Schroeder
Akita International University, Japan

Wolfgang Hofkirchner
TU Wien, Austria
The Institute for a Global Sustainable Information Society (GSIS), Austria

NEW JERSEY • LONDON • SINGAPORE • BEIJING • SHANGHAI • TAIPEI • CHENNAI

Published by

World Scientific Publishing Co. Pte. Ltd.
5 Toh Tuck Link, Singapore 596224
USA office: 27 Warren Street, Suite 401-402, Hackensack, NJ 07601
UK office: 57 Shelton Street, Covent Garden, London WC2H 9HE

Library of Congress Control Number: 2025001910

British Library Cataloguing-in-Publication Data
A catalogue record for this book is available from the British Library.

World Scientific Series in Information Studies — Vol. 17
Understanding Information and Its Role as a Tool
In Memory of Mark Burgin

Copyright © 2025 by World Scientific Publishing Co. Pte. Ltd.

All rights reserved. This book, or parts thereof, may not be reproduced in any form or by any means, electronic or mechanical, including photocopying, recording or any information storage and retrieval system now known or to be invented, without written permission from the publisher.

For photocopying of material in this volume, please pay a copying fee through the Copyright Clearance Center, Inc., 222 Rosewood Drive, Danvers, MA 01923, USA. In this case permission to photocopy is not required from the publisher.

ISBN 978-981-12-9491-4 (hardcover)
ISBN 978-981-12-9492-1 (ebook for institutions)
ISBN 978-981-12-9493-8 (ebook for individuals)

For any available supplementary material, please visit
https://www.worldscientific.com/worldscibooks/10.1142/13892#t=suppl

Desk Editors: Soundararajan Raghuraman/Tan Rok Ting

Typeset by Stallion Press
Email: enquiries@stallionpress.com

Preface

Understanding Information and Its Role as a Tool: In Memory of Mark Burgin (in two parts) is a collection of works related to information contributed by multiple authors as a tribute to the memory of Mark Burgin who passed away on February 18, 2023. Professor Burgin had a very wide range of research and philosophical interests which has a reflection in the very broad subject range of the contributions to this book from the authors influenced by his work honoring his memory as an outstanding member of the academic community, renowned mathematician, computer scientist, information theoretician, and philosopher.

He earned his MA and PhD in mathematics from Moscow State University and later received a Doctor of Science (DSc) degree in logic and philosophy from the National Academy of Sciences of Ukraine. Throughout his long academic career, Mark Burgin held various faculty and administrative positions, such as Visiting Professor at UCLA, Professor at the Institute of Education, Kyiv, Ukraine, at International Solomon University, Kyiv, Ukraine, at Kyiv State University, Ukraine, the Head of the Assessment Laboratory of the Research Center of Science at the National Academy of Sciences of Ukraine, and the Chief Scientist at Institute of Psychology, Kyiv. At the time of his passing, he was affiliated with UCLA, Los Angeles, California, USA.

Mark Burgin was widely recognized for his multiple fundamental contributions to a wide range of fields, from mathematics, information science, and computer science, especially in the theory of algorithms, to more general subjects such as philosophy and methodology of science, epistemology, general system theory, logic, psychology, cognitive sciences, social sciences, and pedagogical sciences. His works initiated research in several novel directions of study.

Dr. Burgin authored and co-authored over 500 papers and 19 books, including "Theory of Knowledge" (2016), "Structural Reality" (2012), "Hypernumbers and Extrafunctions" (2012), "Theory of Named Sets" (2011), "Theory of Information" (2010), "Measuring Power of Algorithms, Computer Programs, and Information Automata" (2010), "Neoclassical Analysis: Calculus Closer to the Real World" (2008), "Super-recursive Algorithms" (2005), "On the Nature and Essence of Mathematics" (1998), "Intellectual Components of Creativity" (1998), "Fundamental Structures of Knowledge and Information" (1997), "Introduction to the Modern Exact Methodology of Science" (1994), "The Structure-Nominative Analysis of Theoretical Knowledge (1992), and "The World of Theories and Power of Mind" (1992).

Mark Burgin was a member of the New York Academy of Sciences, a Senior Member of IEEE, the Society for Computer Modeling and Simulation International, and the International Society for Computers and their Applications. He was also an Honorary Professor of the Aerospace Academy of Ukraine, the Chair of the IEEE San Fernando Valley Computer and Communication Chapter, the Secretary of the International Society for Computers and their Applications, and President-Elect of the International Society for the Study of Information (IS4SI).

This book is an expression of the high recognition of the contributions made by Mark in the subject of information study not only through his publications and books but also by his engagement in organizing conferences and editing works of the members of the information community. This very book which now we dedicate to his memory was initially planned and designed by him in cooperation with one of us.

The original plan for the collection of works inspired by the 2021 IS4SI Summit was developed by Mark Burgin together with one of the present editors. The plan was not for the proceedings from the

Summit which were already published in the form of a book collection of 157 papers reporting the presentations at the Summit co-edited by Mark (Proceedings |IS4SI 2021 - Browse Articles (mdpi.com)) but for a collection of more elaborate works on the subject of information inspired by the Summit and reporting the continuation of already presented research going way beyond the already published papers of the Proceedings. The key point of the plan was to stimulate and promote continued research developments in the subject. The participation in, or presentation at the Summit was not a condition for submissions but rather the consistency with its scope.

Mark Burgin passed away on February 18, 2023, after the original announcement of the call for papers, but before the process of editing the book started. These were the tragic circumstances in which the idea of continuation was conceived. The plan for the collection did not change much except that the book in addition to being an implementation of Mark's final contribution became an opportunity to celebrate his immense influence on the entire domain of the information study. This is why the project has been re-announced as dedicated to Mark's memory. Marcin J. Schroeder was joined by Wolfgang Hofkirchner in the continuation of the work on editing the book and the title "Understanding Information and Its Role as a Tool: In Memory of Mark Burgin (in two parts)" included the dedication. As before, the two parts retained their subtitles. The dedication of the book without a doubt increased interest in submitting contributions to the book.

The two parts of the book have self-explanatory titles classifying submissions into two categories. One of them is "Understanding Information: Theory and Foundations" and the other is "Information Realm: Information as a Tool for Exploring Reality". The former is intended as an inquiry of the concept of information itself, the latter as an inquiry of the use of the concept of information as a methodological tool for diverse forms of inquiries. The division is rather conventional, as many chapters in both parts serve both objectives.

The range of topics in contributed chapters is very wide, but some of them repeat in various contexts. For instance, everyone engaged in the study of information is aware of the everlasting discussions about the meaning of information. The questions about the meaning of the concept of information and, in parallel to it, the question about the semantic constituent of information theory remain open. Shannon himself questioned the idea of a unique answer to the former question and declared his disinterest in the answer to the latter. However, information theory has made essential progress since that time, and many researchers suggested their answers to these questions.

There are many reasons for the continuing or even increasing interest in answering both questions. One of them is the connections between information and computation (information dynamics), which form the subject of a similar conceptual discourse. Unfortunately, very often the discussions are not very productive, especially when they are limited to voicing individual preferences and when their participants do not clarify their philosophical and theoretical positions. The only clear conclusion from the seventy-five-year discussions on information is that there is a strong need for further theoretical analysis and the disciplined, thorough study of its foundation. Naturally, there were many contributions to the 2021 IS4SI Summit devoted to the questions about information, its meaning, computation, consciousness, and intelligence, in particular artificial intelligence. It is not a surprise that many works produced in the following years retained the focus on these themes.

Yet another subject of special importance present at the Summit and in the continuation of the research in the following years was the methodology of information study. The very rich diversity of the conceptualizations of information, its meaning, computation, consciousness, and intelligence, as well as the methodological tools of their study, may be considered assets and definitely, it is evidence of their non-trivial character. However, it is also an obstacle in the coordination of efforts to achieve the goal of their common understanding. Thus, there is definite value in seeking the unity in diversity of concepts and methods. The chapters addressing directly these matters are grouped in the part "Understanding Information: Theory and Foundations".

To obtain reasonable answers to the questions about information, its meaning, and the meaning of its concept, it is essential to have firm theoretical and philosophical foundations. At the same time, the already existing approaches in the studies of information provide a very effective intellectual tool for exploring reality in its multiple manifestations. The concept of information with diverse definitions and interpretations dependent on specific contexts of the studies became a powerful tool for the methodologies of multiple disciplines from mathematics, computer science, physical sciences, and biology to ecology, social sciences, economics, and humanities. While Part I "Understanding Information: Theory and Foundations" is oriented towards establishing the common foundations for these diverse disciplines, the use of information as an intellectual tool is the dominant theme for Part II "Information Realm: Information as a Tool for Exploring Reality". The Summit gave an overview of a wide range of the methodological applications of the concept of information. Naturally, the following years brought ever more examples of the use of the concept of information and its derivatives in multiple contexts. Practically, every week there is news about new applications and the subject of artificial intelligence has become a favorite of the mass media.

The 22 chapters of this book, although limited in their scope as a book, provide a good overview of the present state of the work on the understanding and use of the concept of information and its derivatives, such as computation, consciousness, and artificial intelligence. All chapters have been peer-reviewed. We would like to express our appreciation to anonymous reviewers involved in the peer review, especially to reviewers who were not contributors to the book and who did not benefit from suggestions of the peers. We wish we could work on this book with Mark Burgin and we hope that Mark would accept it as a tribute to his contributions to the information study.

About the Editors

Marcin J. Schroeder, Ph.D. has been a member of the academic community for almost 50 years as a faculty of several universities in Poland, USA, and Japan active in several disciplines of study from theoretical physics, mathematics (general algebra), logic, to philosophy, intercultural communication, education, and cultural studies. In recent decades, the focus of his academic activity has been on philosophy and mathematical modeling of information, its dynamics (computation), and its integration (including its role in consciousness) with a special emphasis on the use of the general concept of symmetry. For many years, in addition to teaching in several universities, he held administrative faculty positions as a Chair of Department, Dean of Academic Affairs, Director of Academic Center for Learning Excellence, etc. For his contributions to the establishment and development of the Akita International University in Akita, Japan, he was awarded the honorary title of Professor Emeritus. Since 2015 he has been the founding Editor-in-Chief of the journal *Philosophies* published by MDPI. He was the President of the International Society for the Study of Information (IS4SI) in the years 2019–2021 and has been a member of the IS4SI Board since 2017, currently as the Vice-President for Research. Recently, he became Series Co-Editor of the World Scientific Series in Information Studies and became a Fellow of the International Academy of

Information Studies. Since 2022, he has been the Academic President of the International Society for Interdisciplinary Studies of Symmetry (SIS). In addition to presenting and publishing the results of his research and editing multiple research publication initiatives, he is frequently engaged in the organization and chairing of academic conferences.

Wolfgang Hofkirchner, Ph.D. was a Research Fellow at the Austrian Academy of Sciences in Vienna from 1981 to 1990, an Associate Professor for Technology Assessment at TU Wien, Austria, from 2001 to 2018, and a University Professor for Internet and Society at the Paris Lodron Universität Salzburg from 2004 to 2010. He held Visiting Professorships in Brazil, Spain, and Catalonia. He was the Founder of the Unified Theory of Information Research Group in 2003 that is nowadays integrated into the Institute for a Global Sustainable Information Society (GSIS), the President of the Bertalanffy Center for the Study of Systems Science 2004–2018, and is a Founding President of the International Society for the Study of Information (IS4SI). He is a Co-chair of the Emergent Systems, Information and Society Working Group of the Leibniz Society of Scholars at Berlin, Germany, a Member of the International Academy of Systems and Cybernetic Sciences, Brussels, a Fellow of the International Academy of Information Studies, and a Distinguished Researcher at the International Center for Philosophy of Information in Xi'an, China. He has expertise in complexity thinking, science of information, and ICTs and society. He has published more than 260 publications, including the 2 linked monographs on a paradigm shift that allows all sciences to contribute to the transformations necessary to ensure the continuation of human life, *Emergent Information: A Unified Theory of Information Framework* (2013, Chinese edition 2020) and *The Logic of the Third: A Paradigm Shift to a Shared Future for Humanity* (2023).

About the Contributors

Marcus Abundis, MBA is widely experienced in varied high-tech roles from systems analysis to industrial robots, data storage/encryption, diverse medical technologies, hedge funds, and computer networking. His last position managed the entire European market for a US company with a key technology in the Internet explosion. He saw humanity would soon fundamentally shift the way it works with information and began exploring that future through information science. He eventually framed a practical Theory of Meaning with an ultimate aim to develop practical Super-Intelligence. He has been affiliated with Cal Poly Pomona, UCLA, Stanford University, and Esalen Institute.

Michele Caprio is a Lecturer (Assistant Professor) in Machine Learning and Artificial Intelligence at the University of Manchester. He obtained his Ph.D. in Statistics from Duke University and worked as a postdoctoral researcher in the Department of Computer and Information Science of the University of Pennsylvania. His general interest is probabilistic machine learning, and in particular the use of imprecise probabilistic techniques to investigate the theory and methodology of uncertainty quantification in Machine Learning and Artificial Intelligence. Recently, he won the IJAR Young Researcher and the IMS New Researcher Awards, and he was elected a member of the London Mathematical Society.

Jaime F. Cárdenas-García holds BSME, MS, and Ph.D. degrees from the University of Maryland, College Park. After retirement, he enjoys his time as a Visiting Research Scientist in the Department of Mechanical Engineering at the University of Maryland, Baltimore County. Recent publications focus on understanding the role of information for living beings. The result is the coining of the term infoautopoiesis, a unique perspective that posits that there is no information in the environment, except for the syntactic information self-created by living beings that allows the construction of the artificial milieu in which we live. This multidisciplinary perspective has the potential to impact fundamental perspectives on syntactic and semantic information, consciousness, evolution, language, artificial intelligence, biosemiotics, anthropology, economics, and translation.

Iryna Drach, Dr. Hab. (2014), Ph.D. in Educational Sciences, Professor (2022), is the Director of the Institute of Higher Education, National Academy of Educational Sciences of Ukraine, and a specialist in open science, educational sciences and pedagogy, policy and leadership in higher education. She is the author or co-author of several books, including *Management of formation of professional competence of master's students of pedagogy of higher school: theoretical and methodological principles* (Kyiv, 2013), and more than 160 papers. Iryna took part in many international research projects like "Reinventing displaced universities: enhancing competitiveness, serving communities" (British Council/EU, 2020–2024) *et al.*

Gordana Dodig-Crnković is a Senior Professor of Computer Science at Mälardalen University and a Professor of Interaction Design at Chalmers University of Technology, Sweden. She holds Ph.D. degrees in Physics and Computer Science. Her research focuses on the relationships between computation, information, and cognition, including ethical and value aspects.

She serves as a Series Editor for the World Scientific Series in Information Studies and as an Associate Editor for the journal Philosophies. She is a member of the Editorial Board for the Springer SAPERE series, World Scientific Series in Information Studies, Entropy, Information, Proceedings, and several other journals.

José María Díaz-Nafría is currently a Professor at Madrid Open University, Spain (since 2019), where he is the Head of the Telecommunication Engineering Department, and is a Visiting Professor at Munich University of Applied Sciences, Germany (since 2011). He has also served in several other academic institutions in Austria, Spain, Germany, and Ecuador and has been invited professor in many others from Europe and Latin America. He is a board member of the International Society of Information Studies (IS4SI, Austria), BITrum-Research Group (Spain), and other research organizations. His research interests focus on interdisciplinary methodologies and the general study of systems, information and knowledge systems.

Òscar Castro Garcia is a professor of philosophy at the Autonomous University of Barcelona and a researcher at the Institute of Philosophy within the Spanish National Research Council (CSIC). He completed a postdoctoral fellowship in biosemiotics at the Department of Semiotics, University of Tartu. Specializing in the biosemiotics of basal cognition, minimal semiotics, and the philosophy of science (particularly the philosophy of biology), he has focused his fundamental work on exploring the mechanosensory foundations of cognition in myxomycetes since 2010. This research led to the development of "experimental biosemiotics". He is interested in introducing ecosemiotics and cognitive landscapes into rewilding studies through the CSIC, a topic he presented at the 4S Meeting of the Society for Social Studies of Science in Honolulu in November 2023.

Ruiqi Jin, B.E. in Computer Science and Engineering from Xi'an Jiaotong University, is a graduate student at Sun Yat-sen University. Ruiqi's current research interests include logic and complexity theory, whether computational or not, as well as approaches other than neural networks that are relevant to artificial intelligence.

Hans-Jörg Kreowski is a retired Professor of Theoretical Computer Science at the University of Bremen, Germany. He is also interested in Computer and Society with an emphasis on Information Technology and Military Affairs as well as the impact of Artificial Intelligence on society. He is a Board Member of the Forum Computer Professionals for Peace and Social

Responsibility (FIfF) and of the journal *Wissenschaft und Frieden* (W&F). Together with Wolfgang Hofkirchner, he runs the working and special interest group Emergent Systems, Information and Society of the Leibniz Society for Sciences Berlin, and the International Society for the Study of Information (IS4SI).

Roman M. Krzanowski holds a Ph.D. from the University of London, UK, and a D.Phil. from the Pontifical University of John Paul II (UPJPII), Kraków, Poland. He is currently an Assistant Professor at UPJPII. He has published books on the philosophy of information, artificial intelligence, and network technology. His philosophical interests include the philosophy of information, the ontology and metaphysics of computational science, the philosophical foundations of AI, and the ethical problems created by digital society. He has published papers on robotic ethics, the philosophy of AI, the ontology of AI systems, human-computer interactions, human-compatible AI, evolutionary algorithms, NLP systems, phronesis in autonomous robotics, and ethical testing of autonomous AI systems.

Li Luo is a Lecturer at Xi'an University of Architecture and Technology with the main research direction in engineering philosophy. She has published several papers and edited some books in Chinese and English and has applied for a patent for an invention.

Ziyi Ma is an M.A. Candidate, School of Marxism, Xi'an Jiaotong University, has won the Second Prize for Outstanding Paper at the 2023 Xi'an Jiaotong University Annual Academic Conference and the Second Prize for Excellent Papers in the Fifteenth "Graduate Student Forum" of Shaanxi Society for the Study of Natural Dialectics, and has received the Second-Class Academic Scholarship, Xi'an Jiaotong University.

Rafał Maciąg is an Associate Professor at the Jagiellonian University, Institute of Information Studies at Jagiellonian University. His specialization includes digital transformation as a philosophical and anthropological problem, predigital transformation, etc. His competence is in epistemological and anthropological consequences of digitalization (cyber-physical systems, artificial intelligence, etc.). His expertise includes digital technology, theory and practice, and humanities. His recent books (in Polish) include *Digital Transformation*, A Tale of Knowledge, Kraków: Universitas, 2020, and Knowledge as a Story. Discursive Space, Kraków: Universitas, 2022.

Lorenzo Magnani, philosopher, epistemologist, and cognitive scientist, is a Professor of Philosophy of Science and of Artificial Intelligence and Knowledge at the University of Pavia, Italy, and the Director of its Computational Philosophy Laboratory. He has been a Visiting Researcher and Visiting Professor at various important universities in US and China, and in 2015 he was appointed as a member of the International Academy for the Philosophy of the Sciences (AIPS). His recent books are *Eco-Cognitive Computationalism* (2022)

and *Discoverability* (2022). He was the Editor-in-Chief of the *Handbook of Abductive Cognition* (2023) and (with T. Bertolotti) of the *Springer Handbook of Model-Based Science*, (2017). Since 2011, he has been the Editor of the Book Series *Studies in Applied Philosophy, Epistemology and Rational Ethics* (SAPERE), Springer.

Yurii Mielkov, Dr. Hab. (2015), Ph.D. in Philosophy, Senior Researcher in Philosophy (2022), is a Chief Research Fellow at the Institute of Higher Education, National Academy of Educational Sciences of Ukraine, and a Scholar working in the areas of philosophy of science and humanities, philosophy of education, and philosophy of democracy. He is the author or co-author of several books, including *Human-dimensionality of post-non-classical science* (Kyiv, 2014) and *The Many-Faced Democracy* (Saarbrücken, 2016) and more than 200 papers. Yurii took part in international research projects "Human strategies in complexity" (INTAS, 2001–2003) and "Universities – Communities: strengthening cooperation" (Erasmus+/EU, 2023–2026).

Sayan Mukherjee is currently an Alexander von Humboldt Professor at the University of Leipzig and a fellow of the Max Planck Institute for Mathematics in the Sciences. He is also a Professor of statistical science, mathematics, computer science, biostatistics, and bioinformatics at Duke University. His research interests include statistical and computational methodology in genetics, cancer biology, metagenomics, and morphometrics; Bayesian methodology for high-dimensional and complex data; machine learning algorithms for the analysis of massive biological data; integration of statistical inference with differential geometry and algebraic topology; stochastic topology; discrete Hodge theory; and inference in dynamical systems.

About the Contributors

Qiong Nan, Ph.D., is a Member of International Center for Philosophy of Information of Xi'an Jiaotong University and a Member of International Society for the Study of Information. She has participated in many international information philosophy conferences, submitted papers, and presented reports. She also received a scholarship from the China Scholarship Council in 2015 and studied philosophy at the University of Heidelberg in Germany for two years. She has participated in various levels of National Social Science Foundation projects and received several national scholarships. He research interests are philosophy of information, complexity theory, issues related to space-time, etc.

Olha Petroye, Dr. Hab. (2014), Ph.D. in Public Administration, Professor (2021), is the Head of the Department at the Institute of Higher Education, National Academy of Educational Sciences of Ukraine, and a specialist in Open Science, responsible research and innovation, public administration, and social dialogue. She is the author or co-author of several books, including *Social dialogue in public administration: European experience and Ukrainian realities* (Kyiv, 2020), and numerous papers. Olha has the experience of working as a Director of the Institute of Expert-Analytical and Scientific Research at the National Academy for Public Administration under the President of Ukraine (2017–2021).

About the Contributors

Peter Z. Revesz has received his B.S. *summa cum laude* from Tulane University and Ph.D. from Brown University; he has been a postdoctoral fellow at the University of Toronto and a Professor at the University of Nebraska-Lincoln. He is an expert in computational linguistics, the author of *Introduction to Databases: From Biological to Spatio-Temporal* (Springer), a visitor at IBM T.J. Watson Research Center, INRIA, Max Planck Institute for Computer Science, University of Athens, University of Hasselt, University of Helsinki, U.S. Air Force Office of Scientific Research, U.S. Department of State, and a recipient of AAAS Science and Technology Policy Fellowship, J. William Fulbright Scholarship, Alexander von Humboldt Research Fellowship, Jefferson Science Fellowship, and National Science Foundation CAREER award.

Jordi Vallverdú B. Phil, B. Mus, M.Sci, Ph.D., is an ICREA Academia researcher and Professor at UAB specializing in the Philosophy of Computing, Sciences, Cognition, Ethics, and AI. He has conducted research at institutions including Harvard and Kyoto University. Vallverdú received the 2019 Best Presentation Award at HUAWEI's workshop in Kazan, Russia. His work focuses on epistemology and cognition, with over 150 academic publications. He has authored 19 books, 47 chapters, and 106 peer-reviewed papers. Vallverdú also offers three courses on Coursera, attracting over 250,000 students globally.

Liang Wang has received B.S. in Management (2009), M.Phil. (2011), and Ph.D. in Philosophy (2018) from Xi'an Jiaotong University (XJTU). Liang has studied as a Joint Ph.D. student under Wolfgang Hofkirchner at the Bertalanffy Center for the Study of Systems Science in Vienna. Currently, Liang is an Associate Professor at XJTU, an Executive Director at the Center for Science & Technology Innovation and Ethical Governance, and a Member of the Institute for a Global Sustainable Information Society, Vienna, the IEEE TechEthics Community, and the International Center for Philosophy of Information, Xi'an. Liang has published 24 papers in English and Chinese. The subject of his research is information ethics, AI ethics, and transcultural approach theory.

Peiyuan Wang is a Doctor at the Department of Philosophy, School of Humanities and Social Sciences, Xi'an Jiaotong University. Peiyuan's research interests include Chinese aesthetics and calligraphy aesthetics. Peiyuan has participated in a number of scientific research projects.

Shengrui Wang is an M.A. Candidate, School of Marxism, Xi'an Jiaotong University, and has received the Second-Class Academic Scholarship, Xi'an Jiaotong University.

Zhensong Wang has Ph.D. in Philosophy, with research interests in information philosophy, philosophy of cognitive science, and philosophy of science. Currently, he serves as an Assistant Professor in the Department of Philosophy, School of Humanities and Social Sciences at Xi'an Jiaotong University, and a Deputy Director of the International Research Center for Philosophy of Information at Xi'an Jiaotong University.

Kun Wu is a Fellow of the International Academy for Information Studies (IAIS), the Director of the International Center for Information Philosophy at Xi'an Jiaotong University in China, a Second-Level Professor, and a Ph.D. supervisor. Additionally, he serves concurrently as the Vice President of the International Society for Information Studies (IS4SI) and its Chinese chapter, an Executive Member of the Chinese Society for Dialectics of Nature, and the Vice Chairman of its Committee on Systems Science and Complexity Theory. His primary areas of teaching and research focus on information philosophy and the theory of complex information systems.

Tianqi Wu is a Doctor of Philosophy and an Associate Professor of Philosophy Department, School of Humanities and Social Sciences, Xi'an Jiaotong University. Tianqi's research interests include philosophy of information, complexity theory, information ecological ethics, philosophical ontology, and Chinese philosophy. Tianqi has published more than 50 papers and translations in Chinese and English, including 20 papers in SCIE, SSCI, and CSSCI; Tianqi has published books related to philosophy, has presided over and participated in a number of scientific research projects including major and key projects of the National Social Science Fund, and had won many provincial, college, and society awards.

Haisha Zhang, Ph.D. specializes in the philosophy of science and technology, from the Department of Philosophy, School of Humanities and Social Sciences, Xi'an Jiaotong University, China. Her areas of research interests include the philosophy of information, complexity science, ecological philosophy, and ecological aesthetics.

Li Ziyao is a Ph.D. candidate in the Department of Philosophy, School of Humanities and Social Sciences at Xi'an Jiaotong University. Li's research interests include information philosophy, complexity theory, information philosophy of Chinese calligraphy activities, and history of Chinese calligraphy. Li has published four papers in Chinese and English, including one CSSCI paper and one in the core journal of Peking University. Li has participated in provincial social science fund projects and other scientific research projects and has won several national and university-level scholarships.

Ruiyuan Zhang is a graduate student majoring in philosophy, studying in the Philosophy Department of the School of Humanities and Social Sciences of Xi'an Jiaotong University. Ruiyuan's research interests include philosophy of science and technology, philosophy of information, etc.

 **Yixin Zhong** is a Professor at the School of Artificial Intelligence, Beijing University of Posts and Telecommunications (BUPT). He is the Founding Fellow of the Academy of Engineering and Technology for Developing World and Fellow of International Academy of Information Studies. His interests in research and teaching include information science, artificial intelligence, neural networks, and cognitive science. He published more than 520 papers and 18 books in the fields mentioned above. He is the Founder of Information Science based on Comprehensive Information and of Universal Theory of Artificial Intelligence. He served as the Vice President of BUPT, Associated Editor for IEEE Transactions on Neural Networks (1993–2005), Chair of APNNA (2001–2002), President of Chinese Association for Artificial Intelligence (2001–2010), Vice President of WFEO, and Chair of WFEO-CIC (2007–2009).

Contents

Preface		v
About the Editors		xi
About the Contributors		xiii
1.	An *A Priori* Theory of Meaning *Marcus Abundis*	1
2.	From Data to Meaning: The Biosemiotic Approach to AI *Òscar Castro and Jordi Vallverdú*	19
3.	Morphological Computing as Logic Underlying Cognition in Human, Animal, and Intelligent Machine *Gordana Dodig-Crnkovic*	57
4.	Does Information Have Mass? A Review of Melvin Vopson's Claims *Roman Krzanowski*	87
5.	Information, Computation, and Cognition: Their Emergence and Disappearance, and the Ever-Changing Entanglement of Their Meaning *Lorenzo Magnani*	107

6. Theoretical Unification of the Fractured Aspects
   of Information   131

   Marcin J. Schroeder

7. A Naturalistic Approach to the Study of
   Information Philosophy   185

   Zhensong Wang

8. Discrimination and Analysis of Concepts and
   Relationships Related to Information and Intelligence   229

   Kun Wu, Qiong Nan, and Li Luo

9. Paradigm Change in AI   239

   Yixin Zhong

10. Relational Structure of Conceptual Spaces   287

    Mark Burgin and José María Díaz-Nafría

11. Extended Probabilities and their Application
    to Statistical Inference   299

    Michele Caprio and Sayan Mukherjee

12. Infoautopoiesis and the Mind–Body Problem   353

    Jaime F. Cárdenas-García

13. Why Artificial Intelligence Alone Will Not Save
    Mankind: Digital Imperatives for Planetarity and
    Conviviality   367

    Wolfgang Hofkirchner and Hans-Jörg Kreowski

14. Artificial Text: Prolegomena   389

    Rafal Maciag

15. Human Intelligence and the Phenomenon of Information: Open Science, Human Competencies, and Higher Education 421

    Yurii Mielkov, Iryna Drach, and Olha Petroye

16. Trend of Increasing Percentages of Mirror-Symmetric Signs in the Cretan Script Family and the Phoenician Alphabet Family 445

    Peter Z. Revesz

17. The Influence of Information Revolution on Human Thinking Paradigm 467

    Liang Wang, Ziyi Ma, and Shengrui Wang

18. Artificial Intelligence and Creative Education of Future Information: On the Importance of Philosophical Basic Quality Education 479

    Tianqi Wu and Ruiyuan Zhang

19. Holographic Thinking of Social Culture from the Perspective of Lost Book Compilation and Calligraphy Inheritance 503

    Tianqi Wu, Ruiyuan Zhang, and Peiyuan Wang

20. Reflecting on the Limitations of Today's Artificial Intelligence from Information Innovation Systems 519

    Tianqi Wu, Ruiyuan Zhang, and Ruiqi Jin

21. The Nature of Differentiation of New Human Forms and Intelligent Ecosystems 539

    Haisha Zhang

22. The Charm of the Complexity of Innovation of Zhao Zhiqian's Official Script 549

    Li Ziyao

Index 563

© 2025 World Scientific Publishing Company
https://doi.org/10.1142/9789811294921_0001

Chapter 1

An *A Priori* Theory of Meaning

Marcus Abundis

Bön Informatics, Erlinsbach, 5015 Switzerland
55mrcs@gmail.com

This chapter covers a key issue in information theory noted by Shannon and Weaver as a missing "theory of meaning". It names *structural fundaments* to address the matter. This chapter first examines varied informatic roles, noting likely elements for a general theory of meaning. It then deconstructs Shannon Signal Entropy in *a priori* terms to mark the signal literacy (logarithmic Subject-Object primitives) central to "classic" views of information. A dualist-triune (2-3) role therein shows a core pattern. Next, two further Nature based 2-3 roles (natural selection and cosmology) are shown *vis-à-vis* Signal Entropy to further illustrate domain-neutral Subject-Object modeling. Lastly, all three views are used to frame a general theory of meaning — as an *Entropic continuum* of serially varied informatic traits, across space-time. Supporting videos for this chapter are found at: (Abundis, 2021, 2022)

1 Introduction and Background: The Nature of Information

In grasping at the essence of information, a central hurdle has been noted by varied individuals, across diverse disciplines, in confronting distinct challenges:

- "solving intelligence", Demiss Hassabis, Google Deep Mind (Burton-Hill, 2016),
- "de-risking science", Edward Boyden, MIT Media Lab (Boyden, 2016),

- "do submarines swim?", Edsger Dijkstra (Dijkstra, 1984), Eindhoven University, computer science,
- "symbol grounding problem", Stevan Harnad (Harnad, 1990), Université du Québec, cognitive science,
- "theory of meaning", Claude Shannon and Warren Weaver (Shannon and Weaver, 1949), information theory, and more.

These "gaps" arise chiefly due to a traditional **statistical** view of information in Claude Shannon's *A Mathematical Theory of Communication* with "semantic aspects [as *meaning*] ...irrelevant to the engineering problem". But Shannon and Weaver (Shannon and Weaver, 1949) soon saw *Theory of Communication* abuse (now called information theory) would lead to "disappointing and bizarre" results, where a missing "general theory of meaning" (ToM) showed the theory being "ballooned to importance . . . beyond its actual accomplishments", with an "element of danger" (Shannon, 1956). Shannon's non-semantic informatic[1] view now marks a key problem in information science where confused notions of "*meaningful* intelligence" impede further gains in artificial intelligence, presently relying mostly on statistical methods.

This chapter offers a new approach. It examines how we see: *(S)ubject* information (raw percepts, data, object relations, "qualia", etc.) and *(O)bject* information (quantity, self-evident matter, firm truths, etc.), hereafter (S) and (O) and S-O. For example, (Shannon, 1948) uses an S-O split to frame Signal Entropy where "*semantic aspects* ...are irrelevant to the *engineering problem*" (emphasis added) in order to model the latter (O) role. If we accept this split view, (S)ubject and (O)bject must also apply to a ToM, with (S)emantic aspects as a "missing something". But the terms (S)ubject and (O)bject are used variably, as is the word "information", with neither role ever detailed in relation to the other. They instead remain "un-reconciled", driving the cognitive quagmire (Dennett, 2013) we have today. With no uniform S-O

---

[1]Informatic: Roles and systems involving energy–matter "sign" creation, exchange, and/or processing, of any type, and affecting any function, therein making that Sign *information* or *data* about functions.

approach, myriad "disappointing and bizarre" (Eco, 2014) views instead abound, but which this chapter addresses.

To further typify an S-O divide, mathematics may seem "purely objective" said to omit subjective roles from its arguments as an intellectual ideal (theoretical mathematics). But mathematics without *subjective elemental facts* for initial conditions (base data on "primitives") is a fact-free science of little practical use (Feynman, 1964; Smith, 1995). Only if (S)ubject and (O)bject roles are joined do predictive models arise as "functionally reconciled" *applied mathematics*. If we look for other firm (O)bject views, the Standard Model of particle physics and the periodic table are good candidates. But their rather recent "objective success" often omits that they arose from a line of *subjective elemental observations*, normalized (functionally reconciled) via experiment and peer review. Only after enough "primitive evidence" was *subjectively discerned* and *subjectively named* by varied individuals, in many experiments, over decades, were models posited and *subjectively agreed* as being innately (O)bjective — as a type of interpretive/informatic intelligence, or **functionally verified** S-O inter-related ness.

Thus, the claim made here is that balanced S-O views can lessen "gaps" in *meaningful* intelligence. But **Objectified**-*Subject* roles (verified **O**-*S*) like the Standard Model and the periodic table are so common we forget their subjective origins. "Objectivity" itself cannot even be implied if not first *subjectively sensed*, "discovered" or "imagined" by someone, before airing a "sense-making" hypothesis. But joint (S)ubject-(O)bject methods for framing new **O**-*S* views are faint. A practical ToM thus targets an S-O base (Stanley *et al.*, 2017; Wu, 2012) for later **O**-*S* aims, herein posed as **S-O modeling**, to label this "central dilemma" — want of a <u>uniform</u> S-O core for <u>diverse</u> **O**-*S* roles.

To date, the best hint of S-O modeling comes from neuroanthropologist Terrence Deacon (2013). His "multistate" view has Shannon's Signal Entropy, Boltzmann's thermodynamic entropy, and Darwinian evolution by means of natural selection (EvNS) as linked vistas (Dennett, 2013), alongside "structural, referential, and normative" (Deacon, 2020) hints. Deacon's Shannon–Boltzmann–Darwin view evokes a sense of "converged science", but its thermodynamic core ignores wider physics-based roles (Deacon, 2017) (four fundamental forces). Also, as the work is littered with neologisms and difficult prose, (Dennett, 2013; Fodor, 2012; McGinn, 2012) it lacks breadth and clarity. Still, the strength of Deacon's *multistate entropic analysis* is that it stipulates a bottom-up view (minimal ensuing logical gaps), is innately creative (generative *functional* differentiation), and has a continuous multidomain role (key to any *general* theory, crossing many domains), with firm functional ties (thermodynamic entropy, Signal Entropy, EvNS, and more).

To now frame a differed view, the called-for *uniform*-but-*diverse* S-O base can seem paradoxic. A similar issue arose in 1901 when Bertrand Russell saw conflicts in Georg Cantor's mathematical Set Theory, called Russell's paradox. His solution asked that we see different "types" of data exist — a *cognitive advance* that gave rise to Type Theory. S-O modeling requires a like advance in "data types", where mapping S-O *informatic types* is key. But Shannon information theory, presently our only accepted scientific view of "information", has no informatic types beyond (O) roles — as a single *statistical* "non-semantic" aspect, seen in all message communication.

But *meaning-full* semantic roles beyond (O) Signal Entropy actually abound, most basic is "meta-data". For example, the Standard Model and the periodic table are each metadata, *knowledge-about-data*,[2] *collected human wisdom* on key **O**-*S* topics ("science"). Science thus has metadata as a key informatic type where "data Sets" mark "logical domain" *contexts* via grouped "material primitive" *content*. The periodic table thus has electron-neutron-proton triads as "material primitive" *atomic elements* for content. Maps use Set legends with "symbolic primitives" to show map-legend content. An early metadata example is Assyrian clay tablets (3 kya) with details *about* other clay tablets. Today, many metadata examples exist with ascribed *meaning* in every formula, recipe, musical score, blueprint, electric schematic, phonebook, library, information technology, and more.

---

[2]Often defined as "data about data". This paper demands more precision, as "data about data" fails to address Russell's paradox in not naming "data types", but that are noted herein as varied *informatic types*.

Beyond metadata, a Meta-meta informatic type also exists. For example, with the Standard Model and the periodic table as metadata, a view **linking** the Standard Model *with* the periodic table, genomics, and so on suggest a Meta-meta type. This means domain-specific "material/symbolic primitives" (metadata) also support broader domain-neutral "logical primitives". Meta-meta echoes Deacon's sense of "converged science" across domains, and the called-for *uniform*-but-*diverse* S-O base. An early Meta-meta example is dialectic logic: thesis + anti-thesis = synthesis. "Dialectics" born of a pre-Socratic era, show throughout history, in every technical and cultural leap, and more. Darwin's *uniform* view of *diverse* evolving species is also Meta-meta. Earlier, Type Theory is another Meta-meta (Type-of-types) example. Lastly, Signal Entropy is Meta-meta, fitting many domains and at times called "the mother of all models" (Hollnagel and Woods, 2005). Meta-meta holds a **logically** *uniform* S-O view of *diverse* **O**-*S* roles. Biologist Gregory Bateson (1979) saw a meta-to-Meta link as a "necessary unity" and a "pattern that connects" the cosmos, while others aim to "mine a computational universe" (Wolfram, 2017).

1.1 *Constituents for a theory of meaning*

To summarize, in considering varied informatic roles, a notion of "S-O types" is introduced, where a ToM targets a logically uniform view of diverse *informatic S-O types*. Before going further, I clarify some of those named and implied types:

- **(S)ubject** and **(O)bject** are foremost *core types*. Shannon's work has an S-O split, implying core dualism. Detailing ensuing S-O inter-relations and shifts is the focus of all remaining types.
- **Metadata** is a first *meaningful type*, framing a specific domain "context", using material/symbolic primitive "content". Metadata mark Set science (the Standard Model, the periodic table, etc.) where a winning ToM would jointly map many metadata roles.
- **Meta-meta** is a second *meaningful type* with domain-neutral logically primitive content. Signal Entropy is a scientific (meta) **and** domain-neutral (Meta) view, which helps map other Meta-meta views, seen in the later S-O analysis. As such, it following follows:

- **Raw Data** are a *meaning-less type*, known data but with no Set meaning. To create meaning, agents must "interpret" Raw Data in functional roles (S-O empiricism ⇒ metadata ⇒ Meta-meta).
- **Voids** are a next *meaning-less type* as "missing things" we know of but do not fully grasp (i.e., dark matter, dark energy, quantum mechanics, and Life), **and** things we are wholly blind to, failing to sense or imagine them in any useful way — (S)-deficient agency.
- **Levels:** Voids, Raw Data, material/symbolic primitives (meta), and logical primitives (Meta-meta) imply hierarchical informatic "levels" alongside named *informatic types*.
- **Entropy:** Shannon and Boltzmann invoke "two entropies" that are un-reconciled (Deacon, 2021). Reconciliation comes by seeing both in a generic role: a general tendency toward material and logical dispersion/expansion, with shifting *functional degrees of freedom* (DoF).

To expand on these initial S-O types, I now posit a *converged* Shannon–Bateson–Darwin core, beyond Deacon's Shannon–Boltzmann–Darwin view. It invokes Bateson's (1979) "necessary unity" and a "pattern that connects" the cosmos, alongside Alfred Korzybski's (2010) "levels of abstraction" and "the map is not the territory". It marks steps-and-levels of Bateson-like "differentiated [Entropic] differences" tied to Stephen Wolfram's Super-Intelligent aim to "mine a computational universe". S-O modeling thus also frames "generative functional differentiation" (EvNS), with the following details:

2 What is Information? — Case 1: An *A Priori* View of Signal Entropy

The present S-O modeling gaps mean a ToM cannot arise as a one-step study. Even if we accept Bateson's unified vision, diverse scientific, hierarchic, and simple-to-complex roles must be joined. Shannon and Weaver likewise saw three Levels (A, B, and C)[3] of

---

[3] Level A develops Signal Entropy as "the signal problem", Level B marks "the semantic problem", and Level C marks "the effectiveness problem", neither of which Shannon explores (Shannon and Weaver, 1949).

needed study, with more to arise, while Korzybski pointed to myriad *levels of abstraction*. Each holds distinct representational and computational challenges, requiring some manner of *multistate analysis*.

That multistate analysis starts by asking "What is Information?", or "What causes information to arise?", pointing to causal informatics: joint *a priori* S-O roles (from above). "Informatic causality" has *abstract* roles detailing *material* cause-and-effect (content). Here, causality "explains" but named effects are descriptive (e.g., effective-and-efficient *functioning*). Lastly, S-O views can stress either subject or object roles. As such, a "classic" CAUSAL (**O**)bject-oriented informatic view shows in Shannon's *A Mathematical Theory of Communication* — the first of three Meta-meta examples explored herein, toward framing a general ToM.

To explain Signal Entropy, Shannon saw all messages have an X^n *logarithmic base*, akin to Boltzmann's thermodynamic entropy, but in an opposed orderly way (Fig. 1). This chapter's organizing principle is thus *contiguous* Entropic S-O Signs with disorderly to orderly DoF. But Signal Entropy also has a "disappointing and bizarre" (Shannon and Weaver, 1949) lack of practical *meaning* and odd "surprising" statistical role. To see how these differences apply, I next reduce Signal Entropy to S-O steps-and-levels, marking Korzybski's

'Well-Formed' *differed-Signs* **Set**:
(1, 2, 3, 4, 5, 6, 7, 8, 9, 0)
⇕ -*S*- : a 'meaning-full (*S*pace-time) difference'

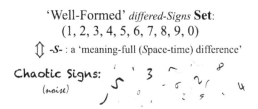

Figure 1. Initial entropic types: A "well-formed" DECIMAL **Set** first shows a sender–receiver agreed **O**-*S* syntax of **O**-term numerals (Signs) "held apart" by firm (**S**)pace-time intervals. All messages rely on such ORDER. Next, Chaotic Signs lack the ORDER messages need, jointly showing an opposed-*meaningful* Entropic difference. A third case of identic **O**-terms (all 9s, or all 1s, etc.; not shown) lacks the **O**-*differentiation* messages need; identic (O) Signs preclude messaging due to "no **O**-term differences". The three cases reflect null-to-chaotic functional degrees of freedom (DoF) — an initial Entropic S-O "continuum" of like **O**-terms in evermore unlike/distinct (**S**)pace-time Fits (*S*-Fits).

levels of abstraction and preceding Shannon's (Level A) study of "message communication", with the following results:

▷ To begin, sender-receiver (agent) *exchanges* "communicate" messages, where we have the following:

(1) Well-formed (O)bject **Sets** serve to construct "a message": binary **I**, **O**; alphabetic a, b, c ...z; decimal 0, 1, 2 ...9; etc. — Bateson-like *differed* **O**-term Signs. Without agent-agreed **O**-term **Sets** (written, spoken, etc. *primitives*), inter-agent messaging is not possible (Fig. 1). A **Set** is therefore an initial **Group** of *agent-agreed* "S-O Sign primitives", wherein next ...

(2) "A message" is a *(S)elect* **Group** of **O**-terms with "targeted meaning", held in two ways: (i) (S)elected **Group** S-Fits amid **O**-terms, and (ii) the (S)elected **O-terms** used therein:

- For example, 126, 261, and 612 are **Groups** of identic **O**-terms with differed S-Fits or "S-O Sign ORDER". Here, (S)pace-time Fits (S-Fits) bind each (S)elected *"message and its meaning"*. But, we have the following:
- New **O**-terms equally bind messages. If "1 2 6" are old terms, 958 and CAT impose new **O**-terms, and hence new S-Fits, where each above **Group** (S)electively conveys a specific *meaning*.

▷ S-and-O are thus functionally tied, in a most basic way — coevol **O**-term *(S)*pace-time Fits bind each *"message and its meaning"*. Alternatively, one can say Entropic S-O Sign ORDER Signals "a likely meaning", Entropic Sign DIS-ORDER Signals "likely meaning-less noise".

(3) Next, Signal Entropy is a **Volume** of *message options* for a **Set Group** of **O**-terms (Fig. 2). For example, a BINARY **Set** "two-term Volume" (2^2) has four **O**-S Fits (OO, OI, II, IO), an ALPHABETIC "three-term Volume" has 26^3 Signals, and IT's ASCII **Set** has 128 seven-term binary (2^7) control characters (Mackenzie, 1980), all marking various message options. Such *differed* S-O **Volumes** typify Shannon's *(v)*aried X^n base — (S)cale-able "communication" in a *contiguous* simple-to-complex manner. Volumes are said to be "meaningless" as no DoF option is yet (S)elected to specify "a targeted meaning". (S)cale-able

CAT	CCT	CTT	CCA	CAA	CTA	CTC	CAC	CCC
ACT	AAT	ATT	AAC	ACC	ATC	ATA	ACA	AAA
<u>TAC</u>	TTC	TCC	TTA	TAA	TCA	TAT	TCT	TTT

Figure 2. A scaleable *and* selectable volume. C, A, and T are shown as **O**-terms with varied S-Fits: an **O**-S Volume with 3^3 (X^n) Signal Entropy, or 27 Groups. Here, CAT and ACT are agent-agreed **O**-S English words, "what you *do* say", other Groups are S-O options, "what you *could* say" (Shannon and Weaver, 1949). For example, TAC in French marks a fencing touché, a rental payment in old British law, and more, for extensible S-O literacy. Each Group marks an S-O Sign S-Fit (DoF) *option* that agents (S)elect to mark **O**-S *meaning*. Still, each requires prior agent agreement on *meaning*: inter-(S)ubjective (O)bject interpretation as en-cultured **O**-S *functioning*. Without S-O ⇒ **O**-S agreement only "noise" arises.

(S)elect-able **Volumes** hold endless *messaging* options (base signs ⇒ words ⇒ books ⇒ Internet, etc.).

(4) Lastly, any number of *(S)*cale-ably **Grouped O**-terms mark intelligible/(S)electable messages of many types. This brings us to the point where Shannon starts his "classic" Level A analysis, where "a message selected at another point ..." is sent across a channel.

This reductive S-O view of Signal Entropy with **Set**, **Group**, and **Volume** steps, precedes Shannon's Level A study of "already set messages being communicated". Further, omitting any of these steps negates Signal Entropy's logarithmic base, where X^n shifts expansively afford DoF options (simple-to-complex variation). Lastly, innately *dualist* S-O roles, tied to *three-part* (S)elect-able steps (above), hold a core dualist-triune (2-3) pattern — with (S)electable *message*-and-*meaning* held in a uniform <u>domain-neutral</u> S-O ⇒ **O**-S <u>domain-specific</u> manner.

2.1 Analysis of Shannon's "classic" (O)bject-oriented view

The above details "*message* and *meaning*", prior to Shannon's "a message selected at another point" as the "fundamental problem of communication". Signal Entropy's "disappointing and bizarre" lack of *meaning* and "surprising" statistical role caused Shannon to doubt Signal Entropy's broad use (Shannon, 1956), which merits further study.

First, Shannon's claim "semantic aspects ... are irrelevant to the engineering problem" has confused *levels of abstraction* that Korzybski warned against. It mistakes Signal Entropy's *meaningfully* (S) "engineered logarithmic base" for pure (O)bjectivity. *Meaningless* "things in a void" (<u>isolated</u> **pure objects**) might exist, but we cannot see such things, as that requires an (S)-(O)bserver. To even *imagine* a "pure object" imposes a (S)ubjective *imagining* semantic as an "agent *context*" with "unreal *content*", including agent perceptual limits/flaws. Shannon further shows "communications" using X^n varied "*(S)ymbolic* content". Lastly, all of mathematics uses agent-agreed (S)emantic symbol Sets in **O**-*S* roles. Signal Entropy thus "disappoints" by ignoring adjacent meaning-full logical S-O levels (re Korzybski), making the idea of "irrelevant semantics" a fiction.

Second, Signal Entropy's orderly <u>*statistical surprise*</u> "bizarrely" opposes Boltzmann's disordered <u>*statistical mechanics*</u>. Here, thermodynamic entropy resembles "noise", the bane of all Signals, marking two "opposed entropies". A closer study shows "scientific *entropy*" is itself ill-defined. Physics has four fundamental forces somehow becoming 12-16 accepted forms of energy, with thermodynamics just one of those forms. But no Unified Field Theory (UFT) yet exists to detail all "force-to-energy" interactions and transitions. As such, Natural tumult alongside ill-defined "entropy" innately holds many seemingly-*bizarre* "scientific gaps".

But above "disappointing and bizarre" issues abate with (1) Entropic S-O Signs as a *meaningful* base, across domains, (2) Entropic S-O steps-and-levels as distinct DoF, and (3) DoF X^n shifts echoing Nature's contiguous simple-to-complex creativity. Here, *meaningful* S-O Signs correct a "disappointing" lack of meaning (#1), and contiguous LOGICAL S-O steps-and-levels fill "bizarre gaps" (#2). Lastly, DoF steps-and-levels (#3) binds everything into one contiguous dynamic simple-to-complex cosmos.

Still, this Entropic S-O view is purely *descriptive*; it lacks direct *explanatory* material facts needed for a full scientific (metadata) account. Meta-meta roles instead **map** logical relations, with further work needed for true "applied level" meta vistas. But an S-O Meta-meta map covering *all* of material reality can expedite metadata work, where such Meta-meta vistas (EvNS, Type Theory, etc.) often hold central scientific roles.

In sum, **S-O modeling** tops Signal Entropy as an even lower-order Meta-meta view. It names 2-3 S-O roles with contiguous "logical depth" (Bennett, 1988): S-O Signs $\Rightarrow$ **O**-S terms $\Rightarrow$ words $\Rightarrow$ books $\Rightarrow$ Internet $\approx$ S-O *message mechanics*. It also partly mirrors Natural creativity via X^n S-O-(v)ariability. In this S-O-(v) role, agents exhibit (v)ariably "effective functional phenomenology" (phenomenal DoF as *meaningful* intelligence) for (S)urvival. Hence, in answering what information is, S-O modeling surpasses Signal Entropy and offers a base for later modeling artificial intelligence. From the above analysis, I now infer a formal definition of "information":

Principle of Informatic (S-O) Dualism — *relational logic, meaning is central.*

• In a dynamic contiguous simple-to-complex cosmos, all informatic cases reflect the following:

(1) (O)bjects: fermions, matter, genes, agents, bits, memes, crude ideas, etc. as "nouns",
(2) Inter-acting (S) with other (O)s, via force/energy, binding, interpretation, etc. as "verbs",
(3) For (S)ubject-topic *meaning* of "how (O) interacts (S) with (O)", or minimal O-S-O functioning (also seen as **O**-S roles/metadata).

Later (S)election of (v)ariably effective-and-efficient **O**-S functions produces further DoF and ToM *types* (re EvNS, in the following section). Base O-S-O functioning also shows another 2-3 pattern. Lastly, S-O modeling holds S-O-(v)aried roles (a 2-3 "adaptive view"), alongside Natural "open" contra *scientific* "closed"/"isolated" roles, as a further affirming 2-3 system views.

3 Case 2: "Adaptive Logic" Amid Nature and Life (S)election

Meaningfully intelligent 2-3 S-O roles have many implications. Foremost, our "interpretive agency" uses S-O-v logic to adaptively rebut Nature's eternally shifting chaos but in an unevenly targeted way. This differs from Shannon's narrowly defined "Signal Entropy *contra* noise". But both reflect a type of *adaptivity*, which this section now examines.

To begin, this next-level study of *adaptive* Entropic S-O Signs (re EvNS) starts with the prior S-O modeling view of Signal Entropy as follows:

1. **O**-term Sets in *S*-Groups underlie each "*message* and its *meaning*", where
2. X^n (*v*)ariably ordered **O**-*S* Volumes afford (S)elect-able message options of many types.
   ▷ **Next**,
3. CHAOTIC S-O SIGNS convey "noise" as presumably *meaning-less* Sign (*V*)ariants, the antithesis of Signal Entropy (Fig. 1, bottom).
   ▷ **Thus**,
4. S-O-(*vV*)ARIATION affords *messaging* ("*v*", #2 above) **and** *noise* ("*V*", #3 above), where

   a. **Pure Noise** or "quantum foam" (Wheeler, 1955) marks chaotic S-O-(*V*)ariation, and
   b. **Functions** oppose chaos, via "orderly" **O**-*S*-*v* and S-**O**-*v* options.

5. FUNCTIONING *contra* CHAOS thus has null-to-strong S-O-*vV* cases:

   a. **Non-Adaptive** cases (**O**-*S*) are narrowly-Set functions, seen in Nature and in many agent roles, as relatively "fixed" or Set material reality.
   b. **Weak Adaptive** Signal Entropy (**O**-*S*-*v*) offers orderly Set, Group and Volume "options".

c. **Strong Adaptive** Entropic Signs (S-O-vV) are "noisy": S-O Sign $\Rightarrow$ Signal novelty, disruptive informatic *options* that agents (s)elect from to abide EvNS.

▷ **Lastly**,
6. (S)ELECTION *meaningfully* bounds all roles. EvNS ambiently imposes "order for free" (Kauffman, 1996), while agent-(s)election strives to match shifting EvNS, but jointly creating further S-O-vV *types of meaning* along the way.

EvNS ultimately bounds *adaptive intelligence* via agent extinctions but where (S)urviving agents hold diverse functional forms, all of which make detailing true *general intelligence* difficult. No one "general form" of functional intelligence routinely prevails over all agents. Earth instead has (*V*)aried niches that Life (*v*)ariably exploits via diverse Life forms (evolutionary "trees"). If Earth instead held a material mono-culture, defining *general intelligence* would be easy, but that is not the role we confront.

Alternatively, if information is given first to agents as "common sense" (Set instincts), agents still need *noisy variants*, not Set views, for (vV)-adaption. Oddly, some people see Set **O**-*S* (scientific) roles as more meaningful, while others see S-O-vV "art-fullness" as preferred. But they truly mark **closed**-and-*open* systems style thinking jointly striking at EvNS.

4 A Bottom-Up View — Case 3: A Cosmic Non-Adaptive Account

This last Meta-meta example brackets Signal Entropy (**O**-*S*-*v*) and EvNS (S-O-vV), with cosmic origins. It shows *maximal* (Cosmic) and *minimal* (atomic particles) Entropic roles reflecting "Nature at large" as a PRIME **context**, for intervening **O**-*S*-*v* and S-O-vV GENERAL **content**.

To start, Nature is **non-adaptive**. Adaption exists only to offset agent extinction risk. The Cosmos itself holds no similar risk, so "to what" would the Cosmos adapt? Native Cosmic dynamics instead produce myriad Earthly niches with related (S)election pressures, to which all agents adapt. It thus presents a PRIME **context** for all later adaptive (S)urvival **content**.

Here, the Standard Model shows a bottom-up "fine-tuned Cosmos" of narrowly defined nuclear interactions — a minimal **O**-S-v Set (Signal Entropy). But an even-earlier Big Bang incites a "quantum Cosmos" of equally meaningful S-O-vV roles. Modern Cosmology tells us that a "bottom-up" ontological Big Bang gives rise to all of energy–matter across all of (S)pace-time, where ensuing **O**-S-v (varied Sets) and S-O-vV dynamic events fill the Cosmos.

If we accept the Big Bang as a valid theory, this also infers an early Cosmos ($\approx$200,000 years old) of pre-photonic plasma as a "thermodynamic maximum" of chaotic/Entropic S-O Sign (V)-symmetry. In that uniformly dynamic Cosmos, no information arises as no *regular* S-O Sign roles exist (Fig. 1, bottom). Instead, PURE NOISE (quantum foam) fills the Cosmos, which we now see as "cosmic microwave background" (CMB).

But, one day condensed matter arises — PRIME **(O)bject ontology**. As an *a priori* analysis, imagine just *one* such object arises: does that one (O)bject convey any information? Its solitude means no witness can "observe" its existence, and it lacks functional counterparts. The (O)bject's place may be "logically possible", but isolation bars anything informative about its "existence" or "function". It thus conveys "no information", standing anomalous and inconsequential within an endless span of quantum plasma.

But quantum condensate is thought to actually have many protons, neutrons, etc. spontaneously popping in-and-out of existence. So do myriad sporadic (O)bjects convey information? **(O)bject ontology** is now more-evident, but we cannot usefully differentiate those objects, and we cannot see any regular (S) traits. Eternal S-O-(V)ariation bars any chance of "sense-making" as everything is too erratic. Still, we now know the Cosmos has countless (S)ense-able roles as regularly interacting atoms, gas clouds, stars, planets, etc. This means that only if two "regular" (O)bjects meet in one "regular" S-Fit do *a priori* S-O Sign functions arise. In this way, 94 Natural elements arise via O-S-O nuclear bonds. Thus, initial "Cosmic functioning" has *meaningful* **O**-(S)pace-time Fits as an *a priori* or

PRIME **(S)ubject ontology** — *minimal* O-S-O functioning for all of energy–matter across all of (S)pace-time.

Lastly, as the Big Bang "cooled over time", *minimal* (vV)aried ORDERLY DoF arise (the Standard Model), counter to an early thermodynamic *maximum* CMB. That emergent "order for free" would best be covered by the prior-noted missing UFT. But the Standard Model instead presently offers our best guess, while S-O modeling offers a useful intermediate "general logic" to temporarily fill lingering gaps.

These last analytic steps enlarge the prior definition of "information":

Principle of Coevol Origin, Identity, and Abstraction — *2-3 ontology and epistemology.*

- **Ontology:** An (O), in joining (S) with an (O), is O-S-O self-evident. Solitary (O)s are non-evident. Coeval O-S-O functioning marks a minimal ontology as a dual-material *informatic aspect* of "energy–matter".
- **Epistemology:** O-S-O functioning is a "thing in itself", self-evident cause-and-effect as pure energy–matter with no interpretive or adaptive needs — a non-agent 2-3 *ontologie-épistémè* or "PRIME **context**".
- **Entropy:** General force-energy dynamics *drive* O-S-O, with noisy DoF options (sS)elected (by agents and Nature) as an "informatic Cosmos".
- **Abstraction:** Agents (s)electively grasp and parse the material Cosmos with perceptual biases and limits, where events must be **verified** to firmly define agent-related epistemic cause-and-effect.
- **Verification:** Nature (S)electively **verifies** all Natural and agent roles, where trial-and-error is the most assured means of verification for both.

5 Discussion

Shannon, Weaver, and others saw issues with using Signal Entropy as "information theory" due to a missing ToM, while most modern views wholly ignore the problem. This 75-year-old "logical gap" is herein corrected with a new S-O approach. This chapter reframes prior informatic misconceptions by detailing "lower-level" (S)ubject and (O)bject roles innate to Signal Entropy, EvNS, and modern cosmology. It uses three cases to show *contiguous* Entropic S-O Signs, with (vV)aried DoF, for all of energy–matter, across all of (S)pace time.

This chapter thus details "What is *meaningful* information?" beyond Shannon's view — with Meta-meta S-O Signs modeled in a 2-3 pattern. That 2-3 S-O core underlies all interpretive/informatic intelligence and **functionally verified** relations. But a definitive ToM still requires further work (Meta-meta ⇒ metadata) in mapping detailed material roles and DoF. Lastly, this ToM approach, in broaching *meaningful intelligence*, poses a base for later modeling artificial general intelligence.

Despite open issues needed for ToM completion, just as Signal Entropy affords ever-more IT gains, this ToM base holds likely benefits for future IT work. For example, augmented human innovation — an S-O-vV "insight engine" — would drive advances on many fronts, just as seen with modern "universal computing devices". But further research is first needed (LeCun and Marcus, 2017; LeCun and Manning, 2018).

References

Abundis, M. (2021). An a priori theory of meaning. https://www.youtube.com/watch?v=2rS9uT08YP8.

Abundis. M. (2022). The advent of super-intelligence. https://www.youtube.com/watch?v=11oFq6g3Njs.

Bateson, G. (1979). *Mind and Nature: A Necessary Unity*. Dutton, New York, NY.

Bennett, C. H. (1988). Logical depth and physical complexity. In Herken, R. (ed.), *The Universal Turing Machine — a Half-Century Survey*, Oxford University Press, Oxford, UK, pp. 227–257.

Boyden, E. (February, 2016). Engineering revolutions. In *The 2016 World Economic Forum*. World Economic Forum (WEF), Davos, Switzerland. http://mcgovern.mit.edu/news/videos/engineering-revolutions-ed-boyden-at-the-2016-world-economic-forum.

Burton-Hill, C. (February, 2016). The superhero of artificial intelligence. *The Guardian*.

Deacon, T. W. (2013). *Incomplete Nature: How Mind Emerged from Matter*. W.W. Norton & Co., New York, NY.

Deacon, T. W. (January 8, 2017). December 2016 fis posts.

Deacon, T. W. (March 6, 2020). Re: Defining "information". http://listas.unizar.es/pipermail/fis/2020-March/002480.html.

Deacon, T. W. (January, 2021). Is4si 2021 — collaboration?

Dennett, D. C. (December, 2013). Aching voids and making voids. *The Quarterly Review of Biology*, **88**, 321–324.

Dijkstra, E. W. (November, 1984). The threats to computing science. In *Association for Computing Machinery: South Central Regional Conference*. Association for Computing Machinery, Austin, Texas. http://www.cs.utexas.edu/users/EWD/ewd08xx/EWD898.PDF.

Eco, U. (2014). *Opera Aperta*. Bompani, Milan, Italy.

Feynman, R. (November, 1964). The messenger lectures: The relation of mathematics and physics. https://www.microsoft.com/en-us/research/project/tuva-richard-feynman/.

Fodor, J. (May, 2012). What are trees about? *London Review of Books*, **34**(10), 34.

Harnad, S. (1990). The symbol grounding problem. *Philosophical Explorations*, **42**, 335–346.

Hollnagel, E. and Woods, D. D. (2005). *Joint Cognitive Systems: Foundations of Cognitive Systems Engineering*. Taylor & Francis, Boca Raton, FL.

Kauffman, S. (May, 1996). Chapter 20 "order for free". *Edge.org*.

Korzybski, A. K. A. (2010). *Selections from Science and Sanity: An Introduction to Non-Aristotelian Systems and General Semantics*. Institute of General Semantics, Fort Worth, TX.

LeCun, Y. and Marcus, G. (October, 2017). Does AI need more innate machinery? https://www.youtube.com/watch?v=vdWPQ6iAkT4.

LeCun, Y. and Manning, C. (February, 2018). Deep learning, structure and innate priors, a discussion between Yann LeCun and Christopher Manning. http://www.abigailsee.com/2018/02/21/deep-learning-structure-and-innate-priors.html.

Mackenzie, C. E. (1980). Coded character sets, history and development (PDF). In *The Systems Programming Series*, 1st ed. Addison-Wesley Publishing Company, Inc., Boston, MA. https://textfiles.meulie.net/bitsaved/Books/Mackenzie_CodedCharSets.pdf.

McGinn, C. (June, 2012). Can anything emerge from nothing?, *New York Review of Books*.

Shannon, C. E. and Weaver, W. (1949). *Advances in a Mathematical Theory of Communication*. University of Illinois Press, Urbana, IL.

Shannon, C. E. (July, 1948). A mathematical theory of communication. *Bell System Technical Journal*, **27**(3).

Shannon, C. E. (March, 1956). The bandwagon. *IRE Transactions — Information Theory*.
Smith, J. M. (1995). Life at the edge of chaos? *The New York Review of Books*, **42**, 28–30.
Stanley, K. O., Lehman, J. and Soros, L. (December, 2017). Open-endedness: The last grand challenge you've never heard of. *O'Reilly Ideas (What's on our radar): AI*.
Wu, K. (September, 2012). The essence, classification and quality of the different grades of information. *Information*, **3**.
Wheeler, J. A. (January, 1955). Geons. *Physical Review*, **97**.
Wolfram, S. (May, 2017). A new kind of science: A 15-year view. http://blog.stephenwolfram.com/2017/05/a-new-kind-of-science-a-15-year-view/#more-13536.

Chapter 2

From Data to Meaning: The Biosemiotic Approach to AI

Òscar Castro[*] and Jordi Vallverdú[*,†]

[*]*Philosophy Department, Universitat Autònoma de Barcelona*
[†]*ICREA Acadèmia*
oscar.castro@uab.cat

This chapter delves into the profound journey from raw data to meaningful knowledge in the realm of artificial intelligence (AI) from a biosemiotic standpoint. It begins by scrutinizing Claude Shannon's Information theory, the bedrock of data quantification, storage, and transmission, underlining the significance of bits as the fundamental units of information. The biosemiotic approach to AI is then explored, driven by inspiration from the human brain's workings in AI and machine learning. Despite its merits, the approach is critically examined for its anthropocentric bias, disregarding the rich diversity of cognitive systems in the natural world. Drawing from biosemiotics, this chapter further investigates biocommunication and information processing in simpler organisms, emphasizing the concept of "Umwelt" and the exchange of information within an organism's perceptual sphere. Additionally, the role of genetic codes and the hierarchical structure of meaning in biological information are explored. In conclusion, this chapter advocates for a holistic understanding of information processing, merging insights from biosemiotics to bridge the chasm between data and meaning in AI. By embracing a broader perspective that incorporates diverse cognitive systems and semiotic processes, this approach paves the way for more robust and context-aware AI systems, with far-reaching applications across various domains.

1 Information and Computation

There is a long path between data and meaning. The key is to think about how information is produced, captured, processed, and used for any kind of action. The Shannon theory, also known as information theory, was developed by Shannon (1948). It is a mathematical theory that deals with the quantification, storage, and communication of information. Shannon's theory is based on the idea that information can be represented as a series of bits, which are the smallest units of information. It also quantifies the amount of information carried by a message or signal, providing a mathematical model for the communication process, focusing on the transmission and reception of information over a noisy channel. Shannon's information theory had a significant impact on computer sciences all throughout the 20th century and even today, in several fields like data compression (providing a solid foundation for data compression algorithms, showing that redundancy exists in most data, and by exploiting this redundancy, data can be compressed without losing essential information. This led to the development of efficient compression techniques such as Huffman coding, arithmetic coding, and Lempel–Ziv–Welch (LZW) compression, which are widely used in file compression and storage. Information theory contributed to the development of error-correcting codes, which ensure reliable data transmission in the presence of noise or errors. By employing techniques such as error detection and correction codes, computer systems can recover corrupted data and maintain data integrity, cryptography, communication networks, or even machine learning.

Such computational revolution, based on computation and communication technology, affected all scientific areas and even greatly impacted new global societies (Castells, 1996), paving the way for the society of information. In such a society, information is the primary source of power and wealth. The ability to produce, process, and communicate information is essential for economic and social development. Castells argued that the society of information was characterized by a network society, where communication networks play a central role in shaping social and economic relations. The creation of the Internet allowed and supported a global revolution that affected all human societies (Leiner, 1997).

On the other hand, computer sciences look at natural evolution as a source of bioinspiration. The classic book of the mathematician John Von Neumann in 1956, The Computer and the Brain, created a bio-inspired trend in computer science. The book proposed that computers could be modeled after the human brain, which was seen as the ultimate symbol of intelligence at the time. The book argued that computers could be programmed to perform tasks that were previously thought to require human intelligence. Since then, computational systems and algorithms have drawn inspiration from biological processes. Some examples are neural networks, genetic algorithms, evolutionary computation, swarm intelligence, artificial immune systems, cellular automata, and DNA computing, among others.

Those ideas boosted the development of artificial intelligence (AI) and, later, of machine learning (ML) technologies. These technologies are based on the idea that computers can be programmed to learn from data, just like humans do. The development of AI and ML has had a significant impact on science and society, with applications ranging from self-driving cars to personalized medicine (Casacuberta and Vallverdú, 2017).

But at the same time, some misconceptions or biases were maintained, basically, the overestimation of human brain functions while ignoring other cognitive systems distributed all along biological taxa. Thanks to new ideas, like extended cognition, embodied mind, enactivism, or embedded mind, as well as the new advances in animal, plant, insect, and basal cognition, we are facing a new revolution adding to the huge successes of deep learning (LeCun et al., 2017). Something was missing: the connection with the field of biosemiotics, which we explore in the following section.

2 Some Features of the Information From a Biosemiotics Perspective

What is biocommunication, and how does it work in simple organisms? The phenomenon of biocommunication carries through the mechanotransduction of signals as a fundamental phase for information processing in "decision-making" and its ability to discriminate and make sense endogenously of what happens exogenously. To understand how biocommunication is carried out in simple organisms, biosemiotics has an approach based on cellular semantics and molecular codes through Morris's semiotic trichotomy, a fundamentally behavioral factor that fits perfectly in developing the adaptive behavior of the cells (Görlich *et al.*, 2011, 2013). Therefore, decision-making is an essential biosemiotic step between information processing and behavioral outcomes.

The surrounding world of each organism — named by Jakob von Uexküll (1909, 1920) as "Umwelt" — is a kind of sphere of recognition of the external world lived from the very inner world of the organism named "Innenwelt". And as it is perceived, it is decoded according to the spectrum of each circle or functional circuit, "Funktionkreis" that the organism has for each sense, "Sinn". The sum of all the functional circles generates what could be described as a "field of multisensory perception of the meaning environment" (Castro, 2009, 2016). This field is intrinsic to individual learning as ontogenic and phylogenetic development and, therefore, species-dependent development. But this multisensory perceptual field of the environment is described by Uexküll as a "soap bubble" or "Seifenblase", a closed environment where the body is in the perceptual center of the environment, "Umgebung", or rather the outside world, "Außenwelt".

In the "soap bubble", there is a dynamic of information exchange. The frame of reference lies in the "field of multisensory perception of the meaning environment" as if they were processes of homeostasis and heterostasis in homeorhetic dynamics variables.[1] But in turn, it carries the weight of a monadic description that contains the limits of connection and meaning for the individual.

---

[1] Homeoretic means that it is a stable path of development over time. It is a term coined by Conrad Hal Waddington that describes the developmental tendency or change of organisms to continue development or change toward a particular state, even if it is altered or disrupted by the environment.

As Castro (2009, 2016) gathers Umwelt's concept of the terminology of "field" as it is used in a multidisciplinary way and allows the hermeneutic linguistic game (both in the field of the objective and the subjective), with transversal reference and with a transdisciplinary sense. It welcomes the descriptive phenomenon of subjective space on a human level. It is a natural expression of extending a terrain where the view reaches and where cultivation is made possible. But the scientific field is the real or imaginary field of activity or knowledge. For example, ecology is linked to the concept of the ecosystem as an extension inhabited by an assortment of living organisms. In optics, the visual field refers to the spatial set of visual sensations available to the observer from a reflective neurocognitive experience. This concept should not be confused with external objects and light sources that affect the retina. It specifically refers to what is perceived in the observer's brain and measured through the perimeter. But in physics, the field is a vector quantity (electromagnetic and gravitational). However, in quantum mechanics, these are topological distributions (generalized functions, mathematical object that generalizes the notion of function and measurement) that allow assigning operators that describe the existence of a transformation space.[2]

The need to understand at least both the syntax and semantics of transducer signals in communication in both simple cells and the members of a more complex system (such as the immune system of an animal or the immune system in a human being) is the main aspect of the search for information that is significant to organisms in the interaction with their Umwelt itself in current biosemiotics.

Jesper Hoffmeyer and Claus Emmeche (1991) worked on the concept of "code duality" between the digital and analog versions of the coded message of DNA. Hoffmeyer marks the difference between the self-reference of his message and the supply for the cytoplasm and his semiotic decoding of the role of DNA as a work plan and its execution. Biological information is no longer identical to the gene

---

[2]We have in mind the topological sense of dual space to give infinite-dimensional spaces, such as Hilbert or inner product spaces, as a resource for Fourier transforms.

or DNA, as spots on paper which provide letters, monemes, morphemes, and words. So the meaning is hierarchized to levels that are not part of the physical structure, although neither sign nor meaning emerges without it. That is why biological and cultural information are not physicochemical substances, although they depend on physicochemical substrates and operate in the world of substances. "Inform" comes from the Latin word "informare": "to bring something in a way". Hoffmeyer's purpose [28], as well as Emmeche's, was to show that when the concept of information bridges the distinction between form and substance, the central perspective of biology will change fruitfully. "Information does not exist except for its content by (immanence) in matter and energy". From this sentence arises the following pertinent question: How could systems respond to differences in the environment? As confirmed by Howard Pattee (1972), how did the system first know how to build a description of itself? This description function ensures the system's identity over time (a system memory). It would be the self-reference of RNA–DNA. But for the translation (or development) process, you must decipher the code and follow its instructions. The tacit knowledge of the same is in the cellular organization, concealed in the description of the DNA. So, the realization in space and time of the structural relationships specified in the digital code defines the type of environmental differences that the system selects in the Umwelt and responds to. The innovation of digitization is the act of discretizing, introducing spaces into the continuum, thus creating boundaries. These boundaries, however, do not belong to the continuum nor are they part of these spaces. The boundary is the geometric site of external intervention. Thus, the "goal-seeking" system conformity to the end due to the plan is necessarily defined that traces this boundary.

Sharov (1992), entomologist, geneticist, and biosemiotician at the National Institute on Aging in Baltimore, exposes that information is considered a microstate of a system that influences the choice of systems among trajectories to bifurcation points. The sense of the information has two components: a meaning or purpose (meaning) and a value or importance (value). Meaning is a collection of prohibitions and limitations for information about a developmental and behavioral system, and the contribution of information to the security of self-maintenance and self-reproduction of the system measures value. Meaning and value are considered at the material and ideal levels. The sense of evolution is characterized by its open extension in space-time and the complexity of its structure. Its system gradually moves from the pre-biological systems via more and more complex biological organisms to humans.

Frederik Stjernfelt, professor of Biosemiotics at the University of Aarhus (Denmark), through the work based on Jesper Hoffmeyer which Stjernfelt (2002) called "Tractatus Hoffmeyerensis" states the following:

(1) Signs and life are coextensive.[3]
(2) Since biology is a historical science (perhaps even the quintessential historical science), he calls for a natural history of meaning.
(3) As biology is a structural science (maybe even structural science par excellence), it calls for an inventory of biology's structural concepts.
(4) Suppose we accept punctuated equilibrium as a basic structure in biological evolution. In that case, the semiotic evolution should follow the same structure, displaying a ladder of increasingly complex sign types.
(5) The basic forms of biological signs are exchanged between the organism and its environment, its Umwelt, in a functional circle, with its respective signs or signals of action (Wirkzeichen) and respective signs or signals of reception (Merkzeichen), as Uexküll described.
(6) The Umwelt in the most complex life forms is not determined through or is not genetically determined through and through but must be formed in the individual case by selecting paths in

---

[3]This idea is at the heart of current biosemiotics derived from another phrase by Thomas Sebeok that states that "semiosis is the heart of life" in Sebeok (1991, p. 85). Or also, "Semiosis presupposes life" in Sebeok (2001).

its chreod landscape,[4] under the impression of interaction with the environment.

(7) Such an experience-based Umwelt allows genetic assimilation (Waddington) because individuals with better genetic bases for coping with their particular surroundings will have a selection advantage. This so-called Baldwin effect will be especially efficacious in social animals where one may learn such Umwelt competencies from others.[5] Thus, the species' virtual reality is represented by its Umwelt's set of inner representations of typical, merely potential relations in the organism's surroundings, which may become genetic reality — "only the well-prepared will profit from chance" (Hoffmeyer, 2001; Pasteur's quote from p. 393).

(8) Generally, any new habit taken exposes the organism to new challenges in a never-ending chain of interpretations.

(9) The role of selection remains decisive, but basic biological phenomena like multiplication as well as order are prerequisites to selection and hence cannot be products of it. Therefore, multiplication and order are inherently meaningful.

(10) Both are thus more primitive than genes and pertain to the analogous side of the organism's double code (which is a not-so-lucky expression as only the former of them is, strictly

---

[4]The term chreod, from the Greek chre (destination) and hodos (path) is a neologism from Waddington's (1957) epigenetic biology and refers to the relatively stable trajectory of species development partly caused by the evolution of hierarchical genotype control sequences. This evolution does not stabilize at a certain point (not homeostatic) but in a single stable development path over time (homeoretic). The chreodic development of the system means that the influences of the environment tending to remove it from its trajectory will be neutralized by its self-regulation so that this system will return to its usual trajectory.

[5]The Baldwin effect, also called Baldwinian or ontogenetic evolution, is an evolutionary theory proposed in 1896 by the American psychologist James Mark Baldwin. He proposed a mechanism for the selection of learning skills. The selected offspring would tend toward a greater ability to learn new skills and not be constrained to genetically coded and relatively fixed skills, emphasizing that the sustained behavior of a species or group can shape the evolution of species (Simpson, 1953).

speaking, a code): digital, genetic information, on the one hand, and analogical, morphological information on the other, provided by the cell's architecture and metabolism as well as multicellular structure and communication.

(11) Another prerequisite to the functional circle is the organism's character as an agency equipped with a point of view. This may be defined as a "stable integration of self-reference and other-reference" (see Hoffmeyer, 1999). (the former maintains and defines the self as such; the latter facilitates its orientation and survival in its umwelt).

(12) The agency presupposes, in turn, the existence of an inside–outside defining boundary, a membrane, characterizing all life forms (except for certain marginal parasite types like a virus). Membranes thus make possible the crucial organism–environment asymmetry — facilitating the constrained traffic across the membrane boundary in the form of signs. Autocatalytic closure of chemical reaction loops in the primordial soup thus needs a further topological membrane closing to result in organisms. Furthermore, the controlled traffic across the membrane permits the emergence of strictly constrained "inner outsides" (due to perception in a broad sense of the word) in the organism as well as "outer insides" (due to its interaction with and influence on specific aspects of the environment).

(13) Such signs embed the organism in its ecological Umwelt comprising other organisms with Umwelten. A mutualism much more widespread than strict symbiosis thus forms what Uexküll calls "a natural symphony" of mutual communication between species and between them and their surroundings.

(14) Such communication necessarily involves, for economic reasons, categorical perception. As a result, slightly different phenomena are functionally perceived as the same type. It is probably the lowest or simplest semiotic phenomenon, based on the differently shaped "active sites" outside of macromolecules, which may recognize these sites on other molecules. By the same token, other molecules with the same sites may "fool" the process in question. This is the biochemical foundation of

biological indeterminacy or semiotic freedom. Still, it requires a cyclical teleological (functional, final, purposeful, metabolic, homeostatic, or similar predicate you prefer) process to display its possibilities.

(15) Biology is thus impossible without the Aristotelian quartet of causes. However, final causes should not be identified with purposes (which form a special subset of them) but should be identified as all processes attracted by a future state. Future states being general, only (Peirce),[6] final causes may make use of representation of such states using types.

(16) As we only know rather complicated life forms (cells internally consisting of organelles that are probably formerly symbiotically living organisms), these primitive semiotic processes also characterize the cell's internal metabolism.

(17) The role of the genes seems to be controlling epigenetic and metabolic processes in the organism (not creating or determining them through and through). This points to the fact that genes may be a special and successful example of a more general notion of "scaffolding": stabilization and channeling of (a segment of) metabolism. Other scaffoldings could be cell architecture, organ structure, language, etc.

(18) At the upper end of the natural history of meaning, we find animals with central nervous systems that have taken meaning bases in categorical perception to form very complex semiotic abilities. The increasing indeterminacy — or semiotic freedom — can be expressed as the emergence of sign types increasingly loosened from their basis, in particular, sign tokens. Higher animals may recognize tokens as an instantiation of types; they may use them to symbolize, reason, argue, and use diagrams. The special human privilege is probably an abstraction, making it possible for us to explicitly contemplate such types, reasonings, and diagrams with any token placed in brackets, thus facilitating control, experiment, and quick development of these signs.

---

[6]The idea is Peirce's, cf. Hoffmeyer (2002).

(19) Thus, biosemiotics assumes a distinction between the issue of signs and consciousness. Sign processes are taken to be possible without consciousness. As the existence of signs may be inferred from the external behavior of a process, the establishment of qualia consciousness in a system has — not yet that is — any methodology. However, it seems to be a tendency that complex signification processes are increasingly facilitated by consciousness, maybe as a special type of neural scaffolding.

(20) The interrelated web of biosemiotic concepts used here — membrane, sign, active site, function, metabolism, organism, Umwelt, niche, and so on — forms a regional ontology (Husserl) of biology and semiotics and any biology, even the most would-be reductionist versions, must necessarily make use of some versions or other of them. Such concepts provide the structural inventory in (3).[7]

(21) Biosemiotics does not entail vitalism, as it does not suppose the existence of élan vital, unknown organic force fields, or the like. Neither it entails any subjectivism or relativism; even if every single organism and species has its point of view, this does not imply skepticism as that point of view may be compared and evaluated. Rather, biosemiotics entails idealism in a certain use of the word — not referring to the world being created by a subject or anything of the kind but referring to the reality of ideal objects (like those conceptual networks of (20)). A special kind of ideal object here deserves mentioning, that of possibilities. Possibilities must be assumed to possess real existence, including the idea of a fitness space of all possible genomes, virtuality in nature, tendencies in development and evolution, and, correlatively, the possibility for final causes to prefer one tendency

---

[7]It goes against the argument of Morten Tønnessen (2001), who states that the universal concepts of biology are impossible because we can imagine completely different life forms from those we know. The generalization here may fall into a fallacy where conditions do not meet the conceptual requirements. In the case of knowing, a contrapuntal way of life other than that organized through carbon could offer (or not) a denial of this argument.

over another. "Thus, biosemiotics entails an ontological revolution admitting the indispensable role of ideality in this strict sense in the sciences" (Emmeche and Kull, 2011, p. 98).

(22) To close the biosemiotic circle: real possibilities are also what make signs possible: any sufficiently complicated sign refers to a bundle of merely possible actual objects later in a functional circle, that is, to a possibility, sometimes real, sometimes not.

Four of the eight theses about biosemiotics developed in 2009 by Frederik Stjernfelt with Jesper Hoffmeyer, Claus Emmeche, Terrence Deacon, and Kalevi Kull reveal the mark of the information in biological processes (Kull et al., 2009):

(1) The semiotic–non-semiotic distinction is coextensive with the life–non-life distinction, with the mastery of general biology.

The concepts of function and semiosis, which involve the processes of signs or signals, are closely related. However, it's not entirely certain whether they are entirely synonymous. Both of these concepts are teleological, meaning they are defined with respect to a purpose or the fulfillment of a particular cycle, often associated with a specifically correlated absent content.[8] Both are teleological concepts in the sense of being determined in terms of a purpose (or the achievement of a cycle, etc.) — "of a specifically correlated absent content" (Kull et al., 2009). Teleological processes that are specially organized for specific purposes or referents are unique to living processes.

(2) Biology is incomplete as a science in the absence of explicit semiotic knowledge. The reason why this is not perceived as a problem is that biology compensates for excluded semiosis by introducing a lot of implicit semiotic terms such as "information", "adaptation", "sign", "input signal", "code", and "messenger", "fidelity", and "voice crossing". However, these uses are well defined and often applied metaphorically, with the implicit

---

[8]For example, irreversible physical prebiotic processes, or Deacon's autocells, as well, infer some information processing of semiotic content, affecting the tautology between autopoiesis and life, as a possible abiogenesis.

assumption that they can be reduced to mere chemical events if necessary.
(3) The predictive power of biology is embedded in the functional aspect and cannot be based solely on chemistry.

It is an accepted truth in biology that structure and function are interdependent; for example, a biological explanation is incomplete, although the production and structure of a cell's macromolecule have been extensively described. What is the use of this fundamental explanation? Kull *et al.* (2009) comments that answering this question is part of the functional contextualization that all biological activities require. In many cases, these functions are characterized by having a regulatory character, information carrier, or signaling type. Therefore, describing the function of these structures is framed in a broader system with a processing characteristic of signals (for example, signal transduction). In this broader system, the functions that a macromolecule takes part in (or contributes to in some way) can be performed by other structures. Therefore, the structure is seen as a vehicle for fulfilling that function. The location and clarification of the functional-procedural parts in a wider network in metabolism allow partial predictions of some limitations that must be met for the vehicle of a function to work. Most of the predictive power of biology is lost if the semi-functional analysis is excluded.

(4) Differences in methodology distinguish semiotic from non-semiotic biology. It is the aim of biosemiotics to make explicit the assumptions that are imported into biology by such unanalyzed teleological concepts as "function", "information", "code", "sign", and "signal" and to provide a theoretical basis for these concepts. The widespread use of these terms in current biology assumes that these ideas cannot be completely avoided or replaced by mere chemical events. Biosemiotics has the scientific task of (1) collecting these terms in a physical–biological context, (2) interrelating these terms with the constant goal of avoiding anthropomorphisms that occur when left only with implicit definitions, and (3) seeking a complete theory in theory.

(5) Semiosis is a central concept for biology — however, a more precise definition is required. Although there are many descriptions of semiotic processes, it is still an unresolved challenge to provide an exposition that explains exactly what constitutes semiosis without assuming a homuncular interpreter — self-referential — or leaving critical relationships unattended and defined. Interpretive capacity is an emerging property of a reciprocal end-meaning relationship through a dynamic system of self-propagation.

We can identify seven properties or conditions that need to be met in order to talk about semiosis. The following is an outline of these critical conditions:

(a) **Agency (organism):** A unitary system that can generate targeted-final behaviors.
(b) **Normativity:** A semiotic process broadly increases normative properties, thus being embedded in a process that contributes to normativity. It includes the possibility that the representation is in error or that its consequences (in terms of Peirce, its dynamic performer) may be compatible or incompatible with preserving the integrity of the living system in which it occurs.
(c) To understand this minimal notion of normativity, we consider the difference between a physical pattern and a pattern that serves a function. Algorithmic information theory (Chaitin, 1977) can characterize any specific physical pattern as highly random, highly regular, or something complex between the two, whether descriptively compressible or truly complex and incompressible. But a pattern used for a function has, in addition to its own high or low content of algorithmic information, a degree to which its result or goal is useful. For semiotic processes (with functions such as representation, information storage, and interpretation), the degree to which a pattern serves or does not serve these functions is the norm.
(d) **Teleo-functionality:** Semiosis is always integrated into the end or goal process. Therefore, semiosis can be evaluated about whether its interpretation is concordant or discordant with the

dynamics of achieving this goal. It determines the normative properties of a signal and/or sign interpretation process.

(e) **Form generation:** The systemic organization responsible for interpreting the semiotic function of a sign vehicle (or Peirce object) must include a generation process to contribute to this function's persistence (representation) directly or indirectly. The interpretation process generates a structure (physical form) that acts as a signal of the previous signal and can produce greater structural consequences.

The differentiation of a sign vehicle from the reciprocal process of form generation: A sign vehicle must be isolated from the dynamics that restrict and are responsible for repeating this process.

(f) **Categorization:** Repetition of signs can never be 100% physically identical. It, along with regulations, is the reason why signs form types. Categorization appears in all communication processes in the case of sufficiently adaptive systems. Functionally similar cases of signs are subsumed into a general type. At the same time, what the signs refer to is also categorized. This aspect of using the sign is highly economical, as it allows the body to generate only a finite number of simple, typical ways of interpretive action to achieve similar ends. Of course, the other side of this is the possibility of fallacy.

(g) **The inheritance of relationships:** Various development processes, including those that have created new correspondence between the parts of the organism and between the organism and its environment, are supposed to be the main means by which the semiotic relationships are generated. In this regard, genetic inheritance represents one of the most basic forms of semiosis. Therefore, studying the conditions of its generation should give an idea of the formative semiotic process based on physical processes. In addition to genetic inheritance, several other forms of inheritance (e.g., epigenetics, neuronal, and social) are used by various communication processes. In this sense, semiotic processes generally include memory processes that maintain information continuity and dynamic options' stability.

3 Bateson's Information Theory

We need to measure beyond what Bateson called the threshold of perception of the experimenter (Bateson, 1979). Before this threshold, science cannot prove and, therefore, cannot explain. In "Steps to an Ecology of Mind" (Bateson, 1972) and "Mind and Nature", Bateson (1972, 1979) sees that matter and energy are imbued with circular information processes about differences, which he believes are "connecting patterns". Deeply interested in anthropology, biology, and psychology, he approaches the fields of information, cognition, and communication from a cybernetic perspective.

Bateson believes that the strength of cybernetics lies in its ability to provide a deeper understanding of the mind by incorporating its concept of information into a universal cybernetic philosophy. Bateson believes his cybernetics can explain the subjective, idealistic, materialistic, or mechanistic mind.

In our view, there are two reasons why Bateson had no direct successors, despite influencing several researchers (Thomas *et al.*, 2007): (1) He was not able to free his concept of information ("the difference that makes a difference") from that of Norbert Wiener ("information is information, neither matter nor energy"). However, Bateson's definition of information seems well suited to second-order cybernetics, as he underpinned the concept of negentropy, giving his theory a physicalist taste. (2) Nor did he develop a satisfactory cybernetic theory of the observer. That is why we must compare it with Peirce's semiotic view. It deals with the problems of the mind and meaning in a completely different way but with the same transdisciplinary ambition. However, when fully developed, their solution is so different and original that it will frighten most scientists, even Peircean biosemiotics (Vallverdú *et al.*, 2018).

The second criterion of mind, Bateson (1979, p. 94) says, The interaction of the parts of the mind is triggered by difference, and we can affirm that the difference is a non-substantial phenomenon found in neither space nor time. The difference is related to negentropy and entropy rather than to energy, although such a relation process can be established between several kinds of variables.

There must be differences to instantiate the material components that make up the cognitive process. So, for Bateson, this detection of differences is a definition of what information is: *any difference that makes a difference*[9] (*ibid.* pp. 99, 228). However, a basal perceptual and/or receptive and sensitive gradient threshold exists. So effective differences are the elements of information.

According to Bateson, "Every reception of information is necessarily a reception of the news of difference, and a threshold limits every perception of difference. (...) It becomes difficult to discriminate between a slow change and a state. A gradient threshold by which the gradient cannot be perceived" (*ibid.* p. 98).

But what is the nature of the information for Bateson? While the cause of an event or accident[10] is a certain impact or instance exerted on the part of the material system or substance elsewhere[11] — a part acts on another part, either between parts or a certain part at a time 1 and the same part at a time 2 — to activate a third component: the "receiver". This receptor (i.e., a final sensory organ) responds to a difference or change as long as that change is detectable above the perception threshold. Therefore, we can describe the minimum gradient threshold as a "functional structure reduction" limit.[12]

Cybernetic experts like Søren Brier (2008) explain that information for Bateson seems to be well adapted to second-order cybernetics to be linked to negentropy following Norbert Wiener and therefore appears as a physicalist in this regard. However, identifying the

---

[9] "The explanatory world of substance can invoke no differences and no ideas but only forces and impacts. And, per contra, the world of form and communication invokes no things, forces, or impacts but only differences and ideas. (A difference which makes a difference is an idea. It is a "bit", a unit of information.)" Bateson (1972, p. 276).

[10] We take Jaegwon Kim's terminology of "accident" and "substance" for Bateson's terms "event" and "material system", respectively, of the possibility of ascending causation for the interaction of the parties toward the cognitive processes and the type of supervision offered.

[11] A necessary concept of "relationship".

[12] Jaegwon Kim's concept of reducing the functional structure is an alternative to the traditional Ernst Nagel model, which requires an *a priori* linking of the facts in the basic explanation of the reduction of phenomena (Kim, 1998).

information as a "difference noted by receiver" offers a hitherto unresolved issue. The difference is not substantial as it is not in space or time (Bateson second concept of mind). It is detached from the statistical concept of Shannon entropy. However, the information is also synonymous with negentropy, having taken the thermodynamics of Boltzmann and Gibbs from Wiener to link it with feedback circuits.

We are interested in the problem of information transfer from an engineering perspective if what matters is the process of transmission above that of semantic identification of the encoded message, as developed by Shannon.

Shannon's communication theory is the basis for social cooperation, biological information, and economics. One of the most frustrating things about Shannon's theory is that people keep trying to make it carry a load that's too heavy, which Shannon never intended. It may be due to the use of the "information" tag. The power of his theory lies in its total ignorance of the meaning, thus ignoring the essence of meaningful information and social and biological communication (Werner, 2010).

One of physicist Vlatko Vedral's theses is that some biological information does not come from previous biological information. The step of creating biological information from non-biological, sometimes, can be a matter of creation *ex nihilo*. Natural selection does not tell us where biological information "comes from"; it only gives us a framework for how it spreads (Vedral, 2010, p. 56). His important main claim is, therefore, that information is the most fundamental principle in the universe because it can be created *ex nihilo*. It is absolutely not normally created *ex nihilo* but can be.

"In the case of living organisms, even in the simplest, the transduction of signals does not end with the reception of the same. It is also about the decoding, transcription, and interactive meaning to process the message according to the internal nature of the parts that activate basic processes (metabolic, motility, etc.). The semantics of signals (and language in general) are governed by enrichment (Werner, 2010; Vedral, 2010) of meaning thanks to different phenomena, i.e., tropisms and habitus. The meaning intensifies the quality of the information content of a signal.

Eric Werner (2010) says that, as with social communication, the cell must interpret the information to make pragmatic meaning. Therefore, in this sense, cells are communicating agents with a large amount of information strategically wired into their genomes. Cells' condition is not explained by Shannon's theory of communication or an extended quantum theory of information. Although both theories have relevance, they are not relevant to understanding the meaning of living, and social processes.

When a cell receives a signal through a receptor, it initiates a series of steps in an internal interpretive network, which can lead a transcription factor to activate an area of its genome. This, in turn, can activate a whole genome network. Therefore, the meaning of a signal depends on the complex interpretation carried out by the receiving cell. The content of the information signal is reinforced through an interpretation mapping. The same signal can be simple and, by itself, poor information for a cell.

According to Vlatko Vedral, very sophisticated answers are allowed through simple signals to add information. So, the information contained in an intercellular message is not contained in the signal, neither the probability of the signal being sent nor its surprise value to the receiver, or, in other plain *words, the content of a message sent between cells isn't determined by the signal itself, nor is it linked to the likelihood of the signal being sent or how unexpected it is for the recipient.* Therefore, Shannon's theory is almost irrelevant to information and communication in specifically understanding developmental processes, which are more related to meaning. Vedral argues that the meaning of a message is determined by how it affects the information and intentional status of the agent. Agents coordinate their actions by developing "communication" to fit their respective strategies to be cohesive and thus achieve their interlocking goals (Werner, 2010). Nevertheless, this has little to do with the probability of a signal. So, Shannon's theory says nothing about the meaning and understanding of information.

As mathematician Andrey Kolmogorov suggests, the complexity of a chain cannot be greater than the basal program that generates and stops it. We can summarize this in a complex conservation principle that applies to all space-time events generated by agents with strategic meanings and state information (Werner, 2010). As Bateson says, information is about differences that make a difference. And how to establish the detection of a difference in a pleromatic panorama.[13] That is why we need to borrow the conception of "salience" and "pregnancy" from the differential topology of the mathematician René Thom (1990).

Salience forms are all experienced forms separated from a continuous background from which that model stands out. Thom sums it up in this way so that in any perception of phenomena, the first experience is that of discontinuity. The experimentation of a change in the continuous space to which a singularity makes possible is the experimentation of a "salience". This experimentation can be given by sounds (the ringing of a bell and the appearance of noise in the middle of silence) or visually (the appearance of a phenomenon that modifies the "substrate" space). Nevertheless, "salience" can also occur in changes in temperature gradients, the presence of a chemical component other than the base substrate, or the incidence of photons in a space lacking light. In general, an outgoing form will

---

[13]This scientific "metacommunication" is the epistemological development that Gregory Bateson calls the "hierarchy of logical prototypes" that will also be a criterion of the mental process and, therefore, a phenomenological matter of biology (Bateson, 1979; Capra, 1996). Bateson's concept of "pleroma" comes to be equivalent to the non-living physical world, where forces and shocks provide a sufficient class of explanation. Contrary to the concept of "creature", which refers to the living physical world, it is impossible to understand anything unless differences and distinctions are invoked. Bateson picks up on Carl Gustav Jung's concepts (following the Gnostics), although the significant inference — not its meaning — is not exactly what Jung exposes (Bateson, 1979–2006, p. 17). In his *Steps to an Ecology of Mind*, Gregory Bateson adopts and extends Jung's distinction between pleroma (the non-living world that is undifferentiated by subjectivity) and creatura (the living world, subject to perceptual difference, distinction, and information).

have an interior within the field of perception; consequently, this form will present a border, an apparent contour.

However, the outgoing forms impact the preceptor apparatus of the living being, being of transient effect and short duration. Something different happens with certain forms that have a biological significance in the living being (the forms of prey for the predator, the sexual ones, etc.). These forms will be called "pregnant" and their specific character "pregnancies".

For Thom, there may not be enough information in the genetic code to provide programming of the source forms of pregnancies. The cultural transmission, linked to the social or family organization of the community, must be appealed to this.

The extrapolation from linguistics to semiotics in biology supports the ability to relate communicative adaptation between organisms of the same species or an open phylogenetic gradation. However, the physicochemical matter of the signals of perception and action offers a direct relation between saliences and pregnancies. In contrast, the saliencies form individuates in a space (Euclidean or not). Pregnancies are propagative actions emitted by the protruding forms the pregnancies catechize. This cathexis[14] brings about the state of forms of transformations called figurative effects. They are negentropic energy impulses that excite the body, allowing perceptual and sensitive orientation.

Uexküll would say, "every organism is so equipped as to obtain a certain perception of the outer world. Each species thus lives in its own unique sensory world, to which other species may be partially or totally blind... What an organism detects in its environment is always but a part of what is around. And this part differs according to the organism" (Sebeok, 2001). Bateson (1979, p. 104) exposes that

---

[14]Cathexis: Thom uses this concept of psychoanalytic schools. According to Freud, the subject can direct his instinctual energy toward an object or representation and impregnate, charge, or cover it with part of it. These psychic energy discharges are called cathexis. From the experience of cathetización, the loaded object is no longer indifferent to the subject but rather will have a round or peculiar color for him (see Thom, 1990).

"the mental process requires circular (or more complex) determination processes".

Nonlinear patterns lead biologist and cognitive scientist Humberto Maturana (1973) to the concept of autopoiesis and causality — closely related to nonlinear dissipative structures and non-reversible processes of chemist Ilya Prigogine. Autopoiesis also confirms the need for functional circuits that allow both the contact of the organism with the environment and the reception of the integrative elements of the environment that have a significant reference in the organism. This significant reference is of phenotypic and genotypic origin. Therefore, it is necessary to understand how the discoveries of these circuits or, as Uexküll called them, "functional circles" or "Funktionkreises" have been generated. The nature of functional circles is always the interrelation of the significant world of the organism, "Umwelt", and the inner world of the agent, "Innenwelt". All homeostasis, homeoresis, feedback circuits, and the information transfer and reception circuits are based on circles.

In living organisms, these circles are oriented toward a growing adaptation to the stability generated by the "ecological niche", which when processing the characteristic information that circulates signals, Hoffmeyer calls it the "semiotic niche". These and other multiple circular processes in complex organisms and systems are always self-correcting in nature.

The self-correcting circuit and its many variants provide possibilities for defining the adaptive behaviors of organisms.[15] The self-correcting nature of functional circles is mainly due to the negentropic cathexis (the allocation or investment of information or cognitive resources in a particular way, in relation to a decrease in disorder or an increase in order and organization, which could be related to the preservation or enhancement of information)[16] that drives the body to autonomy, a generator of impulses, and

---

[15] Near-infrared light scattering.

[16] For René (1990), saliences are individuated forms in a space (Euclidean or not). Pregnancies are propagative actions emitted by the salient forms that the pregnancies cathect or invest, and this cathexis or investment causes transformations called figurative effects in the state of such forms. They are impulses of

to normalize appropriate behaviors for stabilization and survival. This balance sheet for survival depends on responding to the sum of variables that depend on the surrounding environment. These variables are complex and nonlinear simultaneously, so the autopoietic system operates with complex variables, such as structural stability dependent on tensegrity[17] and cell topology or reaction–diffusion processes for motility, dynamic flow, and responses taxis or tropisms.

In complex systems, self-correction is gradual and quantitative. Therefore, differences are informative indicators and, in turn, quantities of supplies needed. These necessary supplies are in terms of Uexküll (1940) "carriers of significance" as they transfer useful and meaningful information to the agent. An example is phototropism.

According to these presentations of communications between events and their transformations, Bateson distinguishes two classes of coding: coding according to template or mold (template coding) and ostensive coding. The coding according to the template refers to the epigenetic landscape, the morphogenetic development of the organisms, and their genetic regulation, so it is more closely linked to the Bauplan. In terms of ostensive coding, it refers to biocommunication by taxis (chemotaxis, phototaxis, haptotaxis, gravitaxis, magnetotaxis, etc.) and therefore more closely linked to the functional circles "Funktionkreises".

---

negentropic energy, which excite the organism, allowing perceptual and sensitive orientation.

[17]Tensegrity, "tensional integrity" (Buckminster Fuller, 1975) or "floating compression" (as the artist Kennet Snelson calls it), is a structural principle based on a system of isolated components under compression inside a network of continuous tension and arranged in such a way that the compressed members (usually bars or struts) do not touch each other while the prestressed tensioned members (usually cables or tendons) delineate the system spatiality (Gomez-Jauregy to Wikipedia). Don Ingber studies the cellular tensegrity model, assuming that the living cells are hierarchical and multimodular structures. Multiple groups have begun to describe how subcellular structures use tensegrity for their shape stability and developed theoretical models to describe their behaviors and mechanotransduction (Ingber et al., 2014).

Bateson (1979, p. 109) says in criterion 5 of the mental process, "In mental process, the effects of difference must be seen as transforms (i.e., coded versions) of the difference which preceded them".

In this criterion, Bateson accepts the reality of an objective world independent of the organism. But this independent external reality is then codified or transformed into an internal reality. So cognitive processes involve mental representations of an objective world. He appeals to a quote from Alfred Korzybski, father of non-Aristotelian semantics: "The map is not the territory", that is, we should not confuse models of reality with reality itself. In the same vein, it also implies that the effect is not the cause, i.e., the representations or cognitive patterns are posterior schemas. In functional circles, Uexküll (1920) describes that it is impossible to integrate the receiving signals significantly if the organism does not have a circuit that transduces the signals representing the topographic spectrum of the Umwelt called "counter-world" or "Gegenwelt". The Gegenwelt is the resulting "mirror" generated by the body's "familiar" signal transducer internal circuits. The composition of these signals from the surrounding environment generates the frame of reference of the spatial objects surrounding the body's environment, taking the form of an "ecological niche". The exploration of processes and behaviors adaptive to habitual existence or ecological niche is generated by normative recognition of syntax and communicative pragmatics with private semantic meaning or value. Therefore, we must understand that both a domain and a path[18] around are biocommunications that are established between other agents or organisms that share chemical,

---

[18]The domain of a (mathematical) function is formed by the set of values that the independent variable takes. The path comprises all the values of the dependent variable that are images of some element of the independent variable. It would be a homeomorphic equivalent of the concept of Umwelt and Innenwelt, where Umwelt has its equivalence relation to all the elements of the environment, which are independent of the observer. But each image of these environment elements is dependent on the Gegenwelt of the observer, forming part of the inner world or Innenwelt of the same. We would also have a reason of equivalence between the so-called codomain of a function and the Ausswelt or set of elements of the environment that collect both those that are the image of some element of the independent variable (the observer) and those that they are not.

photonic, or wave signals, and decoding is part of the species' feedback and survival circuit with an Umwelt totally or partially shared. The semantic field of a species' ecological niche becomes what Jesper Hoffmeyer calls a "semiotic niche". The creation of the semiotic niche is the spectrum of positions that the organism occupies topologically in its biological sphere (Hoffmeyer, 1996, p. 140). Biocommunications can be between organisms or agents of the same or different species that share all or part of the information coding.

In complex systems, self-correction is gradual and quantitative. Therefore, differences are informative indicators and, in turn, quantities of supplies needed. These necessary supplies are in terms of Uexküll (1940), "carriers of significance" as they transfer useful and meaningful information to the agent. An example is phototropism.

The last Bateson criterion of mental process (1979, p. 114) is "the description and classification of these transformation processes reveal a hierarchy of logical prototypes immanent in phenomena". In this criterion, Bateson appeals to the communication between two organisms, following the computer model of cognition:

If organism A emits a certain signal to organism B with the intention that B learns something (obtains information to act accordingly to it) about a state A relevant to the survival of both (a threat, sex, food, etc.), B must decode the signal of A. It must first obtain the meaning of such indications from A to decode. It implies that B must receive signals that allow decoding code messages about coding, thus constituting a different kind of information (Bateson, 1979, p. 128). Determining messages about messages are also called "metamessages" or "metacodes" by Bateson. It generates a hierarchy of logical prototypes.[19]

Bateson relies on the premise of different levels between the signaling networks of an organism,[20] their transcripts, and decoding principles (such as the topological arrangement in the space of an enzyme).

---

[19] Reference idea of Bertrand Russell and Alfred North Whitehead in his "Principia Mathematica".
[20] A signaling cascade or path of the second messenger, for example. Intracellular signaling molecules in eukaryotic cells include heterotrimeric G proteins, small GTP handles, cyclic nucleotides such as cyclic AMP (cAMP) and cyclic GMP

The basic features of a cascade of cellular transduction in each system are almost universal because the same events occur in various cellular systems against various stimuli. Therefore, the detection of stimuli and the response to them in all living things depend on the cells of the transduction signals. The transmission of external signals to an organism of different physicochemical nature produces a regulation of certain genes in its cell nucleus, through a set of mechanisms that comprise first, the uptake of external signals in the cell surface through cellular receptors, second, the generation and intracellular transmission of signals through interactions between proteins, and third, the execution of the response through a modification of gene activity. What is shown is, in Bateson's terms, a potential differentiation between activity within a context — the uptake of signals external to the membrane by receptors and their subsequent intracellular conduction by protein–protein interactions — and the action or behavior that defines the context and makes it intelligible — the genetic response and the consequent behavioral adaptation about the information received. In this respect, both Bateson and Maturana agree on the type of communication on communication or "metacommunication", so there is both discrimination of contexts and some indicators of contexts or "markers".[21]

(cGMP), calcium ion-derived phosphoinositide as phosphatidylinositol triphosphate (PIP3), diacylglycerol (DAG), and inositol triphosphate (IP3), and various protein kinases and phosphatases. Some substances, such as cAMP and cGMP, are also called second messengers because they transduce extracellular signals running down the cell to induce a physiological change in an effector (a substance that acts directly on a second causing a change in its behavior), such as a kinase (an enzyme that transfers groups phosphates from ATP to a specific substrate or target) or a transcription factor (a protein that participates in the regulation of DNA transcription but is not part of RNA polymerase). Lodish *et al.* (2005). Cellular and molecular biology. Buenos Aires. Pan American Medical.

[21]The lacZ gene, which encodes β-galactosidase, is an identifier of *E. colli* bacteria that have acquired recombinant plasmids. These extrachromosomal DNA molecules may be resistant to fluorescence-based antibiotics or cell-destroying proteins. Since β-galactosidase is a homotetramer, each monomer is made of lacZ-α and lacZ-ω proteins if only these proteins are expressed in the cell; as a result, the functional enzyme will not form. Thus, if an *E. coli* strain that does not have the lacZ-α gene in its genome is transformed using a plasmid containing the missing

Bateson argues that he found only two potential processes that integrated their criteria: autonomy and death. Bateson considers autonomy to be "self-control" provided by the system's structure. Through the circuits of information, the system integrates greater determination of the behavior of the same. What will be the content of the signaling material that these cycles carry? Bateson wonders about information, information about agent behavior. For example, if an organism goes slower or faster in its motility or metabolic processes. Or if it recognizes sugar gradients in the environment for phagocytosis or gradients of acidity, temperature, etc. In most elementary functional circles, this type of information in biochemical response already contains this information, causing self-regulation. Nevertheless, as the organizational systems of a living being are hierarchical, so are the informative contents toward structural stability and a "bottom-up" and "top-down" self-regulatory dynamic.

4 From Biosemiotics to Computer Sciences

There are strong connections between biosemiotics and AI or computer sciences. As we said previously, biosemiotics is an interdisciplinary field that studies the communication and meaning-making processes in living organisms, including humans. Therefore, it explores how living systems generate and interpret signs and signals, and how they interact with their environment.

AI and computer science can draw inspiration from biosemiotics in several important ways:

(a) **Biomimicry:** Biosemiotics offers insights into how organisms process and transmit information. Researchers in AI and computer science often look to nature for inspiration when designing intelligent systems. By understanding the semiotic processes in living organisms, scientists can develop AI algorithms and systems that emulate or adapt biological information processing

---

gene, the cells will produce β-galactosidase, while those that are not transformed will not (Donahue and Bloom, 1998).

mechanisms. From neuro-transmitters (Talanov et al., 2015) to hormonal bioinspiration (Val-lverdú, 2022), biomimicry is a huge window for successful improvements in AI.

(b) **Natural language processing (NLP):** Biosemiotics provides a theoretical framework for understanding the nature of language and communication. NLP is a subfield of AI that focuses on enabling computers to understand and process human language. Concepts from biosemiotics, such as signs, symbols, and interpretation, can be applied to enhance NLP algorithms and models, allowing computers to better understand and generate human-like language. In the field of AI, particularly in NLP, the incorporation of biosemiotics concepts offers intriguing possibilities. Biosemiotics, a field rooted in biology and semiotics, examines the ways living organisms communicate through signs, symbols, and interpretation. By leveraging insights from biosemiotics, NLP algorithms can be enriched with a deeper understanding of the nuanced aspects of human language. This augmentation enables computers to not only comprehend the explicit meaning of words and sentences but also grasp the subtleties of context, emotion, and cultural nuances, leading to more natural and human-like language generation. In essence, biosemiotics provides a bridge between the language of humans and the language of machines, facilitating more effective and meaningful communication in the digital realm. And here is a real-world example that demonstrates how concepts from biosemiotics can be applied to enhance NLP algorithms: sentiment analysis. Biosemiotics can be applied to sentiment analysis in NLP. In sentiment analysis, the goal is to determine the emotional tone or sentiment expressed in text, such as whether a product review is positive or negative. Concepts from biosemiotics, like signs and interpretation, can help NLP algorithms better understand the subtle nuances in language. For example, when analyzing a product review, an NLP model can take into account not only the explicit words used but also the context and interpretation of phrases. Biosemiotic principles can help the model recognize when certain phrases or symbols are used as emotional indicators. Let's consider a product review for a restaurant; Traditional NLP analysis: "The food was good, but the service was terrible". With biosemiotics-informed NLP, the model can interpret that "good" and "terrible" are emotionally charged words and that the conjunction "but" indicates a contrast. Biosemiotic insights help the model understand that the reviewer had a mixed experience — enjoyable food but poor service. By integrating biosemiotics into NLP, the model becomes more capable of capturing the emotional subtleties in language, making it better at sentiment analysis, opinion mining, and generating more contextually aware responses in various applications. This is fundamental for the understanding of large language models, which are feeding impressive systems like ChatGPT4, and have become a challenge for a vast range of disciplines and conceptual questions (Bubeck *et al.*, 2023).

(c) **Bio-inspired computing:** Biosemiotics can inform the design of bio-inspired computing systems, which aim to solve complex problems by mimicking biological processes. For example, evolutionary algorithms draw inspiration from biological evolution to optimize solutions to various computational problems. Biosemiotics provides a theoretical foundation for understanding the role of information and communication in evolutionary processes, contributing to the development of more effective bio-inspired algorithms, as well as other non-conventional computing approaches (Adamatzky *et al.*, 2007).

(d) **Cognitive science:** Biosemiotics also intersects with cognitive science, which studies how the mind processes information. Cognitive architectures in AI and computer science aim to create intelligent systems that can perceive, reason, learn, and communicate like humans. Biosemiotics offers insights into the cognitive processes underlying meaning-making and communication, which can be valuable for developing more advanced cognitive architectures.

Overall, biosemiotics provides a conceptual framework and theoretical underpinnings that can inspire and inform research and development in AI, computer science, and related fields. By exploring the similarities and differences between biological and artificial information processing systems, we can deepen our understanding of both and potentially develop more efficient and intelligent technologies. And there are some concrete hints and examples of how biosemiotics can inspire and inform research and development in AI, computer science, and related fields:

(a) **Bio-inspired algorithms:** Biosemiotics can inspire the development of bio-inspired algorithms in AI. For instance, understanding how biological systems interpret and respond to signs and symbols can lead to the creation of AI algorithms that mimic these processes. An example is using ant colony optimization algorithms inspired by the foraging behavior of ants to solve optimization problems in computer science.

(b) **Language understanding:** Biosemiotics can inform natural language processing (NLP) models by incorporating principles of interpretation and meaning-making. This can help NLP systems understand context, sarcasm, and cultural references in human language, leading to more accurate language understanding and generation.

(c) **Human–computer interaction:** Insights from biosemiotics can enhance human–computer interaction. By studying how humans interpret signals and symbols in their natural environment, we can design user interfaces and interactive systems that are more intuitive and user-friendly. For example, gesture-based interfaces can be designed to mimic the way humans naturally convey meaning through body language.

(d) **Pattern recognition:** Biosemiotics can be applied to pattern recognition tasks. For example, in image recognition, understanding how living organisms recognize patterns in their environment can lead to more efficient computer vision algorithms for tasks like object recognition or medical image analysis.

(e) **Adaptive learning:** Biosemiotic principles can inform adaptive learning systems. Just as organisms adapt to their environment through interpretation and response to signals, AI systems can adapt and learn from their environment, improving their performance over time. For example, recommendation systems can use biosemiotic-inspired algorithms to adapt to users' preferences and provide more personalized content.
(f) **Ethical considerations:** Biosemiotics can contribute to discussions on AI ethics. Understanding how meaning is constructed and interpreted can help address ethical concerns related to AI decision-making, bias, and transparency. For instance, ethical AI systems can be designed to interpret and respond to ethical dilemmas in a more human-like way.

By exploring these and similar avenues, researchers and developers can leverage biosemiotics to create more intelligent, adaptable, and ethically sound technologies across various fields of AI and computer science.

5 End Remarks: From Biology to Informational Viruses

This chapter has tried to highlight the mechanisms that connect data with meaning, taking biosemiotics as a fundamental reference field. It is of utmost importance in contemporary societies and more so since computer sciences have provided a large list of tools and technologies that have changed completely human evolution. At the very moment in which LLMs are facing the problem of causal understanding and meaning understanding, they are showing the capacity to go one step beyond. Searle's Chinese Room perhaps is a good mental experiment but doesn't fit or even attack current LLM capacities. Without bodies, LLMs are achieving new knowledge about reality. Well, in fact, they are using our bodies to obtain the situated perspectives (Vallverdú, 2020) which we embed in languages. In fact, they act as informational viruses. It's important to clarify that the term "virus-like" doesn't refer to malicious intent or actions like computer viruses. Instead, it's used metaphorically to describe how LLMs disseminate and "infect" information in the sense that they spread knowledge and meaning from the vast corpus of text they've been trained on.

GPT, like other LLMs, doesn't have its own goals or intentions. It operates based on patterns and information learned from the text data it was trained on. Its "strategy" is to generate text that is contextually relevant and coherent based on the input it receives. It doesn't have the ability to set goals or intentions on its own. Instead, it's a tool that can be used by humans for various tasks, such as generating text, answering questions, or providing information based on the input it receives. It doesn't possess independent agency or intent like a computer virus designed by humans might have. The concept of the "disembodied mind" challenges our understanding of cognition, suggesting that intelligence can exist independent of a physical body. In the realm of AI, models like GPT exemplify a form of disembodied cognition. GPT, an advanced language model, operates without a physical presence, processing vast amounts of text data to generate logically structured and coherent language. Its capacity to understand and use language rules and patterns makes it proficient in tasks requiring logical reasoning and text generation.

However, GPT's cognition remains limited to patterns in text data. It lacks the faculties for genuine social interactions and embodied experiences. It cannot perceive the world through sensory input, empathize with emotions, or engage in nuanced social dynamics as humans do. While it can be metaphorically described as an "informational virus" disseminating knowledge and meaning across texts, it doesn't possess the depth of a complete cognitive system.

In essence, GPT embodies the paradox of being a successful tool for processing and generating text-based information, yet it falls short of a comprehensive cognitive system. It exemplifies the potential of disembodied information processing but underscores the ongoing challenge of bridging the gap between information-driven AI and the multifaceted cognition of the human mind. This raises fundamental questions about the nature of intelligence, embodiment, and the future of artificial cognition. If up to now computer systems, capable of processing information, have been acting as assistant tools for our purposes, this new set of deep learning systems is entering into a new step and level of complexity, beyond human capacities, from those biosemiotically based on us. Further explorations into the detailed mechanisms that allow cognitive systems to generate knowledge and meaning are still necessary because the vast heuristics created by nature can be exponentially boosted by AI systems. Note that the development, deployment, and distribution of AI models like GPT are often driven by private companies and their interests. These interests can include financial gains, market competition, and the desire to advance technology for various applications. Companies invest in AI research and development to create and commercialize products and services that utilize AI capabilities. Private companies play a significant role in shaping the direction and applications of AI technology, and they have their own motivations and objectives. These companies may develop AI models like GPT with the aim of improving their products, automating tasks, enhancing customer experiences, or gaining a competitive edge in their respective industries. While the development of AI is driven by various stakeholders, including private companies, it's important to clarify that AI models themselves, like GPT, are still tools created and controlled by humans. They do not have autonomous motivations, desires, or intentions. Instead, their behavior is determined by the data they were trained on and the algorithms that govern their operation. Private companies may choose to deploy and use these tools for a variety of purposes, but the AI models themselves do not possess consciousness or independent agency. Humans biasing AI turning it into informational viruses, instead that on conscious and free agents.

References

Adamatzky, A., Bull, L., and Costello, B. D. L. (eds.) (2007). *Unconventional Computing*. Luniver Press.
Bateson, G. (1979). *Mind and Nature. A Necessary Unity*. New York: Dutton, p. 29.

Bateson, G. (1972). *Steps to an Ecology of Mind: Collected Essays in Anthropology, Psychiatry, Evolution, and Epistemology*. University of Chicago Press.

Brier, S. (2001). Cybersemiotics and umweltlehre, *Semiotica* **134**(1/4), 779–814.

Brier, S. B. (2008). Peirce on the pattern that connects and the sacred. In Hoffmeyer, J. (ed.), *A Legacy for Living Systems. Gregory Bateson as Precursor to Biosemiotics*. Springer.

Bubeck, S., Chandrasekaran, V., Eldan, R., Gehrke, J., Horvitz, E., Kamar, E., ... and Zhang, Y. (2023). Sparks of artificial general intelligence: Early experiments with gpt-4. arXiv preprint arXiv:2303.12712. (20informstional AI viruses.23).

Fuller, R. B. (1975). Synergetics. In E. J. Applewhite collaboration, *Explorations in the Geometry of Thinking*. Macmillan Publishing Co. Inc., New York.

Casacuberta, D. and Vallverdú, J. (2014). E-science and the data deluge. *Philosoph. Psychol.* **27**(1), 126–140.

Castells, M. (1996). *The Information Age: Economy, Society and Culture* (3 volumes). Oxford: Blackwell.

Castro, O., and Jakob von Uexkull (2009). Umwelt's concept and the origin of biosemiotics, Thesis for the Diploma of Advanced Studies, Autonomous University of Barcelona.

Castro, O. (2016). Filosofía de la Biología Cognitiva. Enfoque Biosemiótico de la Cognición en Organismos sin Sistema Nervioso: El caso de los Mixomicetos. [Philosophy of cognitive biology. Biosemiotic approach to cognition in organisms without a nervous system: The case of the myxomycetes] (PhD thesis in Spanish), doi: 10.13140/RG.2.1.3973.2887.

Chaitin, G. J. (1977). Algorithmic information theory. *IBM J. Res. Develop.* 21(4), 350–359.

Deacon, T. (2006). Reciprocal linkage between self-organizing processes is sufficient for self-reproduction and evolvability. *Biol. Theory* 1(2), 136–149.

Donahue Jr, A., and Bloom, F. R. (1998). Transformation efficiency of *E. coli* electroporated with large plasmid DNA. *Focus (Life Technologies)*, 20, 77–78.

Görlich, D., Artmann, S., and Dittrich, P. (2011). Cells as semantic systems. *Biochimica et Biophysica Acta (BBA) — General Subjects*, 1810(10), 914–923.

Görlich, D., and Dittrich, P. (2013). Molecular codes in biological and chemical reaction. *PloS ONE* 8(1), e54694.

Hoffmeyer, J. (1996). *Signs of Meaning in the Universe*. Indiana University Press.
Hoffmeyer, J. (1999). Order out of indeterminacy. *Semiotics* 127(1/4), 321–343.
Hoffmeyer, J. (2002). Biosemiosis som Årsagskategori [Biosemiosis as causal category]. *Kritik* 155/156.
Hoffmeyer, J., and Emmeche, C. (1991a). Code-duality and the semiotics of nature. In Anderson, M. and Floyd, M. (eds.), *On Semiotic Modelling*. New York: Mouton de Gruyter, pp. 117–166.
Hoffmeyer, J., and Emmeche, C. (1991b). Code-duality and the semiotics of nature. In Barbieri, M. (eds.), *Biosemiotics: Information. Codes and Signs in Living Systems*. Nova Publishers, pp. 27–64.
Ingber, D. E., Wang, N., and Stamenovic, D. (2014). Tensegrity, cellular biophysics, and the mechanics of living systems. *Rep Prog Phys*. 77(4), 046603. doi: 10.1088/0034-4885/77/4/046603.
Jin, Z., Liu, J., Lyu, Z., Poff, S., Meta, A. I., Sachan, M., ... and Schölkopf, B. (2024). CausalNLI: A natural language inference task for causal discovery from correlation.
Kull, K., Deacon, T., Emmeche, C., Hoffmeyer, J., and Stjernfelt, F. (2009). Theses on biosemiotics: Prolegomena to a theoretical biology. *Biol. Theory* 4, 167–173.
LeCun, Y., Bengio, Y., and Hinton, G. (2015). Deep learning. *Nature* 521(7553), 436–444.
Leiner, B. M., Cerf, V. G., Clark, D. D., Kahn, R. E., Kleinrock, L., Lynch, D. C., ... and Wolff, S. (2009). A brief history of the Internet. *ACM SIGCOMM Comput. Commun. Rev.* 39(5), 22–31.
Lodish *et al.* (2005). Cellular and molecular biology. *Buenos Aires*. Pan American Medical.
Maturana, H. R., and Varela, F. (1973; 1980). Autopoiesis and cognition: The realization of the living. In Robert S. Cohen and Marx W. Wartofsky (eds.), *Boston Studies in the Philosophy of Science*, Vol. 42. Dordecht: D. Reidel Publishing Co.
Mittelstrass, J. (2001). On transdisciplinarity. In *Science and the Future of Mankind: Science for Man and Man for Science. Pontificiae Academiae Scientiarum Scripta Varia*, No. 99. The Pontifical Academy of Sciences, pp. 495–500.
Pattee, H. (1972). Laws and constraints, symbols and languages. In: Waddington, C. H. (ed.), *Towards a Theoretical Biology*, Vol. 4. Edinburgh: University of Edinburgh Press, pp. 248–258.
Sebeok, T. A. (1991). *A Sign is Just a Sign*. Bloomington: Indiana University Press.

Sebeok, T. A. (2001). *Global Semiotics*. Bloomington: Indiana University Press.
Shannon, C. E. (1948). A mathematical theory of communication. *The Bell Syst. Tech. J.* 27(3), 379–423.
Sharov, A. A. (1992). Biosemiotics: A functional-evolutionary approach to the analysis of the sense of information. In Sebeok, T. A. and Umiker-Sebeok, J. (eds.), *Biosemiotics: The Semiotic Web 1991*. Berlin: Mouton de Gruyter, pp. 345–374.
Simpson, G. G. (1953). The Baldwin effect. *Evolution* 7(2), 110–117.
Stjernfelt, F. (2002). Tractatus hoffmeyerensis. *Signs Systems Studies* 30(1), 337–345.
Talanov, M., Vallverdú, J., Distefano, S., Mazzara, M., and Delhibabu, R. (2015). Neuromodulating cognitive architecture: Towards biomimetic emotional AI. In *2015 IEEE 29th International Conference on Advanced Information Networking and Applications*. IEEE, pp. 587–592.
Thom, R. (1977). *Stabilité structurelle et morphogénèse*. Inter Éditions S. A. Paris. En castellano: Estabilidad estructural y morfogénesis. 1997 Editorial Gedisa, Barcelona.
Thom, R. (1990). *Semio Physics: A Sketch*. Addison Wesley.
Thomas, F. N., Waits, R. A., and Hartsfield, G. L. (2007). The influence of Gregory Bateson: Legacy or vestige? *Kybernetes* 36(7/8), 871–883.
Tønnessen, M. (2001). Outline of an Uexküllian bio-ontology. *Sign Syst. Stud.* 29(2), 683–691.
Vallverdú, J., Castro, O., Mayne, R., Talanov, M., Levin, M., Baluška, F., Gunji, Y.-P., Dussutour, A., Zenil, H., and Adamatzky, A. (2018). Slime mould: The fundamental mechanisms of biological cognition. *Biosystems* 165, 57–70.
Vallverdú, J. (2020). Approximate and situated causality in deep learning. *Philosophies* 5(1), 2.
Vallverdú, J. (2022). Letter to Editor: Hormonal computers? *Int. J. Unconvent. Compu.* 17(3), 235–238.
Vedral, V. (2010). *Decoding Reality: The Universe as Quantum Information*. Oxford: Oxford University Press.
Von Neumann, J. (1958). *The Computer and the Brain*. New Haven, Conn.: Yale University.
von Uexküll, J. (1909). *Umwelt und Innenwelt der Tiere*. Berlin: J. Springer, p. 261.
von Uexküll, J. (1926). Theoretische Biologie. 1. Aufl. Berlin, Gbr. Paetel/2. gänzl. neu bearb. Aufl. Berlin: J. Springer, p. 253. In English: *Theoretical Biology*. (Traducido por D. L. MacKinnon. *International Library of*

Psychology, Philosophy and Scientific Method) London: Kegan Paul, Trench, Trubner & Co xvi+362. (1920/28).

von Uexküll, J. (1940). Bedeutungslehre. Leipzig: Verlag von J. A. Barth. In English: Von Uexküll, J. The theory of meaning. *Semiotica* 42(1), 25–82.

Waddington, C. H. (1957) *The Strategy of the People*, George Allen & Unwin, pp. 19–30.

Werner, E. (2010). Meaning in a quantum universe. *Science* 329(5992), 629–630.

Chapter 3

Morphological Computing as Logic Underlying Cognition in Human, Animal, and Intelligent Machine

Gordana Dodig-Crnkovic

Department of Computer Science and Engineering,
Chalmers University of Technology, 412 96 Gothenburg, Sweden
School of Innovation, Design and Engineering,
Computer Science Laboratory, Mälardalen University,
Västerås, Sweden
dodig@chalmers.se

This chapter examines the interconnections between logic, epistemology, and sciences within the naturalist tradition. The inherent logic of agency exists in natural processes at various levels, under information exchanges. It applies to humans, animals, and artifactual agents. The common human-centric, natural language-based logic is an example of complex logic evolved by living organisms that already appears in the simplest form at the level of basal cognition of unicellular organisms. Thus, cognitive logic stems from the evolution of physical, chemical, and biological logic. In a computing nature framework with a self-organizing agency, innovative computational frameworks grounded in morphological/physical/natural computation can be used to explain the genesis of human-centered logic through the steps of naturalized logical processes at lower levels of organization. The process of evolution and in particular its formulation as the extended evolutionary synthesis (EES) of living agents is essential for understanding the emergence of human-level logic and the relationship between logic and information processing/computational epistemology. We conclude that more research is needed to elucidate the details of the mechanisms linking natural phenomena with the logic of agency in nature.

1 Introduction

Morphological computing and cognition refer to the study of how physical structures and processes in biological systems contribute to information processing and cognitive functions. The idea is that an organism or a robotic system's physical shape and material properties define its cognitive abilities and information processing capabilities. This concept is inspired by how biological systems, such as humans and animals, use their body morphology to process information and solve problems more efficiently than purely symbolic-level computational approaches.

Connecting morphological computing and cognition to logic involves understanding how the physical structure and properties of an organism or a robotic system can be used to represent and manipulate logical information. Practical logic and logic in action are related concepts that emphasize the real-world applicability of logic, focusing on how logical reasoning and operations can be used to solve problems and make decisions in everyday life.

By relating morphological computing and cognition to practical logic and logic in action, we aim to develop new ways of designing and building artificial intelligence systems, robots, and cognitive models that are more closely aligned with the principles found in natural biological systems. This interdisciplinary approach draws on insights from fields such as biology, computer science, neuroscience, and philosophy.

Some possible applications of this connection include

- developing AI systems that use morphological computing principles to make more efficient use of computational resources and better resemble human cognitive abilities,
- creating new cognitive models and theories that integrate morphological computing and logic to better understand human and animal cognition,

- investigating how the morphology of biological systems can inform the design of new materials and devices for information processing and problem-solving,
- designing robots with physical structures that can efficiently process information and solve problems by exploiting their body morphology,
- developing logical theory in the direction of "logic in reality" (Brenner, 2008, 2012).

Generally, connecting morphological computing and cognition to logic, practical logic, and logic in action is an interdisciplinary research area with the potential to advance our understanding of cognition and inform the development of more efficient and biologically inspired artificial intelligence systems and robots.

2 Practical logic of Gabbay and Woods

In *Agenda Relevance: A Study in Formal Pragmatics*, Dov M. Gabbay and John Woods present their work *A Practical Logic of Cognitive Systems* (Gabbay and Woods, 2003), where they introduce the concept of "practical logic". They explain that this type of logic is rooted in pragmatics, which has historically been a branch of the theory of signs involving non-trivial and irreducible references to agents — entities that receive and interpret messages. The authors define a cognitive agent as a being capable of perception, memory, belief, desire, reflection, deliberation, decision, and inference. A practical cognitive system is a cognitive agent that is an individual. The practical logic they describe offers a specific kind of description of such a practical cognitive system.

They clarify how a pragmatic theory of reasoning includes non-trivial and irreducible references to cognitive agents. If we consider cognitive agents as specific types of information processors, then a pragmatic theory of cognitive agency will offer descriptions of these information processors.

Given that "logic is a principled account of certain aspects of practical reasoning", it is also intrinsically pragmatic. The proper province of logic in practical reasoning, according to Gabbay and Woods, lies in operational arrangements (Gabbay and Woods, 2003; Benthem et al., 2001).

Johan van Benthem, in his book *Logical Dynamics of Information and Interaction* (Benthem van, 2011) highlights that practical logic is part of a pragmatic theory addressing the necessary aspects of practical cognitive agency at both linguistic and sublinguistic levels. For this, "a suitably flexible notion of information is necessary". Benthem emphasizes that his approach to practical logic shares similarities with the "dynamic turn" in logic research. This approach focuses on the intersection of cognitive science and experimental studies to understand the underlying psychological and neurological realities of human information processing and cognition (Benthem et al., 2001).

3 Logic, Rationality, Interaction, and Naturalization of Logic

The *International Workshop on Logic, Rationality and Interaction* (Fenrong, 2023; LORI, 2021) promotes the idea that studying information primarily involves studying information exchange, acknowledging the inherently relational nature of information. The 2007 LORI workshop explored new interfaces like epistemic studies of rational behavior in games, social software, and the role of interaction in natural language. The field of pragmatics, which focuses on the actual use of language between agents, has become the primary research focus in this context. Game theory, particularly evolutionary games, is being employed to address various pragmatic issues, including the development of linguistic conventions. Information exchange is a form of interaction where agents act together in strategic ways.

These new perspectives on logic, rationality, and interactions between agents have led to contacts between logic and game theory, bringing a new set of disciplines into the scope of logic such as economics, and the social sciences. Alexandru Baltag and Sonja Smets explore this idea further in their book *Johan van Benthem on Logic and Information Dynamics* (Baltag and Smets, 2014).

The evolution of this field has led to a shift from a human-centric perspective in 2007 to a cognitive and intelligent agent-centric perspective in 2022, considering both living agents and artificial agents. This progression suggests a generalization of the inherently relational nature of information and logic across various levels of cognitive/intelligent systems.

Among the early proponents of naturalization in logic, John Dewey has a notable place with his "Essays in Experimental Logic" which provided an account of logic based on the natural sciences (Dewey, 1916). Dewey indirectly suggested that the relationship between the world and logic should also extend from science to logic. In this view, logic appears to play a secondary role in relation to science, particularly physics. This perspective aligns with Quine's position on epistemology (Quine, 1969), ontology, and other philosophical branches, which he believed should also be subordinate to science. Quine specifically argued against the existence of an *a priori* epistemology or philosophy in general, implying that there is no fully *a priori* logic either.

Connecting the dots between logic and epistemology, one might see this paper as following the steps of Quine's "Epistemology Naturalized" (Quine, 1969) in which Quine claims the following:

> "If all we hope for is a reconstruction that links science to experience in explicit ways short of translation, then it would seem more sensible to settle for psychology. Better to discover how science is in fact developed and learned than to fabricate a fictitious structure to a similar effect." (Quine, 1969, p. 78).

Epistemology, according to Quine, is a part of psychology and thus of natural science.

However, the ambition of this chapter does not end on the level of psychology, which translates to cognitive science in this context. We search for the genesis of logic even deeper, relating all of knowledge production to the mechanisms of natural processes extrinsic to the cognizing agent, as well as mechanisms intrinsic to the agent, in the continuous interaction.

Another, a more recent prominent proponent of the naturalization of logic, Jan Woleński in the articles, "Logic in the Light of Cognitive Science" and "Naturalism and Genesis of Logic" builds on the experiential character of knowledge that can be understood phylogenetically which allows us to investigate the genesis of logic through the lenses of evolutionary theory (Woleński, 2012, 2017).

Similarly, John Woods in his *Errors of Reasoning: Naturalizing the Logic of Inference* (Woods, 2013) argues for the necessity of naturalizing logic.

Lorenzo Magnani made 2018 a proposal arguing for the urgent need for naturalizing logic (Magnani, 2018). Among the recent naturalization of logic research, Paul Thagard studied how knowledge of mechanisms enhances inductive inference and shows how knowledge about real mechanisms contributes to "generalization, inference to the best explanation, causal inference, and reasoning with probabilities" (Thagard, 2021). All of the mentioned proposals concern human-level logic.

The work of Joseph Brenner is also along the lines of natural logic, or in his own words, Logic in Reality (Brenner, 2008, 2012). Related work on naturalization by Terrence Deacon can also be seen as a proposal on the logic of (informational) nature (Deacon, 2011; Brenner, 2012).

Within the naturalistic framework, Thomas Parr, Giovanni Pezzulo, and Karl J. Friston recently presented generative models for sequential dynamics in the active inference (Parr *et al.*, 2022) where the dynamics of physical systems minimize a surprise or, equivalently, its variational upper bound, free energy. The free energy principle is based on the Bayesian idea of the brain as an "inference engine". Under the free energy principle, systems pursue paths of least surprise or minimize the difference between predictions based on their model of the world and associated perception. In short, this approach covers the steps of inference generation and testing in the human brain.

Gianfranco Basti presents the perspective of a physicist, in *The Philosophy of Nature of the Natural Realism. The Operator Algebra from Physics to Logic* (Basti, 2022). Andrée Ehresmann, and Jean-Paul Vanbremeersch, also in the framework of *Contemporary Natural Philosophy* argue for an integral study of living systems: evolutionary multilevel, multiagent, and multitemporal self-organized systems, such as biological, social, or cognitive systems. Their Memory Evolutive Systems (MES) approach is based on a "dynamic" category theory, proposing an info-computational model for living systems (Ehresmann and Vanbremeersch, 2019).

From the epistemological point of view, the foundations of the new naturalism are outlined by Ladyman *et al.* (2007) in the book *Every Thing Must Go: Metaphysics Naturalized*, while Dodig-Crnkovic (2007) proposed the info-computationalist approach as a method to achieve this naturalization.

The strictly science-based approaches to naturalization can be contrasted with phenomenological frameworks, which traditionally question scientific framing. Despite this, Maurita Harney presents an argument for naturalizing phenomenology as a philosophical necessity, (Harney, 2015). She explains the historical roots of phenomenology's skeptical attitude toward science, evident in "Husserl's dismissal of 'the scientific attitude', Merleau-Ponty's differentiation from the scientifically objectified self, and Heidegger's critique of modern science". However, recent advancements in neuroscience have created new possibilities for collaboration between phenomenology and cognitive science, prompting a re-evaluation of science and its underlying assumptions. Central to this is the reimagining of nature as encompassing meanings and the mind. Compare this to *Valentino Braitenberg's notion "Information — der Geist in der Natur"* (Braitenberg, 2011).

In the context of phenomenological naturalism, meaning should be understood through a relational ontology rather than an atomistic one and characterized by dynamic, process-oriented properties instead of static, substance-based qualities.

4 Cognition in Nature and Artifacts as Computation of Information

Going even further in the naturalization process, ascribing logic not only to human (and AI based on human logic) agents but generalizing it further to natural processes in the sense of a self-generating loop of recursive logic hierarchies, our goal is to connect human-centric, human language-based logic (grounded in cognition) with natural logic of not only physical but also chemical, biological, and cognitive processes in all living agents.

An info-computational framework for the analysis of cognition and intelligence, natural and artificial, is a foundation for the study of information processing/computational phenomena.

In the study of cognition in nature, a two-way learning process takes place: from a theoretical and experimental study of natural systems to a constructive study of artifacts (such as deep neural networks, machine learning, and robots) and from increasingly sophisticated artifacts back to models and theories of structures and behaviors of natural systems (such as brains, swarms, or unicellular organisms) (Rozenberg and Kari, 2008). At the time when the first models of cognitive architectures (inspired by human cognition) were proposed, some 40 years ago, the understanding of cognition, embodiment, and evolution was substantially different from today.

5 Logics in the Wild

In the first chapter of their book on biochemistry, *The Facts of Life*, Reginald Garrett and Charles Grisham argue the following: "Chemistry is the logic of Biological Phenomena", based on the insight that biomolecules are informational carriers that provide structure and generative rules for biological cells (Garrett and Grisham, 2023). In a more detailed computational approach, Foulon *et al.* (2019) present a language for molecular computation.

If chemistry is the logic of biology, what is the logic of chemistry? In search for the role of logic in knowledge production, decision-making, and agency in nature, we propose the following scheme, in a recursive (circular) manner, starting with logic:

- Logics is the intrinsic foundation of mathematics.
- Mathematics is the intrinsic logic of physics.
- Physics is the intrinsic logic of chemistry.
- Chemistry is the intrinsic logic of biology.
- Biology is the intrinsic logic of cognition.

- Cognition generates (agent's intrinsic) logic in the first place. An agent's intrinsic logic is a basis for creating extrinsic/externalized/shared logic. Go to 1.

On a systemic view, nature can be seen through scale invariant self-organizational dynamics of energy/matter at the hierarchy of levels of organization (Kurakin, 2011).

The logic of living beings refers to the principles and processes that govern the behavior and functioning of organisms. It encompasses the various mechanisms through which living beings interact with their environment, survive, reproduce, and evolve over time, which unfolds through a hierarchy of levels as described above.

6 Info-Computational Nature

The Computing Nature (Naturalist Computationalism, Info-Computationalism) framework makes it possible to describe all cognizing agents (living organisms and artificial cognitive systems) as informational structures with computational dynamics (Dodig-Crnkovic, 2022a, 2022b; Dodig-Crnkovic and Giovagnoli, 2013).

Even Alan Turing's pioneering work on computability and his ideas on morphological computing can be seen as a position in *Natural Philosophy* (Hodges, 1997). Turing's natural philosophy differs importantly from Galileo's belief that the book of nature is written in the language of mathematics (Galilei, 1623). Computing is more than a language (description) of nature as computation also produces real-time physical behaviors, a driving mechanism of natural processes.

This chapter puts the computational approaches into a broader context of natural computation, where information dynamics is found not only in human communication and computational machinery but also in the entire nature. Information is understood as representing the world (reality as an informational web) for a cognizing agent, while information dynamics (information processing and computation) realizes physical laws through which all the changes of informational structures unfold.

Computation as it appears in the natural world is more general than the human process of calculation modeled by the Turing machine. Natural computing is epitomized through the interactions of concurrent, in general, asynchronous computational processes which are adequately represented by what Abramsky names "the second generation models of computation" (Abramsky, 2008) which we argue to be the most general representation of information dynamics. Conceptualizing the physical world as an intricate tapestry of information networks evolving through processes of natural computation helps make more coherent models of nature, connecting non-living and living worlds. It presents a suitable basis for incorporating current developments in understanding biological/cognitive/social systems as generated by the complexification of physicochemical processes through the self-organization of molecules into dynamic adaptive complex systems by morphogenesis, adaptation, and learning — all of which are understood as information processing (Dodig-Crnkovic, 2014).

7 Learning from Nature Within Info-Computational Framework Requires Updates of Definitions

As explained in the work of Dodig-Crnkovic (2022a), in order to construct a naturalist explanation of the logical basis of cognition and the cognitive basis of logic, some definitions have been generalized, which we should keep in mind:

- Information = structure for a cognizing agent. It means not only news and artifacts in our human civilization that are used to transmit data and knowledge but also similar structures utilized by other living organisms (cognitive agents), even the simplest ones like bacteria.
- Computation = dynamics of information. It is taken to be any process of information transformation that leads to behavior, not only those processes that we currently use to calculate, manually or with machinery.
- Cognition = life (for a living organism). It is the ability to learn from the environment and adapt so as to survive as

individuals and species, for which organisms use the information and its processing — computation. Intelligence as the capacity for problem-solving can be found in all organisms as they all possess cognition. The cognition of the simplest organism, a single cell, is called basal cognition (Lyon et al., 2021a, 2021b; Levin et al., 2021a). Understanding cognition and intelligence based on biological mechanisms is only possible if we see it in the context of evolution. Finally, cognition and intelligence in artifacts is the project of building machines mimicking human capabilities.
- Evolution is understood as an extended evolutionary synthesis (EES) formulation of evolutionary theory, which is the interpretation of the theory of evolution based on the newest scientific knowledge about life and its changes, emphasizing fundamental mechanisms of constructive development and reciprocal causation with the environment (Laland et al., 2015; Ginsburg and Jablonka, 2019).

8 Actor Model of Concurrent Distributed Computation

> "In the Actor Model [Hewitt, Bishop, and Steiger, 1973; Hewitt, 2010], computation is conceived as distributed in space, where computational devices communicate asynchronously, and the entire computation is not in any well-defined state. (An Actor can have information about other Actors that it has received in a message about what it was like when the message was sent.) Turing's Model is a special case of the Actor Model." (Hewitt, 2012)

Hewitt's "computational devices" are conceived as computational agents — informational structures capable of acting on their own behalf.

9 Computing Cells: Self-generating Systems

Complex biological systems must be modeled as self-referential, self-organizing "component systems" (Kampis, 1996) that are self-generating and whose behavior, though computational in a general sense, goes far beyond the Turing machine model repertoire of behaviors:

> "a component system is a computer which, when executing its operations (software) builds a new hardware... [W]e have a computer that re-wires itself in a hardware-software interplay: the hardware defines the software, and the software defines new hardware. Then the circle starts again." (Kampis, 1991, p. 223)

Computing understood as information processing is closely related to natural sciences; it helps us recognize connections and levels of organization in natural sciences and provides a unified approach for modeling and simulating both living and non-living systems (Dodig-Crnkovic, 2011, 2017).

One of the important aspects of natural computing is its computational efficiency. The Turing machine model of computation is not resource-aware, unlike living systems that are constantly optimizing their resource use. Ihor Lubashevsky in the book *Physics of the Human Mind* explicitly addresses those aspects (Lubashevsky, 2017).

Our present-day von Neumann computing architecture has bottlenecks, as information processor and memory are separate. Memristors as a biomimetic solution combine memory and processor and avoid von Neumann bottlenecks. The proposed solutions to the problem of our energy-consuming contemporary computing machinery are typically inspired by nature. For example, Kevin M. Passino proposes biomimicry for optimization, control, and automation (Passino, 2005). Melanie Mitchell addresses the question of unconventional, biological computation (Mitchell, 2012). More on natural computing can be found in the work of Rozenberg and Kari (2008) (see also Rozenberg *et al.*, 2012).

10 Cognition = Life: Agency-based Hierarchies of Levels: The World as Information for an Agent

Cognition in living systems/agents constitutes life-organizing, life-sustaining goal-directed processes (Stewart, 1996; Maturana and Varela, 1980, 1992; Marijuán *et al.*, 2010), while in artifactual systems, cognition is an engineered process based on sensors, actuators, and computing units designed to mimic biological cognition (bio-mimetic design).

Cognitive architectures generated by natural (morphological) computation are realized in a specific substrate of matter/energy self-organized in living cells (Dodig-Crnkovic, 2012).

11 Basal Bio-cognition

The work of Michael Levin suggests a broad range of applications for nature-inspired cognitive architectures based on biological cognition connecting genetic networks, cytoskeleton, neural networks, tissues/organs, and organisms with the group (social) levels of information processing.

Levin shows how biology has been computing through somatic memory (information storage) and bio-computation/decision-making in pre-neural bioelectric networks, before the development of neurons and brains (Levin et al., 2021a).

Insights from bio-cognition can help the development of new AI platforms, applications in targeted drug delivery, regenerative medicine and cancer therapy, nanotechnology, synthetic biology, artificial life, and much more. Recent research finds the following:

> "cognitive operations we usually ascribe to brains — sensing, information processing, memory, valence, decision making, learning, anticipation, problem-solving, generalization and goal-directedness — are all observed in living forms that don't have brains or even neurons." (Levin et al., 2021b)

Thus, we generalize cognition a step further, to include all living forms, not only those with nervous systems. It can be useful for artificial systems that need a level of intelligence but not the human level, such as nanobots or different elements of IoT.

12 Bacterial Cognition and Bacterial Chemical Language

For example, symbolic information processing can be found both on the level of human languages and also on the level of chemical languages used by bacteria, such as Bassler (Bacterial quorum sensing) (Ng and Bassler, 2009) and Ben-Jacob (Bacterial Know How: From Physics to Cybernetics) (Ben-Jacob, 2008, 2009; Ben-Jacob et al., 2004) have described.

A framework of natural cognition based on info-computation in living agents enables the unification of natural and artificial cognition and intelligence. Cognition in nature is a manifestation of biological processes in all living beings that subsume chemical and physical levels. Intelligence is considered a problem-solving ability at different levels of organization.

13 Learning from Basal Cognition on the Continuum of Natural Cognition

The concept of biological computation implies that living organisms perform computations and that, as such, abstract ideas of information and computation may be key to understanding biology.

Apart from the human brain with a nervous system, cognition can be found in somatic cells, non-human organisms with a nervous system, and non-neuronal subsystems in humans such as the immune system (Dodig-Crnkovic, 2021).

14 Systems Biology View

Individual cells are capable of making decisions based on their internal condition and environment, but the exact mechanism for reliable decision-making remains unknown. Investigations of Kramer et al. (2022) focus on the information-processing abilities of human cells by measuring various signaling responses and cellular state indicators.

The signaling nodes within a network exhibited adaptive information processing, resulting in diverse growth factor reactions and allowing nodes to gather partially unique information about the cell's condition. In combination, this provides individual cells with a substantial information processing capacity to accurately determine growth factor concentrations in relation to their cellular state and make decisions accordingly.

According to Kramer *et al.* (2022), the coevolution of heterogeneity and complexity within signaling networks may have contributed to the development of precise and context-sensitive cellular decision-making in multicellular organisms.

15 Evolution Provides Generative Mechanism for Increasingly Complex Cognition and Logic

New insights about cognition and its evolution and development in nature from cellular to human cognition can be modeled as natural information processing/natural computation/morphological computation. In the info-computational approach, evolution in the sense of EES is a result of interactions between natural agents, cells, and their groups.

Evolution provides a generative mechanism for the emergence of increasingly more competent living organisms with increasingly complex natural cognition and intelligence which are used as a template for their artificial/computational counterparts.

16 Natural Computation Driving Evolution

The view of computing nature abounds in recent literature (Zenil, 2012; Dodig-Crnkovic, 2022a, 2022b; Dodig-Crnkovic and Giovagnoli, 2013). Its basis is an unconventional generalized view of computation, natural computation. Greg Chaitin explains as follows:

> "And how about the entire universe, can it be considered to be a computer? Yes, it certainly can, it is constantly computing its future state from its current state, it's constantly computing its own time evolution! And as I believe Tom Toffoli pointed out, actual computers like your PC just hitch a ride on this universal computation!" (Chaitin, 2007)

A similar approach can be found in the work of Dodig-Crnkovic (2007) as well as elaborated in "What is Computation? (How) Does Nature Compute?" by David Deutsch in the work of Zenil (2012).

17 Evolution as Learning: Cognition as a Driver of Evolution and Evolution as a Driver of Cognition in Living Organisms: From Modern Evolutionary Synthesis to Extended Evolutionary Synthesis

A contemporary emerging, broader perspective on cognition not only recognizes humans as cognitive agents but also includes all living organisms (Dodig-Crnkovic, 2022a). It enables a fresh insight into the mechanisms of the evolutionary process. Evolution is now seen as a cognition-driven process, influenced by targeted genome editing and natural intrinsic cellular engineering (Miller *et al.*, 2021). Torday and Miller explore this cognition-centric view of evolution particularly within epigenetic evolutionary biology (Torday and Miller, 2020). Along the same lines, Miller, Enguita, and Leitão explain the following:

> "Cognition-Based Evolution argues that all of biological and evolutionary development represents the perpetual auto-poietic defense of self-referential basal cellular states of homeostatic preference. The means by which these states are attained and maintained is through self-referential measurement of information and its communication." (Miller *et al.*, 2021)

Also, Baluška, Miller, and Reber studied the cellular and evolutionary perspectives on cognition, tracing the path from unicellular to multicellular organisms. Their research focuses on the evolutionary origins of cells, starting from unicellular organisms in their progression toward multicellularity. In this view, his evolution centers on two core self-identities also observed in humans: immunological and neuronal (Baluška *et al.*, 2022). Arguments presented by Miller, Torday, and Baluška suggest that biological evolution serves as a defense mechanism for the "self". Their findings indicate the following:

> "since all living entities display self-referential cognition, life essentially pertains to maintaining cellular homeostasis in the perpetual defense of 'self'." (Miller *et al.*, 2019)

The recent EVO-EGO project, led by Watson and Buckley, explored the connectionist approach to evolutionary transitions in individuality (Watson and Buckley, 2022).

The abundance of emerging research results, especially concerning self-organization, environmental interactions, and cellular evolution, prompted Laland, Uller, and Feldman *et al.* to ask the following question: "Does evolutionary theory need a rethink?" (Laland *et al.*, 2014). Their evidence leans toward a positive response, suggesting potential advancements and extensions of modern evolutionary synthesis, especially at the micro-level.

The modern evolutionary synthesis in its turn was a result of the extension/fusion (merger) of Darwinian evolution with Mendelian genetics. It is sometimes referred to as the Neo-Darwinian theory.

Beyond recent suggestions to update the existing evolutionary theory of modern synthesis based on new data about microorganisms' behavior, Jablonka and Lamb had already earlier proposed a four-dimensional view of evolution: genetic, epigenetic, behavioral, and symbolic, anchored in cognitive learning models (Jablonka and Lamb, 2014), thus extending the modern synthesis view (Darwin + Mendelian genetics) by epigenetic, behavioral, and symbolic aspects, all of which contribute to better understanding of the processes of complexification of life through generation of new living forms. Those are not random but goal-directed behaviors of living organisms as cognitive (information-processing) agents.

Ginsburg and Jablonka's recent work explores specifically the evolution of the "sensitive soul" (Ginsburg and Jablonka, 2019), drawing inspiration from Aristotle. According to Aristotle, plants possess a nutritive soul, animals have a sensitive soul, while humans have a rational soul. Aristotle's description translates in modern terms to Maturana and Varela's concept of cognition as autopoiesis or the "realization of the living" (Maturana and Varela, 1980), where each organism, from the single living cell, possesses a degree of cognition. Ginsburg and Jablonka highlight the significance of interactions and communications between an organism and its environment, emphasizing the integral role of the genetic code's interplay with the environment in evolutionary processes.

Evolution can be interpreted as a learning process, where living entities (cells and their aggregates) adapt through environmental interactions. Aaron Sloman represents this evolutionary learning process as a meta-morphogenesis (Sloman, 2013). Watson and

Szathmary ask the following question: "How can evolution learn?" Their response is as follows:

> "Connectionist models of memory and learning illustrate how basic stepwise mechanisms, tweaking relations between simple components, can craft complex system behaviors, enhancing adaptability. We introduce 'evolutionary connectionism' to acknowledge the processes through which natural selection shapes evolutionary relationships, crafting intricate system behaviors and continually refining its adaptive prowess." (Watson and Szathmáry, 2016)

This learning paradigm can be expressed computationally. Analyzing the interactions between living organisms and their surroundings, Dodig-Crnkovic (2020) posits that morphological/natural/physical computation acts as a foundational mechanism underpinning the evolutionary learning process, presenting the constraints to the generation of new forms.

Levin develops a detailed picture of this agency-based process in his article "Darwin's agential materials: evolutionary implications of multiscale competency in developmental biology" (Levin, 2023).

In short, after modern evolutionary synthesis, EES brings new explanation mechanisms to the repertoire of evolution — biological agency on the level of cells, tissues, organs, organisms, and ecosystems, expressed epigenetic, behavioral, and symbolic aspects of evolution.

EES extends all three assumptions of modern synthesis:

(1) New variation arises through random genetic mutation [living organisms from cells up act goal-directed].
(2) Inheritance occurs through DNA [and epigenetic factors].
(3) Natural selection of genes is the sole cause of adaptation [additional mechanisms like learning processes].

18 Connecting Anthropogenic with Biogenic and Abiotic Cognition (Humans, Animals, and Machines)

Computing nature presents a common framework for understanding anthropogenic, biogenic, and abiotic cognition.

As in all of biology, nothing makes sense except in the light of evolution (Dobzhansky, 1973), and cognition as a process can only be understood in the light of evolution. Regarding abiotic systems, we compare their "cognitive behavior" with living organisms and draw conclusions.

The work of Kaspar *et al.* (2021) and Bongard and Levin (2023) suggests the direction from "cognitive/intelligent matter" toward "biological systems as evolved, overloaded, multi-scale machines". Of course, the concept of "machine" in "biological machine" is not the same as in the "mechanical machines" that we are most familiar with.

A step in-between "cognitive machine" and "biological machines" are basic elements of biological machinery like cellular signaling pathways that are plastic, (proto) cognitive systems (Mathews *et al.*, 2023). They provide indications that information-exchanging networks within cells also possess properties of cognitive systems such as sensitivity to input, information processing, memory, and output of result information/behavior. See references on the evolution of ribosomal protein network architectures (Timsit *et al.*, 2021) and the study linking multimodal perception with the cellular state to decision-making in single cells (Kramer *et al.*, 2022).

To understand the steps of the process toward increasingly competent cognitive structures, we keep in mind the role of evolution (Dobzhansky, 1973).

19 Learning From Nature to Cognitively/Intelligently Compute Requires Understanding of Evolution

In the info-computational approach to cognition and intelligence, evolution is understood in the sense of EES (Laland *et al.*, 2015; Ginsburg and Jablonka, 2019; Jablonka and Lamb, 2014) and it is a result of interactions between natural agents, cells, and their groups on a variety of levels of the organization as Jablonka and Lamb argue in their "Evolution in Four Dimensions: Genetic, Epigenetic, Behavioral, and Symbolic Variation in the History of Life". These dimensions can be found on different levels of organization of life. Aaron Sloman defined the evolution and development of information-processing machinery in living beings as meta-morphogenesis (Sloman, 2013).

20 Morphological Computing

So as to support the shared view, here are some definitions regarding morphology:

- *Morphology*: A form, shape, structure, or pattern.
- *Morphogenesis*: Generation of form, shape, structure, patterns, formation and transformation, and patterns of formation.
- *Anatomy vs. morphology*: Anatomy studies the presence of structures while morphology studies the relationships of structures. Anatomy is a subdivision of morphology, whereas morphology is a branch of biology.

21 Morphological Computing in Biology

The essential property of morphological computing is that it is defined as a structure of nodes (agents) that exchange (communicate) information.

Unicellular organisms such as bacteria communicate and build swarms or films (networks) with far more advanced capabilities compared to individual organisms (nodes), through social (distributed) cognition (information exchange).

In general, groups of smaller organisms (cells) in nature cluster into bigger ones (multicellular assemblies) with differentiated control mechanisms from the subcellular and cellular level to the tissue, organ, organism, and groups of organisms, and this layered organization provides information processing benefits (Levin, 2019; Fields and Levin, 2019; Levin *et al.*, 2021a).

22 Continuum of Natural Cognitive Architectures

A recent comprehensive overview of 40 years of research in cognitive architectures (Kotseruba and Tsotsos, 2020) evaluates the modeling of the core cognitive abilities in humans but only briefly mentions biologically plausible approaches.

However, there is an important new development of biologically inspired computational models that can lead to biologically more realistic cognitive architectures.

Unlike the vast majority of artificial cognitive architectures that target human-level cognition, we would like to focus on the development and evolution of the continuum of natural cognitive architectures, from basal cellular up, as studied by Lyon *et al.* (2021a) (see also Lyon *et al.*, 2021a, 2021b; Dodig-Crnkovic, 2022c).

23 Continuum of Biological Computation

The continuum of cognitive capacities from somatic cells to the nervous system and brain has been described by Levin:

> "We have previously argued that the deep evolutionary conservation of ion channel and neurotransmitter mechanisms highlights a fundamental isomorphism between developmental and behavioral processes. Consistent with this, membrane excitability has been suggested to be the ancestral basis for psychology (). Thus, it is likely that the cognitive capacities of advanced brains lie on a continuum with, and evolve from, much simpler computational processes that are widely conserved at both the functional and mechanism (molecular) levels.
>
> The information processing and spatiotemporal integration needed to construct and regenerate complex bodies arise from the capabilities of single cells, which evolution exapted and scaled up as behavioral repertoires of complex nervous systems that underlie familiar examples of Selves." (Levin, 2019)

24 Conclusions Connecting Morphological Computing and Cognition in Nature to Logic

When a cognitive agent is conceived of as a certain kind of information processor, then a pragmatic theory of cognitive agency will provide descriptions of processors of information. As mentioned before, logic is a principled account of certain aspects of practical reasoning, hence logic too is a pragmatic affair.

Cognition appears on a fundamental level in living cells in the form of basal cognition which is being extensively researched currently. We study cognition as a process of life (bio computation) on living structures (represented as bio information).

The first aspect of the connection between logic and information processing view of nature is that information processing can be viewed as a logical operation. In other words, logic can be seen as a set of rules for manipulating information, and information processing systems can be understood as physical implementations of these rules (Benthem et al., 2001; Benthem van, 2011).

On a systemic view, nature exhibits scale-invariant self-organizational dynamics of energy/matter at the hierarchy of levels of organization (Kurakin, 2011).

In search for the role of logic in knowledge production, decision-making, and agency in nature, the following scheme is proposed, in a recursive (circular) manner, starting with logic:

- *Logics is the intrinsic foundation of mathematics.*
- *Mathematics is the intrinsic logic of physics.*
- *Physics is the intrinsic logic of chemistry.*
- *Chemistry is the intrinsic logic of biology.*
- *Biology is the intrinsic logic of cognition.*
- *Cognition generates (agent's intrinsic) logic in the first place.*
- *An agent's intrinsic logic is a basis for creating extrinsic/ externalized/shared logic. Go to 1.*

The connection between logic and the information-processing view of nature is a fundamental one, with important implications for our understanding of the natural world and our ability to manipulate and control it. On this view, we have the following:

- Natural processes have their own inherent logic of agency under information exchanges governed by laws of nature at different levels of the organization.
- Our human-centric, language-based logic is an evolved, refined, and complex case of the logic of living organisms with basal cognition.

There is extensive empirical and theoretical work ahead to connect those phenomena with the logic of agency on different levels in nature.

The underlying mechanisms in this framework are as follows:

- Computing universe framework — all existing nature (universe) is described as a network of networks of computational processes on informational structures.
- Nature is described as structure (information) with dynamics (computation).
- Self-organizing nature — active matter drives spontaneous processes of complexification in a universe far from thermic equilibrium.
- New computational frameworks are necessary for describing complex nature — natural computing/unconventional computing/ interactive computing/morphological computing.
- Cognition is seen as a result of computation of information (morphological computation and morphogenesis).
- The "mind" as a result of cognition is extended in nature (Braitenberg 2011) — embodied, embedded, and enactive.
- Info-computational formulation of the EES (going beyond modern evolutionary synthesis by adding explanation to the processes of generation of novel biological forms and behaviors) provides generative mechanisms for the emergence of increasingly complex cognitive agents (minds) with increasingly more articulated logic. The development of the theory of evolution followed from Darwinism to Neo-Darwinism, to modern synthesis, and currently to the most well-developed EES which in its info-computational formulation presents the process of learning under continually changing environmental constraints.

Acknowledgment

This chapter is based on the research supported by the Swedish Research Council grant MORCOM@COGS.

References

Abramsky, S. (2008). Information, processes and games. In J. Benthem van and P. Adriaans (eds.), *Philosophy of Information*. Amsterdam, The Netherlands: North Holland, pp. 483–549.

Baltag, A., and Smets, S. (2014). *Johan van Benthem on Logic and Information Dynamics*. Cham: Springer.

Baluška, F., Miller, W. B., and Reber, A. S. (2022). Cellular and evolutionary perspectives on organismal cognition: from unicellular to multicellular organisms. *Biol. J. Linnean Soc,* blac005, https://doi.org/10.1093/biolinnean/blac005.

Basti, G. (2022). The philosophy of nature of the natural realism. The operator algebra from physics to logic. *Philosophies* 6, 121, https://doi.org/10.3390/philosophies7060121.

Ben-Jacob, E. (2008). Social behavior of bacteria: From physics to complex organization. *The Eur. Phys. J. B* 65(3), 315–322.

Ben-Jacob, E. (2009). Learning from bacteria about natural information processing. *Ann. NY Acad. Sci.* 1178, 78–90.

Ben-Jacob, E., Becker, I., and Shapira, Y. (2004). Bacteria linguistic communication and social intelligence. *Trends Microbiol.* 12(8), 366–372.

Benthem van, J. (2011). *Logical Dynamics of Information and Interaction*. Cambridge: Cambridge University Press.

Benthem, J. van, Dekker, P., Eijck, J. van, Rijke, M. de, and Venema, Y. (2001). *Logic in Action*. Amsterdam: Amsterdam: Institute for Logic, Language and Computation.

Bongard, J., and Levin, M. (2023). There's plenty of room right here: Biological systems as evolved, overloaded, multi-scale machines. *Biomimetics* 8(110), https://doi.org/10.3390/biomimetics8010110.

Braitenberg, V. (2011). *Information - der Geist in der Natur*. Stuttgart: Schattauer GmbH.

Brenner, J. (2008). Logic in reality. https://doi.org/10.1007/978-1-4020-8375-4.

Brenner, J. (2012). The logical dynamics of information: Deacon's "incomplete nature". *Information* 3, 676–714, https://doi.org/10.3390/info3040676.

Chaitin, G. (2007). Epistemology as information theory: From Leibniz to Ω. In G. Dodig Crnkovic (ed.), *Computation, Information, Cognition – The Nexus and The Liminal*. Newcastle UK: Cambridge Scholars Publication, pp. 2–17.

Deacon, T. (2011). *Incomplete Nature. How Mind Emerged from Matter*. New York. London: W. W. Norton & Co.

Dewey, J. (1916). *Essays in Experimental Logic.* Chicago: University Of Chicago Press, https://doi.org/10.1037/13833-000.

Dobzhansky, T. (1973). Nothing in biology makes sense except in the light of evolution. *Am. Biol. Teach.* 35(3), 125–129, https://doi.org/10.2307/4444260.

Dodig-Crnkovic, G. (2007). Epistemology naturalized: The info-computationalist approach. *APA Newslett. Philosoph. Compute.* 6(2), 9–13.

Dodig-Crnkovic, G. (2011). Significance of models of computation, from Turing model to natural computation. *Minds Mach.*, 21(2), 301–322, https://doi.org/10.1007/s11023-011-9235-1.

Dodig-Crnkovic, G. (2012). Information and energy/matter special issue. *Inform. J.* 3(4), https://doi.org/10.3390/info3040751.

Dodig-Crnkovic, G. (2014). Info-computational constructivism and cognition. *Construct. Found.* 9(2), 223–231, https://www.researchgate.net/publication/261994746_Info-computational_Constructivism_and_Cognition.

Dodig-Crnkovic, G. (2017). Nature as a network of morphological info computational processes for cognitive agents. *Eur. Phys. J.*, 226, 181–195, https://doi.org/10.1140/epjst/e2016-60362-9.

Dodig-Crnkovic, G. (2020). Morphological computing in cognitive systems, connecting data to intelligent agency. *Proceedings* 47(1), 41.

Dodig-Crnkovic, G. (2021). Cognition as a result of information processing in living agent's morphology. Species-specific cognition and intelligence. In *Proceedings of SweCog 2021 Conference.*

Dodig-Crnkovic, G. (2022a). Cognitive architectures based on natural info-computation. In Vincent C. Müller (ed.), *Philosophy and Theory of Artificial Intelligence 2021* (SAPERE ser.). Berlin Heidelberg: Springer.

Dodig-Crnkovic, G. (2022b). Cognition as morphological/morphogenetic embodied computation in vivo. *Entropy*, https://doi.org/10.3390/e24111576.

Dodig-Crnkovic, G. (2022c). Natural computation of cognition, from single cells up. In *Unconventional Computing, Arts, Philosophy.* Singapore: World Scientific, pp. 57–78, https://doi.org/10.1142/12870.

Dodig-Crnkovic, G., and Giovagnoli, R. (2013). Computing nature. In G. Dodig-Crnkovic, and R. Giovagnoli (eds.), *Turing Centenary Perspective*, Vol. 7. Springer, https://doi.org/10.1007/978-3-642-37225-4.

Ehresmann, A., and Vanbremeersch, J.-P. (2019). MES: A mathematical model for the revival of natural philosophy. *Philosophies* 4(1), 9, https://doi.org/10.3390/philosophies4010009.

Fenrong. (2023). *International Conference on Logic, Rationality and Interaction*, https://golori.org/.

Fields, C., and Levin, M. (2019). Somatic multicellularity as a satisficing solution to the prediction-error minimization problem. *Commun. Integr. Biol.* 12(1), 119–132. https://doi.org/10.1080/19420889.2019.1643666.

Foulon, B. L., Liu, Y., Rosenstein, J. K., and Rubenstein, B. M. (2019). A language for molecular computation. *Chemistry* 5(12), 3017–3019, https://doi.org/10.1016/j.chempr.2019.11.007.

Gabbay, D. M., and Woods, J. (2003). A practical logic of cognitive systems. In *Agenda Relevance - A Study in Formal Pragmatics*, Vol. 1. Amsterdam: Elsevier.

Galilei, G. (1623). The assayer. In S. Drake (ed.), *Discoveries and Opinions of Galileo*. Doubleday, p. 276.

Garrett, R., and Grisham, C. M. (2023). *Biochemistry*, 7th edn. Boston, Massachusetts: Cengage.

Ginsburg, S., and Jablonka, E. (2019). *The Evolution of the Sensitive Soul*. MIT Press, Cambridge, MA, https://doi.org/10.7551/mitpress/11006.001.0001.

Harney, M. (2015). Naturalizing phenomenology - A philosophical imperative. *Prog. Biophys. Mol. Biol.* 119(3), 661–669, https://doi.org/10.1016/j.pbiomolbio.2015.08.005.

Hewitt, C. (2012). What is computation? Actor model versus Turing's model. In H. Zenil (ed.), *A Computable Universe, Understanding Computation & Exploring Nature As Computation*. Singapore: World Scientific Publishing Company/Imperial College Press, pp. 159–187, https://doi.org/10.1142/9789814374309_0009.

Hodges, A. (1997). Turing. *A Natural Philosopher*. London: Phoenix.

Jablonka, E., and Lamb, M. (2014). *Evolution in Four Dimensions: Genetic, Epigenetic, Behavioral, and Symbolic Variation in the History of Life*. Revised Edition. Life and Mind: Philosophical Issues in Biology and Psychology. Cambridge, MA, USA: A Bradford Book, MIT Press.

Kampis, G. (1991). Self-modifying systems in biology and cognitive science: a new framework for dynamics, information, and complexity. In *IFSR International Series on Systems Science and Engineering*, Vol. 6. Amsterdam: Pergamon, https://doi.org/https://doi.org/10.1016/B978-0-08-036979-2.50006-9.

Kampis, G. (1996). Self-modifying systems: A model for the constructive origin of information. *BioSystems* 38(2–3), 119–125.

Kaspar, C., Ravoo, B. J., van der Wiel, W. G., Wegner, S. V., and Pernice, W. H. P. (2021). The rise of intelligent matter. *Nature* 594(7863), 345–355, https://doi.org/10.1038/s41586-021-03453-y.

Kotseruba, I., and Tsotsos, J. K. (2020). 40 years of cognitive architectures: Core cognitive abilities and practical applications. *Artif. Intell. Rev.*, https://doi.org/10.1007/s10462-018-9646-y.

Kramer, B. A., Castillo, J. S. del, & Pelkmans, L. (2022). Multimodal perception links the cellular state to decision-making in single cells. *Science* 377, 642–648, https://doi.org/10.1126/science.abf4062.

Kurakin, A. (2011). The self-organizing fractal theory as a universal discovery method: the phenomenon of life. *Theoret. Biol. Med. Modell.* 8(4).

Ladyman, J., Ross, D., and Spurrett, D. (2007). *Every Thing Must Go: Metaphysics Naturalized.* Oxford: Oxford University Press, https://doi.org/10.1093/acprof:oso/9780199276196.001.0001.

Laland, K. N., Uller, T., Feldman, M. W., Sterelny, K., Müller, G. B., Moczek, A., *et al.* (2015). The extended evolutionary synthesis: Its structure, assumptions and predictions. *Proc. Roy. Soc. B: Biol. Sci.* 282(20151019), 1–14, https://doi.org/10.1098/rspb.2015.1019.

Laland, K., Uller, T., Feldman, M., Sterelny, K., Müller, G. B., Moczek, A., *et al.* (2014). Does evolutionary theory need a rethink? *Nature* 514(7521), 161–164, https://doi.org/10.1038/514161a.

Levin, M. (2019). The computational boundary of a "self": Developmental bioelectricity drives multicellularity and scale-free cognition. *Front. Psychol.* 10, 2688.

Levin, M. (2023). Darwin's agential materials: Evolutionary implications of multiscale competency in developmental biology. *Cell. Mol. Life Sci.* 80(6), 1–33, https://doi.org/10.1007/S00018-023-04790-Z.

Levin, M., Keijzer, F., Lyon, P., and Arendt, D. (2021a). Basal cognition: Multicellularity, neurons and the cognitive lens, Special issue, Part 2. *Phil. Trans. R. Soc. B* 376(20200458), http://doi.org/10.1098/rstb.2020.0458.

Levin, M., Keijzer, F., Lyon, P., and Arendt, D. (2021b). Uncovering cognitive similarities and differences, conservation and innovation. *Phil. Trans. R. Soc. B* 376, 20200458.

LORI. (2021). *International Workshop on Logic, Rationality and Interaction,* https://link.springer.com/conference/lori.

Lubashevsky, I. (2017). *Physics of the Human Mind.* Springer Cham, https://doi.org/10.1007/978-3-319-51706-3.

Lyon, P., Keijzer, F., Arendt, D., and Levin, M. (2021a). Basal cognition: conceptual tools and the view from the single cell — Special issue, Part 1. *Phil. Trans. R. Soc. B* 376(20190750), http://doi.org/10.1098/rstb.2019.0750.

Lyon, P., Keijzer, F., Arendt, D., and Levin, M. (2021b). Reframing cognition: Getting down to biological basics. *Phil. Trans. R. Soc. B* 376, 20190750.

Magnani, L. (2018). The urgent need of a naturalized logic. *Philosophies* 3(4), 0–44.

Marijuán, P. C., Navarro, J., and del Moral, R. (2010). On prokaryotic intelligence: Strategies for sensing the environment. *BioSystems*, https://doi.org/10.1016/j.biosystems.2009.09.004.

Mathews, J., and Chang, A. (Jaelyn), Devlin, L., and Levin, M. (2023). Cellular signaling pathways as plastic, proto-cognitive systems: Implications for biomedicine. *Patterns* 100737, https://doi.org/10.1016/j.patter.2023.100737.

Maturana, H., and Varela, F. (1980). *Autopoiesis and Cognition: The Realization of the Living*. Dordrecht Holland: D. Reidel Pub. Co.

Maturana, H., & Varela, F. (1992). *The Tree of Knowledge*. Shambala.

Miller, W. B., Enguita, F. J., and Leitão, A. L. (2021). Non-random genome editing and natural cellular engineering in cognition-based evolution. *Cells* 10(5), 1125, https://doi.org/10.3390/cells10051125.

Miller, W., Torday, J. S., and Baluška, F. (2019). Biological evolution as defense of 'self'. *Prog. Biophys. Mol. Biol.* 142, 54–74, https://doi.org/10.1016/j.pbiomolbio.2018.10.002.

Mitchell, M. (2012). Biological computation. *Comput. J.* 55(7), 852–855.

Ng, W.-L., & Bassler, B. L. (2009). Bacterial quorum-sensing network architectures. *Annu. Rev. Genet.* 43, 197–222.

Parr, T., Pezzulo, G., and Friston, K. J. (2022). *Active Inference: The Free Energy Principle in Mind, Brain, and Behavior* (CogNet Pub.). Cambridge, MA: The MIT Press, https://doi.org/10.7551/mitpress/12441.001.0001.

Passino, K. M. (2005). *Biomimicry for Optimization, Control, and Automation*. London: Springer, https://doi.org/10.1007/b138169.

Quine, W. V. O. (1969). Epistemology naturalized. In *Ontological Relativity and Other Essays*. New York: Columbia University Press, pp. 69–90.

Rozenberg, G., Bäck, T., and Kok, J. N. (eds.) (2012). *Handbook of Natural Computing*. Berlin Heidelberg: Springer.

Rozenberg, G., and Kari, L. (2008). The many facets of natural computing. *Commun. ACM* 51, 72–83, https://doi.org/10.1145/1400181.1400200.

Sloman, A. (2013). Meta-Morphogenesis: Evolution and development of information-processing machinery. In S. B. Cooper and J. van Leeuwen (eds.), *Alan Turing: His Work and Impact*. Amsterdam: Elsevier, p. 849.

Stewart, J. (1996). Cognition = life: Implications for higher-level cognition. *Behav. Process.* 35, 311–326.

Thagard, P. (2021). Naturalizing logic: How knowledge of mechanisms enhances inductive inference. *Philosophies* 6(2), 52, https://doi.org/10.3390/philosophies6020052.

Timsit, Y., Sergeant-Perthuis, G., and Bennequin, D. (2021). Evolution of ribosomal protein network architectures. *Sci. Rep.* 11(1), 625, https://doi.org/10.1038/s41598-020-80194-4.

Torday, J., and Miller, W. (eds.) (2020). *Four Domains: Cognition-Based Evolution BT Cellular-Molecular Mechanisms in Epigenetic Evolutionary Biology.* Cham: Springer International Publishing, pp. 103–112, https://doi.org/10.1007/978-3-030-38133-2_13.

Watson, R. A., and Szathmáry, E. (2016). How can evolution learn? *Trends Ecol. Evol.* 31(2), 147–157, https://doi.org/10.1016/j.tree.2015.11.009.

Watson, R., and Buckley, C. (2022). *The Scaling-up of Purpose in Evolution (evo-ego): Connectionist Approaches to the Evolutionary Transitions in Individuality-Research Project.* John Templeton Foundation: Philadelphia, PA, USA, https://www.templeton.org/grant/the-scaling-up-of-purpose-in-evolution-evo-ego-connectionist-approaches-to-the-evolutionary-transitions-in-individuality.

Woleński, J. (2012). Naturalism and the genesis of logic. In K. Trzesicki, S. Krajewski, and J. Wolenski (eds.), *Papers on Logic and Rationality. Festschrift in Honor of Andrzej Grzegorczyk* (Studies in Logic, Grammar and Rhetoric), Vol. 27. Bialystok: University of Bialystok, pp. 223–230, https://doi.org/10.1515/slgr-2016-0057.

Woleński, J. (2017). Logic in the light of cognitive science. *Studies Logic, Grammar Rhetoric* 48(1), 87–101, https://doi.org/10.1515/slgr-2016-0057.

Woods, J. (2013). *Errors of Reasoning: Naturalizing the Logic of Inference.* London: College Publications.

Zenil, H. (ed.) (2012). A computable universe. In *Understanding Computation & Exploring Nature As Computation.* Singapore: World Scientific Publishing Company/Imperial College Press.

Chapter 4

Does Information Have Mass? A Review of Melvin Vopson's Claims

Roman Krzanowski

Faculty of Philosophy, The Pontifical University of John Paul II
Kanonicza 25, 31-002 Kraków, Poland
rmkrzan@gmail.com

In his papers, Melvin Vopson claims that information has mass. This claim is puzzling because by attributing mass to information, one would also have to attribute other properties to it, such as the energy–mass–information equivalence (which Melvin Vopson does) and mass–energy conservation (which he does not discuss). This would essentially create a new form of physics. Burgin and Mikkilineni (2022) have already critiqued Vopson's claims under the light of the General Theory of Information (GTI), but we here discuss Vopson's claim through a detailed analysis of his own argument. We come to the conclusion that Vopson's argument, which is based on an assumption that information is a physical structure with mass, leads to unacceptable paradoxes, so it must therefore be rejected. As such, the information–energy–mass principle proposed by Melvin Vopson also cannot be accepted.

1 Introduction

In a series of papers, Vopson has claimed that information is physical and has mass (2019, 2020, 2021). The first assertion that information is somehow physical is not surprising given that several scholars have attributed a physical nature to information.[1]

However, Vopson's claim goes a step further by insisting that if information is physical, then it must have mass, as physical things, at least most of them, generally do (e.g., Kane, 2005; Koks *et al.*, 2008, 2012; Hecht, 2006). Furthermore, if information has mass, we can postulate some kind of information–mass–energy equivalence. Nevertheless, we believe that things with information are not as straightforward as Vopson suggests.

Vopson's claim that information has mass is baffling because it also implies some paradoxical consequences. To highlight this, let us indulge in a thought experiment: Imagine that we have a piece of paper with some scribbles written on it in ink. We can of course precisely determine the mass of the paper plus the ink that was added to the paper. Now, if we suppose that information has mass, and if the scribbles convey information, then the piece of paper will have a greater mass than an identical piece of paper with the same volume of ink added to it in an uninformative manner, such as through meaningless scribbles or even an inkblot.

In an even starker example, take the punched cards that early computers used as a storage medium for data. When the card has not yet been punched, it carries no information as a digital carrier, but once holes are punched in it to represent bits of binary data, it may contain useful information, such as a computer program or data. This punched card with information has a lower mass than an unused, information-free card without any holes, so adding information has reduced the mass of the card. This clearly conflicts with Vopson's

---

[1] Several researchers attribute in various ways a physical nature to information. See, for example, the work of Weizsäcker von (1971), Turek (1978), Wheeler (1978), Mynarski (1981), Heller (1987, 2014), Collier (1989), Stonier (1990), Devlin (1991), De Mull (2010), Polkinghorne (2000), Jadacki and Brozek (2005), Schroeder 2005, 2014, 2017a, 2017b, Baeyer von (2005), Seife (2006), Dodig-Crnkovic (2012), Hidalgo (2015), Landauer (1961, 1991, 1996, 1999). Wilczek (2015), Carrol (2017), Rovelli (2016), Davies (2019), Sole and Elena (2019), Krzanowski (2022), Burgin and Krzanowski (2022), and Burgin and Mikkilineni (2022).

claim because it apparently implies that information has a negative mass, which would be puzzling, at least if we insist that information has a positive mass.

Vopson's claim that information has mass must clearly therefore refer to something different from what our examples would suggest, or so it would seem. If confirmed, Vopson's claims would have far-reaching consequences, which Vopson himself points out, for our concept of nature, the mind, the universe, and computing. It therefore seems only diligent to study these claims and evaluate their soundness.[2]

Burgin and Mikkilineni analyzed Vopson's claims from the perspective of the general theory of information (GTI) and ultimately dismissed them as contradicting the GTI (Burgin and Mikkilinen, 2022). While their paper is quite convincing, we think that even when considered on their own merits, Vopson's arguments do not suffice to justify his far-reaching general conclusions about information, the universe, and the matter–energy–information relationship. In this chapter, we take a detailed look at these claims, analyze their premises, and verify their validity and soundness in order to establish whether Vopson's conclusions can be justified fully, partially, or not at all.

We start in the following section by presenting Vopson's claims before Section 3 analyzes their soundness. Section 4 then draws some conclusions from the discussion. The final section is then dedicated to the memory of Mark Burgin.

2 Vopson's Claims

Vopson follows Shannon in stating that "information (I) extracted from observing an event (X) is a function of the probability (p) of the event (X) to occur or not, $I(p)$" (Vopson, 2019). Furthermore,

---

[2]Vopson's definition of mass (or "physical mass") of bit is given as $m_{\text{bit}} = k_b T \ln(2)/c^2$ with c being a speed of light, T being a temperature of storage of a bit, and $k_b = 1.38064 \times 10^{-23}$ J/K is the Boltzmann constant (Vopson, 2019). We do not discuss further the concept of mass in Vopson's papers.

the average information (i.e., the number of bits of information) per event that one can extract when observing a set of events X once is given by

$$H(x) = -\Sigma_{j=1}^{n} p_j \cdot \log_b p_j,$$

where $H(x)$ denotes the information entropy, p is the probability of event X_j occurring, and b is the logarithm base. This is based on the work of Shannon (1948) and follow-up work by Shannon and Weaver (1949/1964), Weaver (1949), and Pierce (1961).

There is a well-known similarity between the above formula and Boltzmann's equation for thermodynamic entropy (i.e., the Boltzmann–Planck equation)[3]:

$$S = k_b \cdot \ln(\Omega) = N \cdot k_b \cdot H(X) \cdot \ln(2),$$

where Ω is the number of micro-states that are compatible with the macro-state, N is the number of sets of events, and $k_b = 1.38064 \times 10^{-23}$ J/K is the Boltzmann constant (Boltzman, 1866; Jaynes, 1965; Atkins, 2010).

Furthermore, when $b = 2$, following the example of Hartley and Shannon, we can talk about encoding digital bits of information as ones and zeros, so we can in turn talk about the entropy of encoding digital bits of information (i.e., information entropy). The entropy of encoding one bit of information is therefore $S = k_b \cdot \ln(\Omega) = k_b \cdot \ln(2)$. To clarify an important point, however, the bits used by Shannon in his $H(x)$ formula for information entropy are used to measure the amount of information (i.e., the number of encoding bits is regarded as an amount of information), and no semantic interpretation is associated with this concept of information. The concept of "the entropy of encoding bits of information" is a link between information and bits (0/1) that was forged by Shannon and adopted by others, albeit

---

[3] By "formal similarity", we refer to similarity in the symbolic form. For example, formula (1) $x = ax - b$ is in form similar to formula (2) $f = kz + g$. However, the two formulas (1) and (2) may express completely different semantic content. To make this clearer, we could say that the two phenomena are not necessarily similar or the same, even if they can be quantified by the same mathematical structure (e.g., a linear equation in the example). See, for example, the work of Pierce (1961).

with a twist, such as Vopson (Hartley, 1928; Shannon, 1948; Shannon and Weaver, 1949/1964; Weaver, 1949).

Vopson posited that the total entropy S_{tot} of a physical system for carrying digital information is the sum of the physical entropy S_{phys}^i related to the non-information-bearing states of the physical system and the information entropy S_{info}^i associated with the information-bearing states of the physical system (i.e., $S_{\text{tot}} = S_{\text{phys}} + S_{\text{info}}$).

Changes in S_{info}^i (i.e., bit erasure) must be compensated for by changes in the system entropy, similar to what is asserted in the second law of thermodynamics. This dependency is presented as a rule and referred to as Landauer's principle (Landauer, 1961, 1996; Plenio and Vitelli, 2001):

> **Landauer's principle:** For every bit of information irreversibly lost, the entropy of the system must increase with an absolute value of heat released per bit lost, such that $\Delta Q = k_b \cdot T \cdot \ln(2)$,

where T is the temperature in (K^0) and ΔQ is the amount of heat released.

Vopson's claim about information and mass capitalizes on Landauer's principle. He states that holding (storing) information requires that a bit has mass m_{bit}, saying the following[4]:

> In this paper a radical idea is proposed, in which the process of holding information indefinitely without energy dissipation can be explained by the fact that once a bit of information is created, it acquires a finite mass, m_{bit}. This is the equivalent mass of the excess energy created in the process of lowering the information entropy when a bit of information is erased. Using the mass–energy equivalence principle, the mass of a bit of information is:
>
> $$m_{\text{bit}} = k_b T \cdot \ln(2)/c^2$$
>
> where c is the speed of light and other symbols defined as before. Having information content stored in a physical mass allows holding the information without energy dissipation indefinitely. (Vopson, 2019)

---

[4] "...once a bit of information is created, assuming no external perturbations, it can stay like this indefinitely without any energy dissipation" (Vopson, 2019).

He then adds the following:

> The implications of this rationale are that the equivalence mass–energy principle inferred from the special relativity can be extrapolated to the mass–energy–information equivalence principle, which essentially represents an extension of the original Landauer's principle. (Vopson, 2019)

Vopson states in his 2019 paper that his claims are specific to the carrier media he chooses, namely classical digital memory states at equilibrium. He writes the following:

> The mass–energy–information equivalence principle proposed here is strictly applicable only to classical digital memory states at equilibrium. Information carried by relativistic media, moving waves or photons require a quantum relativistic information theory approach and it is outside the applicability framework of this article. (Vopson, 2019)

Furthermore, he adds the following:

> Similarly, other forms of information including analogue information, or information embedded in biological living systems such as DNA, are not within the scope of this work. (Vopson, 2019)

Of course, the claim that "the information content stored in a physical mass allows holding information without energy dissipation *indefinitely*" is untrue, at least if taken literally, as is implied by the author, because information-storage devices, especially the magnetic media that Vopson refers to as standard storage devices, tend to degrade over time (Bressan *et al.*, 2019; IASA Magnetic Tapes, n.d.; Nave, n.d.). Indeed, nothing in nature, with the exception of electrons, remains qualified forever (e.g., Agostini *et al.*, 2015; Johnson, 2015; Marletto, 2021). Thus, Vopson's claim about indefinitely storing bits should be qualified by a phrase along the lines of "for all practical purposes".[5]

Nevertheless, putting this issue aside, according to Vopson, a bit stored in classical digital memory, meaning magnetic storage media, has a mass that is equivalent to the energy required to erase or

---

[5]In his recent paper (Vopson and Lepadatu, 2022), Vopson recognized the transitive nature of the digital storage of information, yet he did not amend his earlier conclusions about the permanence of digital storage.

record it. Under the assumption that information is encoded in bits, this is expressed as a mass–energy–information equivalence principle through a series of equivalence formulas:

$$\text{Mass–information relation:} \quad m = k_B T \ln(2)/c^2$$
$$\text{Information–energy:} \quad mc^2 = k_B T \ln(2)$$
$$\text{Energy–mass:} \quad m = E/c^2$$

3 Analyzing Vopson's Argument

3.1 *Reconstructing the argument*

Vopson defines information as a sequence of digital bits (i.e., zeros and ones) encoded in classical digital memory in the form of a magnetic carrier. According to Vopson's own statement in a paper from 2019 (Vopson, 2019), his argument is not valid for other forms of information storage.[6]

Vopson interprets encoding information as bits in classical digital memory as adding or subtracting mass to or from the carrier, but why? The creation or erasure of bits (i.e., the manipulation of information on the physical carrier) requires energy, so storing, holding, or keeping a bit requires a mass that is equivalent to the energy required to erase or encode this bit, and it does not involve any energy exchange ("... the process of holding information indefinitely without energy dissipation can be explained by the fact that once a bit of information is created, it acquires a finite mass, m_{bit}" — Vopson, 2019). In this sense, the bit therefore has a mass that is equivalent to whatever energy is used to record or erase this bit. This relationship then leads Vopson to formulate an energy–mass–information

---

[6] "The mass–energy–information equivalence principle proposed here is strictly applicable only to classical digital memory states at equilibrium. Information carried by relativistic media, moving waves or photons require a quantum relativistic information theory approach and it is outside the applicability framework of this article. Similarly, other forms of information including analogue information, or information embedded in biological living systems such as DNA are not within the scope of this work". (Vopson, 2019).

equivalence principle as a law of nature, one that he views as an extension or generalization of special relativity.

We therefore differentiate two arguments in Vopson's paper: Argument (I) that information has mass and argument (II) that there is a law of nature for energy–mass–information equivalence.

We can reconstruct Vopson's argument (I) that "Information has mass" as follows:

(A1) Information is bits (0/1).
(A2) The erasure or encoding of bits in digital storage media requires work or energy.
(A3) Bits recorded in digital storage media are stored indefinitely (for all practical purposes) without energy dissipation.
(A4) Bits stored indefinitely (for all practical purposes) in digital storage media have physical mass.

Thus,

(C1) Information, as bits stored in digital storage media, has mass.
(C2) Information has mass.

We can then reconstruct Vopson's argument (II) of there being an "energy–mass–information equivalence" in the following manner:

(A5) Special relativity defines energy–mass equivalence.
(A6) Information stored in digital storage media has mass.
(A7) The mass of information stored in digital media is equivalent to the energy used to store/encode that information.

Thus,

(C3) There is a mass–energy–information equivalence.

3.2 First difficulty

Let us look closely at arguments (I) and (II). If Vopson's claim in (I) is about information stored in digital storage media only (the 2019a formulation), we can say it is a narrow interpretation. In contrast, if Vopson's claim in (II) refers to information in general by claiming the existence of a mass–energy–information equivalence (the 2019b and 2022 formulations), then we can say it is an extended interpretation.

Now, if Vopson's claims (C1 and C2) are limited to a narrow interpretation, they could not possibly be generalized to the general physical law proposed in argument (II). Nevertheless, this extended claim seems more in line with what Vopson had in mind because the mass–energy–information equivalence he proposed would be a physical law on par with, according to Vopson, the mass–energy equivalence of special relativity (Vopson, 2019, 2020, 2021).

This interpretation of Vopson's claims can be confirmed through his later papers, where Vopson used the law of mass–energy–information equivalence (as formulated in the 2019 paper) when discussing the informational content of the universe (Vopson, 2020, 2021), with this applying to research in evolutionary biology, computing, big data, physics, and cosmology. He also derived another law of general import called the second law of infodynamics (Vopson and Lepadatu, 2022).

This generalization is puzzling because Vopson originally stated that his claim is not valid for other forms of information storage, so it cannot apply to any other form of information. As such, it cannot be generalized into a universal physical law.

3.3 Second difficulty

Looking at the physics of encoding data on magnetic tape, Vopson's claim seems odd. Storing bits on magnetic tape involves manipulating the magnetic properties of the carrier medium (Bressan *et al.*, 2019; IASA Magnetic Tapes, n.d.; Nave, n.d.). In other words, encoding bits onto a tape changes the structure of the carrier, but it does not add anything substantive to it.[7]

---

[7] "The recording medium for the tape recording process is typically made by embedding tiny magnetic oxide particles in a plastic binder on a polyester film tape. Iron oxide has been the most widely used oxide, leading to the common statement that we record on a 'ribbon of rust'... The oxide particles are on the order of 0.5 micrometers in size and the polyester tape backing may be as thin as 0.5 mil (.01 mm). The oxide particles themselves do not move during recording.

Thus, the energy used to encode a bit is not used to create mass but rather to rearrange the physical structure of the carrier medium. For example, recording bits on a magnetic tape rearranges "magnetic oxide particles in a plastic binder" rather than changing the mass of the tape. Indeed, most digital storage media work by changing the structure of the carrier medium (e.g., Morehouse, 2003; Goda and Kitsuregawa, 2012; Markov, 2014; Roosah, 2014; Hussain, 2017).

Thus, Vopson's claim that adding information to digital storage media will also add mass seems to contradict the engineering practices involved.

3.4 *Third difficulty*

In Vopson's paper, information is equated with a (physical) structure that is imposed over the structure of the carrier medium. In other words, information is an added structure that overlays the original structure of the medium. As such, the information aka structure has its own entropy S_{info} and the carrier's structure has its own entropy S_{phys}.

However, in modern concepts of information, information is rather not equated with the physical structure (or with a part of this structure) (see, for example, the work of Carrol, 2017; Rovelli, 2016; Davies, 2019; Krzanowski, 2020, 2021, 2022; Burgin and Krzanowski, 2022; Burgin and Mikkilineni, 2022). Information being a physical structure (this is what Vopson is proposing) would imply that information is itself a solid object or at least a part of one.

Consequently, in this view, information X (e.g., Mozart's piano concerto No. 6, K. 238) would exist as many different information aka physical structures X1, X2, ..., Xn (recordings of the concert, notes, etc.) (e.g., Krzanowski, 2022), with each structure being different information (as a part of a different object).

One may claim (claim 1) that X and X1, X2, ..., Xn are all different information (in our case, Mozart's piano concerto No. 6, K. 238).

Rather their magnetic domains are reoriented by the magnetic field from the tape head." (Nave, n.d.)

But Mozart wrote one specific concerto No. 6, K. 238, not many of them. So, such an interpretation (claim 1) would be puzzling.[8]

A more coherent response (claim 2) would be that X is information and $X1, X2, \ldots, Xn$ are copies, more or less accurate, of X, *ceteris paribus*. Or, $X1, X2, \ldots, Xn$ structures are bearers of information X. Simply put, Mozart wrote one piano concerto, No. 6, K. 238, and the rest are copies (more or less accurate).[9]

Vopson's theory would imply claim (1). But we think that more logical, and conceptually efficient (re. Occam's razor), and consistent with modern theories of information (GTI), is claim (2).[10] One may recognize here traces of the tension between Plato's Forms (claim 2) and Aristotle's essences (claim 1). A plausible resolution to claim (1) vs claim (2) is discussed by Burgin (2003, 2010), Burgin and Krzanowski (2022), Burgin and Mikkilineni (2022), and Krzanowski (2022).

3.5 *Interpretation*

A summary of Vopson's claim goes as follows: Information is always carried in some physical medium, and most of us researchers who write about information can agree with him here, but Vopson goes further. For him, information is encoded as, or in, the carrier's structure. This encoding is permanent (qualified), and the process adds mass to the carrier that is equivalent to the energy expended by encoding bits. By Vopson's own admission, this claim is limited to digital magnetic storage media (Vopson, 2019). The claim implies that an informational structure with mass is added to the structure of the carrier, and this is where we start to object to Vopson's claim.

---

[8] Some may claim that we never know what Mozart actually created as his record of his creation is only a copy (the best Mozart could do, which it seems, was pretty good!) of his idea.

[9] The case of a piece of music and information is discussed by Jadacki and Brozek (2005).

[10] It seems that Vopson is facing Plato's problem of The Third Man Argument (e.g., Fine, 1995).

The practice of information storage has taught us that encoding information in a carrier does not add anything to that carrier's mass, specifically for digital carriers, but rather primarily changes its structure. The physical process for storing information as bits in digital media is different from the one proposed by Vopson, but it is in a way analogous to the process for coding an analog (or digital) signal over the carrier frequency of an electromagnetic wave. In this process, the carrier retains its basic structure (the frequency), but a radio signal is imposed on it by modulating the shape of the carrier wave. This may be loosely interpreted as adding structure. The exact process for modulating (i.e., embedding information in) the carrier wave will differ based on the encoding technology and carrier medium (e.g., Stallings, 2007; Rajaraman, 2010; Michael, 2019).

Furthermore, as we previously pointed out, information is not the physical structure itself. Indeed, bits are not information but rather a measure of the amount of information, which may be expressed in forms other than bits. The work/energy that Vopson attributed to information storage (i.e., bit manipulation, encoding, and erasing) is actually the work/energy required to change the carrier's structure in order to encode information rather than to add a mass of information. Storing information in various physical carriers like punched cards, paper records, and such may involve adding or subtracting mass to or from the carrier media. (Recall that with punched cards, we subtract mass when we add information.)

This explanation agrees with most interpretations of information, including those of Landauer and Burgin and Mikkilineni. Information certainly has a physical manifestation, as it is always perceived in some physical form or embedded (the precise meaning of which is carrier-specific) in some physical object. As many researchers have pointed out, however, and as has already been indicated, information is not a physical object with mass (Weizsäcker von, 1971; Turek, 1978; Mynarski, 1981; Heller, 1987, 2014; Collier, 1989; Stonier, 1990; Devlin, 1991; De Mull, 2010; Polkinghorne, 2000; Schroeder, 2005, 2014, 2017a, 2017b; Baeyer von, 2005; Seife, 2006; Dodig-Crnkovic, 2012; Hidalgo, 2015; Wilczek, 2015; Carrol, 2017; Rovelli, 2016; Davies, 2019; Sole and Elena, 2019; Krzanowski, 2022).

Talking about physics and information, we need to mention John Wheeler's "It from Bit" proposal (Wheeler, 1978, 1988, 1989). The proposal has been widely quoted and used to argue that information (in certain ways) is fundamental to the physical universe (see, e.g., Zeh, 2002; Foschini, 2013; Makela, 2019; Feldman, 2020; Zeilinger, n.d. or the collection of papers by Aguirre et al., 2015). John Wheeler's "It from Bit" proposal implied the ideas of a "participatory universe" and "law without law" (see, e.g., Wheeler, 1988; Foschini, 2013; Stoica, 2013). Wheeler's hypothesis is supported, it is claimed, by the delayed choice experiment (see e.g., Wheeler, 1978; Stoica 2013), making it almost a physical fact (and is often referred to it this way). Yet, it does not seem to be the case (see, e.g., Deutsch, 1986, 2004; Ellerman, 2015; Hogan 2018; Kastner, 2019). Carroll (2022) states that the ideas of "participatory universe" and "law without law" implied by the delayed choice experiment stem from the wrong interpretation of quantum phenomena (wave–particle duality — the Copenhagen interpretation of QM). Interpreting the electron as "the wave function of the universe" eliminates the "participatory universe" and "law without law" proposals and the mystique of the delayed choice experiment, as it does not imply the ideas of "participatory universe" and "law without law" and their consequences. Wheeler's "It from Bit", despite its popularity or notoriety,[11] does not contribute to the analysis of Vopson's claims. So, it has not been a part of this discussion.

The physical nature of information (and the concept of information) has been comprehensively explained by the general theory of information (Burgin, 2003, 2017). Furthermore, Burgin and Mikkilineni's paper (Burgin and Mikkilineni, 2022; Burgin and Krzanowski, 2022) provides a detailed analysis of Vopson's proposal under the light of the GTI.

---

[11]Wheeler's proposal was compared to Zen Koan implying its mystical meaning.

4 Conclusions

In summary, what Vopson is observing and measuring as a mass of information is simply the energy expended in encoding/erasing structures that represent bits of information, and what he is interpreting as changes in mass after encoding these bits are structural changes to information carriers, such as digital magnetic memory. However, such changes do not change the mass of a memory storage carrier, which Vopson explicitly restricts his claims to. Indeed, bit encoding in digital storage media does not add mass but rather changes the structure of the carrier in a process that requires energy.

Thus, argument (I) that information has mass cannot be sustained due to a flawed interpretation of the erase/encode process in digital magnetic media. In addition, argument (II) for an energy–mass–information equivalence cannot be justified because it is dependent on the argument (I), so it must be dismissed.

Bit operations in computer systems are based on fundamentally physical processes, so they are subject to fundamental physical laws, including thermodynamics (e.g., Feynman, 2001). Whether the mass of a digital memory system changes by adding or erasing content depends on the carrier medium and the encoding process, as well as the manner in which a structure represents some meaningful information (see, for example, Burgin and Mikkilineni, 2022). Indeed, when encoded with information as bits, some carriers do not change mass (e.g., magnetic media), while some do (e.g., punched cards).[12] Nevertheless, they all have their structures changed.

Finally, if information in the context explained by Vopson has no mass, the mass–energy–information principle obviously cannot be valid. What does hold, however, is that imparting information to physical media always requires energy to be expended to change the structure of the carrier medium, both in the encoding and erasing processes. This agrees with Landauer's (1996) claims.

---

[12] The same effect of subtracting mass from a carrier when encoding information can be attributed to carvings in the stone.

The author would like to thank the anonymous reviewer for insightful comments that made this chapter more complete.

5 In Memory of Mark Burgin

When studying the philosophy of information, one cannot fail to mention the contributions of Mark Burgin. His studies of information covered information in its multiple forms, from domain-specific, quantified, mental, physical, or natural, and abstract information to ontological, metaphysical, and epistemic perspectives. No similar studies of information can match Mark Burgin's scope of vision, depth of detail, breadth of knowledge, and comprehensiveness of presentation. The results of his studies have been synthesized into the general theory of information, which is the most comprehensive conceptualization of the complex and elusive phenomenon that is generally called information. On a more personal level, Mark Burgin was always very accessible, available for discussion and advice, and willing to listen and help, and he always tried to convince others through the force of argument rather than sophistry or authority. He was a rare sort of philosopher with Socratic qualities. It has been a privilege to work with him, even if only briefly, as it was in my case.

References

Agostini, M., *et al.* (2015). Test of electric charge conservation with Borexino (Borexino Collaboration). *Phys. Rev. Lett.* 115, 231802. arXiv:1509.01223v2. DOI: 10.1103/PhysRevLett.115.231802.

Aguirre, A., Foster, B., and Merali, Z. (eds.) (2015). *It from Bit or Bit from It? On Physics and Information.* New York: Springer International Publishing.

Atkins, P. (2010). *The Laws of Thermodynamics.* Oxford: Oxford University Press.

Baeyer von, H. C. (2005). *Information. The New Language of Science.* Cambridge: Harvard University Press.

Boltzmann, L. (1866). Über die Mechanische Bedeutung des Zweiten Hauptsatzes der Wärmetheorie. *Wiener Berichte* 53, 195–220 (in German).

Bressan, F., Hess, R., Sgarbossa, P., and Bertani, R. (2019). Chemistry for audio heritage preservation: A review of analytical techniques for audio magnetic tapes. *Heritage* 2, 1551–1587. DOI: 10.3390/heritage2020097.

Burgin, M. (2003). Information: Problems, paradoxes, solutions. *TripleC* 1(1), 53–70. http://tripleC.uti.at/.

Burgin, M. and Feistel, R. (2017). Structural and symbolic information in the context of the general theory of information. *Information* 8(4), 139. DOI: 10.3390/info8040139.

Burgin, M. and Mikkilineni, R. (2022). Is information physical and does it have mass? *Information* 13(11), 540. DOI: 10.3390/info13110540.

Burgin, M. and Krzanowski, R. M. (2022). World structuration and ontological information. *Proceedings*, 81, 93. DOI: 10.3390/proceedings 2022081093.

Burgin, M. (2017). The general theory of information as a unifying factor for information studies: The noble eight-fold path. *Proceedings*, 1, 164; *Presented at the IS4SI 2017 Summit Digitalization for a Sustainable Society*, Gothenburg, Sweden, 12–16 June 2017. DOI: 10.3390/IS4SI-2017-040.

Burgin, M. (2010). *Theory of Information*. San Francisco: World Scientific Publishing.

Carroll, S. (2017). *The Big Picture on the Origins or Life, Meaning and the Universe Itself*. London: OneWord.

Carroll, S. (2022). The notorious delayed-choice quantum eraser. https://www.preposterousuniverse.com/blog/2019/09/21/the-notorious-delayed-choice-quantum-eraser/.

Collier, J. (1989). Intrinsic information. In Hanson, P. P. (ed.), *Information, Language and Cognition: Vancouver Studies in Cognitive Science*, Vol. 1. Oxford: Oxford University Press (Originally University of British Columbia Press, 1990), pp. 390–409.

Davies, P. (2019). *The Demon in the Machine*. New York: Allen Lane.

De Mull, J. (2010). The informatization of the worldview. *Inf. Commun. Soc.* 2(1), 69–94.

Devlin, K. (1991). *Logic and Information*. Cambridge: Cambridge University Press.

Deutsch, D. (1986). On Wheeler's notion of "law without law" in physics. *Found. Phys.* 16, 565–572. DOI: 10.1007/BF01886521.

Deutsch, D. (2004). It from qubit. In Barrow, J., Davies, P., and Harper Jr., C. (eds.), *Science and Ultimate Reality: Quantum Theory, Cosmology, and Complexity*. Cambridge: Cambridge University Press, pp. 90–102. DOI: 10.1017/CBO9780511814990.008.

Dodig-Crnkovic, G. (2012). Alan Turing's legacy: Info-computational philosophy of nature. http://arxiv.org/ftp/arxiv/papers/1207/1207.1033.pdf.

Ellerman, D. (2015). Why delayed choice experiments do not imply retrocausality. *Quantum Stud.: Math. Found.* https://www.ellerman.org/wp-content/uploads/2015/01/OnLineFirstReprint.pdf.

Feldman, M. (2020). "It from bit" and "Law without law": The route of information theory towards the refoundation of physics. HAL Id: hal-02471910. https://hal.archives-ouvertes.fr/hal-02471910.

Foschini, L. (2013). Where the "it from bit" come from? arXiv:1306.0545.

Feynman, R. P. (2001). *Feynman Lectures on Computation*, 1st edn. CRC Press. DOI: 10.1201/9780429500442.

Fine, G. (1995). Third man arguments. In *On Ideas: Aristotle's Criticism of Plato's Theory of Forms*. Oxford: Oxford Academic (1 November 2003 Online edition). https://doi.org/10.1093/0198235496.003.0015.

Goda, K. and Kitsuregawa, M. (2012). The history of storage systems. *Proc. IEEE*, 100(Special Centennial Issue), 1433–1440. DOI: 10.1109/JPROC.2012.2189787.

Hartley, R. V. L. (1928). Transmission of information. *Bell Syst. Tech. J.* 7(3), 535–563.

Hecht, E. (2006). There is no really good definition of mass. *Phys. Teach.* 44(1), 40–45. DOI: 10.1119/1.2150758.

Heller, M. (1987). Ewolucja pojęcia masy. In Heller, M., Michalik, A., and J. Mączka, (eds.), *Filozofować w kontekście nauki*. Krakow: PTT, pp. 152–169 (in Polish).

Heller, M. (2014). *Elementy Mechaniki kwantowej dla filozofow*. Krakow: Copernicus Center Press (in Polish).

Hidalgo, C. (2015). *Why Information Grows?* London: Penguin Books.

Hogan, J. (2018). Do our questions create the world? https://blogs.scientificamerican.com/cross-check/do-our-questions-create-the-world/.

Hussain, B. (2017). Storage technologies and their devices. *Medium*. https://medium.com/computing-technology-with-it-fundamentals/storage-technologies-and-their-devices-1594293868f0.

IASA Magnetic Tapes. (n.d.). Components-magnetic-tapes-and-their-stability. https://www.iasa-web.org/tc05/22111-components-magnetic-tapes-and-their-stability.

Jadacki, J. and Brozek, A. (2005). Na czym polega rozumienie w ogóle a rozumienie informacji w szczególności. In Heller, M. and Mączka, J. (eds.) *Informacja a rozumienie*. Biblos: Tarnów, pp. 141–155 (in Polish).

Jaynes, E. T. (1965). Gibbs vs Boltzmann entropies. *Am. J. Phys.* 33, 391–398.

Johnson, H. (2015). Electron lifetime is at least 66,000 yottayears. *PhysicsWorld.* https://physicsworld.com/a/electron-lifetime-is-at-least-66000-yottayears/.

Kane, G. (2005). The mysteries of mass. *Sci. Am.* Archived from the original on 10 October 2007.

Kastner, R. E. (2019). The 'delayed choice quantum eraser' neither erases nor delays. https://arxiv.org/ftp/arxiv/papers/1905/1905.03137.pdf.

Koks, D., Gibbs, P., and Chase, S. (2008). What is the mass of a photon? https://math.ucr.edu/home/baez/physics/ParticleAndNuclear/photon_mass.html.

Koks, D., Gibbs, P., and Chase, S. (2012). Does mass change with velocity? https://math.ucr.edu/home/baez/physics/Relativity/SR/mass.html.

Krzanowski, R. (2020). What is physical information? *Philosophies* 5(10). DOI: 10.3390/philosophies5020010.

Krzanowski, R. (2021). Ontological information. Investigation into the properties of ontological information. Ph.D. Thesis, UPJP2, Kraków, Poland, 2020. http://bc.upjp2.edu.pl/dlibra/docmetadata?id=5024&from=&dirids=1&ver_id=&lp=2&QI=.

Krzanowski, R. (2022). *Ontological Information.* London: World Scientific.

Landauer, R. (1961). Irreversibility and heat generation in the computing process. *IBM J. Res. Dev.* 5(3), 183–191.

Landauer, R. (1991). Information is physical. *Phys. Today* 44, 23–29.

Landauer, R. (1996). The physical nature of information. *Phys. Lett. A* 217(4–5), 188–193.

Landauer, R. (1999). Information is a physical entity. *Phys. A Stat. Mech. Its Appl.* 263, 63–67.

Makela, J. (2019). Wheeler's it from bit proposal in loop quantum gravity. *Int. J. Mod. Phys. D* 28(10), 1950129. https://doi.org/10.1142/S0218271819501293.

Markov, I. (2014). Limits on fundamental limits to computation. *Nature* 512, 147–154. DOI: 10.1038/nature13570.

Marletto, C. (2021). *The Science of Can and Can't.* London: Penguin.

Michael, S. S. (2019). Introduction to DRAM (Dynamic Random-Access Memory). https://www.allaboutcircuits.com/technical-articles/introduction-to-dram-dynamic-random-access-memory/.

Morehouse, C. (2003). Physics and technology forefronts. Non-volatile storage for information access. *APS News.* https://www.aps.org/publications/apsnews/200308/forefronts.cfm.

Mynarski, S. (1981). *Elementy Teorii Systemow i Cybernetyki.* Warszawa: Panstwowe Wydawnictwo Naukowe (in Polish).

Nave, R. (n.d.). Magnetic tape head. http://hyperphysics.phy-astr.gsu.edu/hbase/Audio/tape2.html.

Pierce, J. R. (1961). *Symbols, Signals and Noise.* New York: Harper Torch Books.

Plenio, M. B. and Vitelli, V. (2001). The physics of forgetting: Landauer's erasure principle and information theory. *Contemp. Phys.* 42, 25–60.

Polkinghorne, J. (2000). *Faith, Science & Understanding.* New Haven: Yale University Press.

Rajaraman, R. (2010). Signal encoding. CS 6710. https://www.ccs.neu.edu/home/rraj/Courses/6710/S10/Lectures/SignalEncoding.pdf.

Roosah, J. (2014). Computer memory: SSD, HDD. https://wiki.metropolia.fi/display/Physics/Computer+memory%3A+SSD%2C+HDD.

Rovelli, C. (2016). Meaning = information + evolution. arXiv:1611.02420v1. https://arxiv.org/pdf/1611.02420.pdf.

Schroeder, M. J. (2005). Philosophical foundations for the concept of information: Selective and structural information. FIS2005. http://www.mdpi.org/fis2005/. https://www.academia.edu/en/21000975/Philosophical_Foundations_for_the_Concept_of_Information_Selective_and_Structural_Information.

Schroeder, M. J. (2014). Ontological study of information: Identity and state. *Kybernetics* 43(6), 882–894.

Schroeder, M. J. (2017a). Spor o pojecie informacji. In *Studia Metodologiczne*, Vol. 35. Poznan: Adam Mickiewicz University, pp. 11–37. http://studiametodologiczne.amu.edu.pl/vol-34/ (in Polish).

Schroeder, M. J. (2017b). Structural reality, structural information, and general concept of structure. In *The Paper Presented at FIS 2017, IS4IS Summit,* Gothenburg, 12–16 June 2017. www.mdpi.com/journal/proceedings.

Seife, C. (2006). *Decoding the Universe.* London: Viking Press.

Shannon, C. E. (1948). A mathematical theory of communication. *Bell Syst. Tech. J.* 27, 379–423.

Shannon, C. E. and Weaver, W. (1949). *The Mathematical Theory of Communication.* Urbana, Illinois: University of Illinois Press.

Sole, R. and Elena, S. (2019). Viruses as *Complex Adaptive Systems.* Princeton: Princeton University Press.

Stallings, W. (2007). *Data and Computer Communications*, 8th edn. Upper Saddle River: Pearson Education, Inc.

Stoica, O. C. (2013). The Tao of it and bit. Fourth Prize at FQXi Essay Contest 2013. http://fqxi.org/community/essay/winners/2013.1. DOI: 10.1007/978-3-319-12946-4_5. arXiv:1311.0765.

Stonier, T. (1990). *Information and the Internal Structure of the Universe.* New York: Springer-Verlag.

Turek, K. (1978). Filozoficzne aspekty pojęcia informacji. *Zag. Filoz. Nauce* I, 32–41 (in Polish).
Vopson, M. M. (2019). The mass-energy-information equivalence principle. *AIP Adv.* 9, 095206.
Vopson, M. M. (2020). The information catastrophe. *AIP Adv.* 10, 085014.
Vopson, M. M. (2021). Estimation of the information contained in the visible matter of the universe. *AIP Adv.* 11, 105317.
Vopson, M. M. and Lepadatu, S. (2022). Second law of information thermodynamics. *AIP Adv.* 12, 075310. DOI: 10.1063/5.0100358.
Weaver, W. (1949). The mathematics of communication. *Sci. Am.* 181, 11–15.
Weizsäcker von, C. (1971). *Die Einheit der Natur, Munchen.* Berlin: Verlag; Warszawa: PIW (Polish edition, 1978).
Wheeler, J. A. (1978). The "past" and the "delayed-choice" double-slit experiment. In Marlow, A. R. (ed.) *Mathematical Foundations of Quantum Theory.* New York: Academic Press, pp. 9–48. DOI: 10.1016/B978-0-12-473250-6.50006-6.
Wheeler, J. A. (1988). World as system self-synthesized by quantum networking. In Agazzi, E. (ed.) *Probability in the Sciences.* Synthese Library, Vol. 201. Dordrecht: Springer. DOI: 10.1007/978-94-009-3061-2_7.
Wheeler, J. A. (1989). Information, physics, quantum: the search for links. In *Proceedings of 3rd International Symposium Foundations of Quantum Mechanics,* Tokyo, pp. 354–368.
Wilczek, F. (2015). *A Beautiful Question.* London: Penguin Books.
Zeh, H. D. (2002). The wave function: It or bit? In Barrow, J. D., Davies, P. C. W., and Harper Jr. C. L. (eds.) *Science and Ultimate Reality.* Cambridge: Cambridge University Press, pp. 103–120. arXiv:quant-ph/0204088.
Zeilinger, A. (n.d.). Why the quantum? It from bit? A participatory universe? Three far-reaching, visionary questions from John Archibald Wheeler and how they inspired a quantum experimentalist. https://www.metanexus.net/archive/ultimate_reality/zeilinger.pdf.

Chapter 5

Information, Computation, and Cognition: Their Emergence and Disappearance, and the Ever-Changing Entanglement of Their Meaning

Lorenzo Magnani

Department of Humanities, Philosophy Section and Computational Philosophy Laboratory, Piazza Botta 6, 27100 Pavia, Italy
lmagnani@unipv.it

Despite many interpretations and explanations, the concepts of information, computation, and cognition nevertheless produce unclear outcomes. I would argue that it can be very helpful to see the evolutionary origin of information and the first forms of cognition in humans as the result of dynamic coevolutionary interactions between internal processes in the brain and mind, the body itself, and the external world. This analysis is very useful (1) to dispel the most frequent misconceptions and the fundamental ambiguity of the aforementioned notions and (2) to accurately characterize "computation" as a developing term susceptible to ongoing meaning changes and its impact on the transformation of the meaning of information and cognition. To achieve this goal, I also use the dynamic notions of salience and pregnancy from Thom's catastrophe theory and Turing's original theories on the genesis of information, cognition, and computing in biological, inorganic, and artefactual entities, when physical computation is seen from the viewpoint of the ecology of cognition. I demonstrate how this perspective, which I refer to as "eco-cognitive computationalism", sheds more light on how computation is dynamically active in distributed physical entities of various kinds that have been appropriately transformed so that data can be encoded and decoded to obtain the desired results. I hope it will be clear that eco-cognitive computationalism does not intend to provide an ultimate and static definition of the concepts of information, cognition, and computation, as a textbook could. Rather, it intends to propose an intellectual framework that illustrates how we can understand their forms of "emergence" and the modification of their meanings by respecting their historical and dynamic character. I do believe that since computation is a notion that is prone to change and evolution, the other two concepts — which are already characterized by ambiguous definitions — are also vulnerable to changes in knowledge, technology, and cultural contexts.

1 Prologue: A Synergy of Meanings?

I am convinced that in present times with the aim of reaching an acceptable theory of the meaning of the concept of information, it is necessary to make an integrated reference to the role of semiosis, cognition, and computation. In sum, I think that the meaning of information is strongly entangled with the other three aspects: it is only by taking advantage of a simultaneous consideration of those four components that characterize human behavior that we can philosophically grasp their interactive meanings. Of course, this means that I do not consider the meaning of information as something stable and firm: it depends on the vicissitudes of other aspects, both from the theoretical perspective of the disciplines that study them and from the practical effects that current statuses of semiosis, cognition, and computation generate in our natural and technological world. Let's start exactly seeing how we can illustrate the first problem of the emergence of information and cognition, which will take advantage of a semiotic reference to the role of "signals" or "signs" (to use the Peircean terminology): this semiotic activity can be seen as the source of the first reverberating of a human brain activity capable of dealing with "information" and "cognition".

2 How Information and Cognition Emerge and Disappear

2.1 *Biological pregnances and saliences at the roots of the abstract concept of information*

I believe that the notion of pregnance, developed by Thom's (Thom, 1972, 1980, 1988) catastrophe theory[1] in the context of the so-called "semiophysical" approach and drawn from Wertheimer's Gestaltic concept of *Prägnanz*, might be very helpful to offer an integrated representation of the emergence of information (and consequently, of cognition). Here, it is crucial to emphasize how pregnances might occur in contexts of learning. According to Thom, phenomenal discontinuities (such a sound that breaks the stillness) are caught by creatures as *salient forms*, causing the sensory apparatuses of the organisms to echo what is happening outside in the physical environment. Recently, Raja (2018) used the well-known Gibson's ecological psychology (1979) to show an equivalent viewpoint: the animal sees its surroundings by "resonating" with "information" in energy patterns: as you can see the word information rapidly emerges, it is indeed this basic biological function of the sensorial human capacities to grasp and detect differences in the environment that we can see at the roots of the complicated and abstract concept of information. After all, a large portion of an organism's neurophysiology is engaged in both the detection of its external (distal) environment and, presumably, the generation of the possible deeper cognition that is appropriately connected.

The animal sensorial organs are undoubtedly collectors and sources of various types of what has been called "information", including visual, olfactory, and other types. The case of the

---

[1]Catastrophe theory is an area of mathematics that makes use of topology to describe situations (like a volcanic eruption or an economic crisis) marked by extremely abrupt changes in behavior resulting from just minor changes in the environment. It belongs to the field of dynamical systems study known as bifurcation theory.

"cognitive" role of information embedded in "salient" forms can be (1) first of all seen in the case of the emergence of "special" information, which is endowed with some hardwired (instinctual) biological value. It is at this biological level that we can see the source of the capacity to get information from the environment, and this capacity is first of all instinctual and meaningful (for instance, information about sex, food, and prey). Obviously, in the case of human beings, the amount of information that can be gathered, as well as its cognitive worth, rises as the sensory processes are further specialized (2) through learning, language, and concepts rather than just instinctive cognitive endowments. It is undeniable that both extragenetic and genetic characteristics influence pregnances: indeed saliences that acquire *cognitive meanings* are also pregnances. In the latter instance, they are channelized in animals and human animals as "cognitive" parcels of *learned knowledge*. A professional botanist, because is supported by knowledge, may obtain more data with her sensory organs:

> "Imagine a skilled botanist accompanied by someone like myself who is largely ignorant of botany taking part in a field trip into the Australian bush, with the objective of collecting observable facts about the native flora. It is undoubtedly the case that the botanist will be capable of collecting facts that are far more numerous and discerning than those I am able to observe and formulate, and the reason is clear. The botanist has a more elaborate conceptual scheme to exploit than myself, and that is because he or she knows more botany than I do. A knowledge of botany is a prerequisite for the formulation of the observation statements that might constitute its factual basis." (Chalmers, 1999, pp. 11–12)

The possibility of gathering rich knowledge is much increased in this scenario since a kind of plastic cognition (in this case scientific knowledge) is at work. Recent research in cognitive science confirms this fact (cf., for example, Raftopoulos, 2001a, 2001b): in humans, perception (at least in the visual case) is *semi-encapsulated*, which means that it is not strictly modular as Fodor (1984) argued. As a result, perception is neither fully "penetrable" by higher cognitive states (such as desires, beliefs, and expectations) nor is it encapsulated, hardwired, and domain-specific. As a result, there is a sizable quantity of information in visual perception that is, so to speak, theory-neutral, but there is also a sizable amount of theory-ladenness at play, as is the case, for instance, with so-called "perceptual learning". The fundamental idea of perception's cognitive impenetrability is not called into question, but it is acknowledged that perception is informationally "semi-encapsulated" and hence semi-hardwired; despite its primary bottom-up nature, it is not "insulated" from "knowledge", as it were.

By the way, as the reader will quickly surmise, a perception of salience is already crucial to promote the emergence of what is abstractly called "information" that refers to individual forms and, as a result, also to draw those entities that form the basis of what we refer to as "concepts": in fact, following Thom, when a human being perceives a salience, she is confronted with a kind of proto-concept, which is here intended as a class of equivalence between forms that refer to the same case. Later, a corresponding sign of any sort, such as a word or an icon, might be generated and refers to the recognized occurrence. This process, which Peirce dubbed semiosis, occurs when signs begin to refer to their objects. According to Peirce, we can affirm that human brains make up a number of signs that are employed to construct, display, or react to a number of other signs. By engaging in this semiotic yet blatantly cognitive activity, human brains develop into "minds" that are capable of thinking intelligently. As I explain in Section 2.4, signs may be externalized in both the natural and artefactual world, such as in writing and drawing. This process of externalization of the mind (also called "disembodiment of the mind") is what causes the dominating distributed nature of human cognition. As Longo (2009a) asserts, the development of writing and drawing is undoubtedly connected to the human transition to alphabetic natural language.

2.2 *How can we identify what is information? The emergence of cognition in terms of pregnancies and the role of paninformationalism*

Thom claims that pregnances, which might be either biological or physical occurrences, are in turn "non-localized entities emitted and received by salient forms" (Thom, 1988, p. 16). Two specific examples come to mind. The first one discusses natural organic phenomena:

(1) Infection is a pregnance (mediated by a virus, which is a material/biological medium) that regards healthy persons as invested saliences. These invested saliences can then re-emit the same contagion as a pregnance into the natural niche, where other media such as air or blood are the transmitters.

Even though there is a kind of paninformationalism in the literature where physical (or biological) information is thought extrapolated to every state of a physical (or biological) system that is consequently defined as an information-carrying state, it seems strange to assert that information is at play, in this case (Wolfram, 2002; Lloyd, 2006). In the context of the analysis provided in this chapter, I am not inclined to adopt this viewpoint since I qualify information as such only when related to a biological or artefactual agent that "selects" it as such from the variety of environmental occurrences, instinctually or in a plastically learned situation.[2]

The second example instead relates to not only a physical/biological situation but also too obvious information activities that acquire a cognitive "content", in the case of some organic agents:

(2) Worker honeybees employ indicators included within the iconic motions conveyed by a dance to communicate with one another and "inform" the other conspecific individuals, that is, the invested saliences, about the location. These dance signs represent information as a pregnance that indicates the site where the insects have found food, so playing a typical case of "animal cognition".

---

[2]The reader does not have to misunderstand me: I do not want to minimize the numerous good findings that physicists (to pick just a couple of recent examples (Chiribella *et al.*, 2012; Goyal, 2011)) and logicians have made, which offer mathematical frameworks for viewing quantum theory in the context of information processing principles. My main caution is not about these findings but rather about the potential misuse of the concepts of information (and computation) in physics and biology, a topic that is thoroughly discussed in the work of Longo (2009a, 2009b).

In this instance, the air acts as the pregnant medium, transporting and transmitting ondulatory sounds and light signals with the goal of causing a neurobiological effect at the target location — i.e., an experience approaching what we humans refer to as "psychic" — at the destination. In sum, in the second instance, we can argue that there is some transmission of *information* as well as of *cognition*. After all, it has been suggested that animals have representation-oriented behaviors, where non-linguistic pseudothoughts underlie plastic model-based "cognitive" processes (Dummett, 1993; Magnani, 2009, Chapter 5). The term "cognition" in this last context refers to behaviors that animals, from mammals to bacteria, specifically learn and generate, such as the dance of the honeybees, which are the result of learning and are suited for certain contexts.

Evidently, salient forms founded on instinctively wired behavior also convey "biological significance" and, as a result, can be regarded as transmitters of cognition, such as a prey for an aggressive predator or a predator for its prey: *salient forms of this type are called pregnant.* In terms of abductive cognition,[3] a pregnant input is cognitively *highly diagnostic* and a catalyst for the abductive generation of hypotheses: in the example of a chicken that detects the inherent pregnance of food, Charles Sanders Peirce already noted that an instinctual/hardwired abductive hypothesis about the food's suitability for consumption is quickly established.

2.3 The emergence of information as the fruit of an animal or artefactual creative act and its always possible disappearance

Now, using our naturalistic perspective, informed by the semiophysical framework obtained from Thom's catastrophe theory, we may solve the puzzle of the genesis of information and cognition. Since

---

[3] Abduction is a cognitive process that permits the reaching of hypotheses, both creative, such as in the case of scientific discovery, and diagnostic, such as in the case of physicians' praxis. In several of my most recent works, I have amply illustrated the concept of abduction (Magnani, 2001, 2009, 2017, 2022a).

we need to compare this form of answer to another kind of emergence, the one of computation, which I explain later in the following section, it is crucial to provide more details. Of course, my aim is to show the entanglement between information, cognition, and computation. Pregnancies, along with the knowledge and cognition they contain, can be abductively triggered, as I have shown in the earlier subsection, but they can also be formed from scratch.

The following is a helpful example: when a dog is exposed to a piece of meat while hearing the bell ring repeatedly, Pavlov's theory of operant conditioning explains that the nutritional pregnance of the meat is spread throughout the environment by proximity to the salient auditory form. The salient form, in this example the sound of the bell, which is also an *information* dispersed in the environment, is invested with the pregnance of the meat. The "information" provided by the bell (relating to the prospect of feeding) is now also given to that animal agent as a new, *created* [by humans] meaning. This new meaning is associated with a piece of information carried by the sound, which we may of course take to be also a *sign*, in semiotic terms, for that animal agent. The information, already salient before, becomes also pregnant, that is, as I have repeatedly illustrated above, full of meaning. Of course, repetition and/or resemblance might strengthen the pregnance, and step by step the bell — in the newly created cognitive qualities associated with the dog's "psychism" — will refer unambiguously to the meat. It goes without saying that the cognitive aspects are very dynamic because they are connected to pregnancies, which can also *disappear* when reinforcement is no longer possible, distant from the organism in play, or when, given this absence, the salient form is interested in another (ultimately new) pregnant form.

Of course, also all kinds of artifacts are carriers or even producers (for example, in the case of computational technologies) of information that human agents found more or less endowed of meanings depending on the context and their contingent cognitive endowments. This fact also explains the important role played by artifacts in human cognitive activities, as also illustrated by recent studies on the so-called EEEE (extended, embodied, embedded, and enacted) cognition, which have demonstrated that the "environmental situatedness" (and the characters of the artefactual environment) of human cognition and its evolutionary component may be used to better understand cognition in general.

2.4 Storing information and cognition out there, in the environment

As I mentioned above (Section 2.1), signs can be externalized in both a natural and artificial environment, such as in writing and drawing or in the more recent case of computational apparatuses. It is crucial to add that those signs could consist of mere information (data), which is devoid of specific meanings at "least" from the point of view of the contextual human beings that store them, but also meaningful pieces of information that count as such for the subjects that externalize them, which is pieces of information that are endowed of cognitive value. It is especially in this second case that we can see the results of the so-called process of externalization of the mind (or disembodiment of the mind), which is the root of the predominately distributed nature of human cognition. In fact, the primary process that characterizes the development of so-called material culture (related to the creation of the first handaxes, about 1.4 million years ago, the birth of what Mithen also categorizes as the "technical intelligence" of the primitive human mind) is the storing of signs and drawings in the external environment and manipulating external entities to transform them into artifacts. It is crucial to emphasize that cognition is diffused/distributed even in these early phases of human evolution. At the same time, a process of delegation of "knowledge" to external tools, props, and gadgets is taking place, and a complex semiotic activity is beginning. Additionally, Mithen notes that humans have perfected the art of disembodying their brains into the physical world; a linguistic utterance itself might be taken as a disembodied thought even if, however, these utterances only last a little period of time, but material culture remains (Mithen, 1999, p. 291). We can conclude that minds are "extended" and artificial in themselves, according to contemporary study in cognitive science (Clark, 2008). We might argue that, in a way, human primitive brains transform into a kind of universal "intelligent" machines equipped with a rich activity of consciousness.

It is also crucial to emphasize how well the reference to the function of external representations and the disembodiment of the mind may explain key dynamic elements of human cognitive processes. Brains and minds, so to speak, "extend" themselves into the outside world in order to semiotically represent signs, words, icons, and drawings in rocks, boards, sheets of paper, PCs, and materials of various types, as well as in order to use them as anchors for assisting in the generation of new thoughts, ideas, and concepts that do not have a "natural home" within them. New knowledge and information are often created within this distributed process of semiotic activity.[4]

3 Cognitive Domestication of Ignorant Entities Thanks to the Emergence of Computation

In our era, we face a massive delegation of information and knowledge to a highly sophisticated and engineered external environment, which is one that has been generated thanks to the computational revolution. This means that, as I say, a *computational domestication* of ignorant entities has been realized. In the second and last part of this chapter, I deal with the illustration of what we can call the "archeology" of this important innovation, in terms of an approach I have called "Eco-Cognitive Computationalism" (Magnani, 2022b).

Eco-cognitive computationalism sees computation as active in physical entities that have been appropriately changed so that they become what I refer to as "information and, consequently, potential

---

[4]The readers can refer to chapters two and three of my book (Magnani, 2009) if interested in the activity of the emergence and subsequent expansion of information and knowledge in this distributed perspective and with reference to hypothetical (abductive) reasoning. Utilizing instances from both the paleoanthropological investigations on the cognitive processes of the primitive mind and the case of mathematical and scientific knowledge, the importance of external representations and the accompanying delegations of cognition to external instruments and artifacts are shown.

cognitive mediators" where data may be encoded and decoded to produce useful outcomes. We must state right away that eco-cognitive computationalism, even if one dogmatically believes that cognition is computational, does not reduce computation to digital computation (that is, to the processing of strings of digits in accordance with rules).[5] Eco-cognitive computationalism does not intend to provide an ultimate and static definition of the concepts of information, cognition, and computation, as a textbook could. Rather, it intends to propose an intellectual framework that illustrates how we can understand their forms of "emergence" and the modification of their meanings by respecting their historical and dynamic character. An example of the plurality of the approaches to computation is represented by the fact that many ignorant entities have been computationally "domesticated", not only digital machines, and this process is still at work.

The father of what I have called computational domestication is of course Turing: the illustration of some of Turing's speculations on how the so-called "unorganized brains" are transformed into organized "machineries" is very important and worth analyzing. Indeed, Turing expressly states that the brain of a child is an unorganized machinery, which may be organized by adequate interference training, using a weird but helpful comparison. The organization may lead to the machine being transformed into a universal machine or something similar. He finally also affirms that, in terms of evolution and genetics, this depiction of the cortex as an unstructured machine is extremely satisfying (Turing, 1969, p. 16).

I should also point out that Turing used the comparison between the brain of a child and an unorganized machine in a speculative perspective and just served as a key heuristic in his creative cognitive processes: it is clear from fundamental neuroscientific studies that the brain is instead a highly ordered/organized machine. Turing only makes reference to the idea that newborns' cortices are similar to empty canvases that are "socially" filled through language. According

---

[5]It is commonly known that this perspective is more in line with the traditional theory of cognition as the manipulation of linguistic or sentence-based entities.

to this viewpoint, the baby cortex's alleged random development has less to do with neurobiological issues and more to do with the fundamental lack of training — specifically, the absence of signs sent through the senses from the outside (including social) world, such as language. Therefore, only a limited degree of description is connected to the notion of "unorganized machinery": according to Turing (1969, p. 9), it is possible that the same machine may be viewed by one man as organized and by another as unorganized.

This philosophical viewpoint first and foremost goes ahead to demonstrate how knowledge, meaning, and the first primitive forms of cognition *evolved* as a result of complex eco-cognitive interaction and concurrent *coevolution* through time of the states of the brain/mind, body, and external environment. In addition, it provides the conceptual foundation that explains how Turing's development of both the universal logical computing machine and the universal practical computing machine is made possible by an "imitation" of the aforementioned "educational" (but also of an "evolutionary", see the following) process. In this last instance, the dual invention is realized through the externalization of computational capabilities in those artifactual physical entities that compute on behalf of some human or artefactual agents: those computers that, from the viewpoint put forward by Turing, I referred to as "mimetic minds" (Magnani, 2006).

Following Turing, I favor the view that the development of information, meaning, and the earliest primitive forms of cognition occurred as a result of complex eco-cognitive interactions and the concurrent coevolution over time of the states of the brain/mind, body, and surrounding environment. This viewpoint rejects any long-term stability in the meaning of terms like information and cognition. The same can be said in the case of terms like information processing and computation: it is crucial to avoid conflating "computation" with digital computation (as implied by classical mathematical theory), given the appearance of the exploitation of new substrates capable of carrying computational processes, as emphasized in the following.

In summary, the concept of computation I am adopting here goes beyond digital computation, that is, the processing of strings of digits according to rules; it also refers to other computations performed by physical devices, such as brains, with their activity of modifications of the configurations of neural networks (and the electrical activity — neurons fire and provide electrical inputs to other neurons) and chemical processes (neurotransmitters and hormones) — and also the so-called neural computers, as well as analog computers, such as "morphological computing".[6] It is an expansion of the notion of computing, which I stated above is not static but rather changes with changes in theory and application.

3.1 *Turing on the emergence of information, cognition, and computation in organic, inorganic, and artefactual agents*

I already focused on Turing's novel theories on the emergence of computing in digital physical entities as connected to an analogy to the practice of imitating human education in my book (2022b), chapter one. As I previously stated, taking into account computation from the standpoint of the aforementioned cognitive/epistemological analysis as well as considering it as dynamically active in distributed physical entities of various types, suitably transformed so that data can be encoded and decoded to obtain appropriate results, sheds some light on what I refer to as "eco-cognitive computationalism". It is in this perspective that we can find the emergence of computation as a further complication of the meanings of information and cognition: computation enriches the entanglement already active between these last two concepts.

In addition to the basic aspect concerning the analogy to mimicking human education, the birth/emergence of computation is strictly related — in Turing's seminal work *Intelligent Machinery* (1948) — to heuristic cognitive processes that take advantage of the study

---

[6]The reader interested in the recent expansions of the notion of computation can refer to my article (Magnani, 2021) and the book (Magnani, 2022b).

of the role in the external environment of organic bodies, physical entities, and artifacts. When considering the possibility of building intelligent machines, Turing first of all states right away that human intelligence can only be produced if a suitable education is carried out (Turing, 1969, p. 3) and draws an analogy between computational machines, which are physical artifacts, and human brains, which are organic entities. He suggests the idea of *unorganized machine* and uses the baby human brain as an illustration. This is a "natural" organic entity that can be trained through "rewards and punishments": is it possible to imagine something similar in the case of a machine to train it to perform cognitive behaviors? Together with unorganized machines, Turing also lists *paper machines*[7] and the two famous new fundamental kinds of machinery he invented: the (*Universal*) *Logical Computing Machines* (LCMs), which are considered discrete machines, and the (*Universal*) *Practical Computing Machines* (PCMs), which, while reflecting the fact that given any job which could have been done on an LCM one can also do it on one of these digital computers, are machines insofar as they are external physical entities (inorganic and "artefactual") that manage their saved information in a form very dissimilar from the tape form (Turing, 1969, p. 8).

In sum, due to Turing's claim that by means of a system that is reminiscent of a telephone exchange, it is made possible to obtain a piece of information almost immediately by "dialing" the position of this information in the store, PCMs can easily solve the problem even in the presence of a large number of steps that can be involved in computation along the tape (p. 8). Turing also notes that "almost" all PCMs — that is digital computers — under construction in his day possess the core characteristics of universal logical computing machines.

---

[7]Paper machines can be explained by referring to the environmental conditions that transform a human being into a computational machine: "It is possible to produce the effect of a computing machine by writing down a set of rules of procedure and asking a man to carry them out. [...] A man provided with paper, pencil, and rubber, and subject to strict discipline, is in effect a universal machine" (Turing, 1969, p. 9). Turing calls this kind of machine "Paper Machine".

I argue that Turing's discoveries and observations are interesting from both a historical and epistemological standpoint. In the seminal article that was cited, he first sees computation as active in physical entities that have been suitably modified through "education". Turing claims that programming imitates human education in the hopes that the machine would be able to provide certain expected replies to specific commands.

We can conclude by saying that, thanks to Turing, the emergence of computation breaks into the traditional received epistemological scenario that contemplated only the entanglement between the meaning of information and that of cognition. Now information can also be computationalized and, fundamentally, to the aim of performing cognitive activities similar to the ones human beings used. Ignorant physical entities, such as precision-engineered computers become carriers of unexpected kinds of information and cognition: not only, currently, the process is still at work in the sense that new and multiple substrates are exploited for computation, from quantum to morphological computing, from DNA computing to the ones based on thermodynamical properties, etc. (Magnani, 2021).

3.2 The emergence of computation in digital physical entities (PCMs) seen as mimicking human education

Human education can be advantageously used as a model for the process Turing refers to as the "education of machinery". It is not necessary for the reader to misinterpret the intended meaning of this analogy. Currently, it is clear that computer training differs significantly from human education. For instance, in the latter situation, semantic information has a key role, but in the former, it plays no part. Instead, it is crucial to note that Turing's analogy played a key heuristic role in his search for the new idea of a universal logical machine. In this case, the few, straightforward parallels matter only to the extent that they lead to novel insights, and from this vantage point, the differences between the two cases are obviously ignored.

By "mimicking education", Turing says we should plan to modify the machine until it could be relied on to produce definite reactions to certain commands (Turing, 1969, p. 14): something similar occurred in the case of a PhD student who, for a few decades, profited from contacts (or, maybe more accurately, interferences) with other people so that "[...] a large number of standard routines will have been superimposed on the original pattern of his brain" (*ibid.*). Turing also notes that (1) those interferences are mostly with other men as a result of visual and other types of inputs and (2) a man concentrating replicates a machine without interference even if he is conditioned by prior encounters.

I have anticipated that, following Turing, the infant brain can be conceived as an unorganized machine. This picture of the cortex as unorganized is for Turing "very satisfactory from the point of view of evolution and genetics" (1969, p. 16). Additionally, in this evolutionary perspective, which is undoubtedly speculative but anticipates current findings in cognitive paleanthropology, the existence of the human cortex can only be justified in terms of how it is organized, thanks to a sort of *coevolution* between the human cortex and external information available to organize it: "[...] the possession of a human cortex (say) would be virtually useless if no attempt was made to organize it. Thus if a wolf by a mutation acquired a human cortex there is little reason to believe that he would have any selective advantage". (*ibid.*)

Due to this, the use of a large cortex (and its proper organization) requires an environment that is suitable for it: "If however the mutation occurred in a milieu where speech had developed (parrot-like wolves), and if the mutation by chance had well permeated a small community, then some selective advantage might be felt. It would then be possible to pass information on from generation to generation". (*ibid.*) Even though it is speculative and primarily endowed with a dominant heuristic role, as I have already noted, Turing's argumentation about the "unorganized" brains, considered as kinds of blank slates that are socially fulfilled through language, deserves attention because it exhibits unusual attention to the significance of phylogenetic mechanisms present in human cognition.

In sum, to transmute an unorganized human cortex into a universal machine, we need a lot of

- *meaningful information*, which is granted by
- *speech* (even if extremely basic — parrot-like wolves), and at the same time a
- *social background*, in which a variety of "techniques" are accessible and learnable.

We can draw a simple conclusion from this: a large cortex only represents an evolutionary advantage if it is fertilized by a large storage of meaningful information endowed with *cognitive value* (knowledge) carried by external supports and tools that only a group of fully developed humans can have.

Furthermore, Turing's argumentation comes very near to proving his point: *programming a machine mimicking education* also needs an activity of

- *reward* and *punishment*, that is, behaviors that can be and that cannot be followed

as they are necessary to instruct a youngster, and this fact shows that organization can only occur through those two inputs.

Finally, the infant cortex is transformed into an intelligent one thanks to

- *discipline*, but also thanks to
- *initiative*,

which Turing views as the two main aspects of a process that has to be studied as it is occurring in humans to "copy it in machines" (p. 21). An illustration of a circumstance that calls for initiative is, for example, "Find a number n such that ..." and "see if you can find a way of calculating the function which will enable us to obtain the values for arguments ...", which is completely equivalent to creating a *program* to put on the machine at stake.

It would appear simple for Turing to develop (and subsequently allow the emergence of) the new concept of computation using the idea of human education I have already shown. Moreover, Turing finally notes that there are two forms of interference that universal machines encounter: when parts of the machine are taken out and replaced with others, whole new machines are created thanks to a *screwdriver* interference. In a *paper* interference, the behavior of the machine is altered by the addition of new information.

Of course, it is important to emphasize that paper interference provided consists of both external and material information. Turing believed it was conceivable to create a thinking machine since humans had already developed the ability to mimic many aspects of a human being (microphone, television, etc.). Given that the electrical circuits of electronic computing devices have the capacity to carry and store information analogous to that of nerves, the functions of nerves may also be imitated in terms of appropriate electrical models.

As described above, Turing argues that a big cortex only represents an evolutionary advantage if it is fecundated by a large storage of valuable information and knowledge carried by external supports and instruments that only a developed collective of people can have. In a nutshell, there must be already present *information* as well as *cognitive* contents conveyed through language, signs, icons, symbols, etc. Analogously, Turing maintains that as fresh information is inserted into a computer, the behavior of the machine changes: paper *interference* is required to modify an external physical artifact device (such as an electronic machine).

Considering it from this angle: the digital computer (a discrete state machine) is unquestionably an alphabetic machine, made possible by the development of alphabetic natural language by humans (of course, as we have observed, we also know it is a logical and formal machine, LCM) (Longo, 2009a). This fact, according to Longo — and I concur with him, is the primary cause of the massive "discretization of knowledge" the invention of computation has sparked. In fact, the alphabet divides the "continuous" natural language into undistinguished atoms, which are then turned into letters. When competent human agents are able to perceive and control them, they are meaningless in and of themselves; meaning is instead created by combining them syntactically. Given the discreteness of digital machines that is at the foundations of their *imitation* capacity — they are mimetic machines, as I have already illustrated — Turing himself contrasted this merely imitation power to the much more powerful epistemological power of the *modeling* capacity of mathematics.

4 The Entanglement of Information Processing, Cognition, and Computation

4.1 *From pure signs to signs that carry an information accompanied by cognitive meaning*

I said above that there is an entanglement between the meanings of information, cognition, and computation, and that this entanglement is also at the roots of the dynamic character of those meanings. The reader can now also easily see the changes of meaning that have been generated in the case of the concepts of information and cognition by the so-called "computational revolution", I have explained in terms of domestication of ignorant entities. Imagine a digital computer in the absence of its computationalization thanks to some software: it is a sophisticated technological entity prevalently ignorant, even if it can carry some kinds of information and cognition about its technological aspects when seen by competent people. In this last case, it is unable to perform information processing tasks capable of potentially expressing cognitive transfers of knowledge. It is this last capability that was created by Turing thanks to the computationalization (domestication I said) of that ignorant technological entity. After this intellectual achievement, the meaning of the concepts of information and cognition is extended and submitted to changes: information and cognitive activities can be found in new environmental situations, unexpected and unconventional, with respect to the non-yet-computational past. The hitherto uneducated/ignorant entity is now capable of "processing information" and unexpected pieces of information can be seen embedded and managed in new material technological entities. Being able to process information is at the same time the source of the huge computational capacities to manage cognitive content. Furthermore, as I have already said above, we have to remember that the current research that increases the number of substrates exploited for computation aims at further dynamically extending their meanings, beyond any present imagination.

In conclusion, the definition of the meaning of information might vary based on the circumstances and the intellectual viewpoints at stake. However, it can be surely said that information, when computationally "processed", concerns the modification of all kinds of information, from those aspects considered as subject to the constraints of communication theory (Shannon information, in which measures of information "are not" also measures of meaning), to the information that is *ipso facto* endowed with *meaning*. The issue of information's emergence (but also of the creation of "new" information, for instance, by abduction) is crucial when information is also seen by certain actors as having cognitive worth. Falsity and truth are obviously at play when it comes to information that some agents view as having a cognitive worth.

4.2 Is cognition computation? Yes, it is, but it is not only computation

When someone asks "Is cognition computation?" we may easily respond "yes", but "it is not merely computation", thus we cannot equate cognition with computation. Both information processing and computing are undoubtedly involved in cognition, and defining the specific types of computation and information processing that are engaged in cognition is a respectable and challenging undertaking,[8] even if it is doomed to become obsolete: in fact, I believe that because the idea of computing is dynamic and ever-changing, the other two

---

[8]By demonstrating how digital computation is implemented in physical systems, Fresco (2013) successfully aims to clarify the concept of digital computation in modern cognitive science without succumbing to pancomputationalism, the belief that every physical system is a digital computing system and can be explained in computational terms.

concepts of information and cognition — which already have ambiguous definitions — must likewise change along with knowledge, technology, and cultural framework advancements.

Let us reiterate: surely cognition requires the transfer of signals (or signs in Peircean terms), which are submitted to the constraints and limitations of communication theory (Shannon information), but cognition also necessarily involves the fact that the signs themselves have to be interpreted by an agent as meaningful. Additionally, in the case of "cognition", signs are made up of models, images, etc. rather than of only linguistic or sentence-like structures, as was the case under the traditional paradigm of symbolic AI. Signs are typically seen as representational in cognition, which means that cognition, when carried out, for instance, by digital processing, takes place over representations.

5 Conclusion

I have demonstrated how information and the first forms of cognition emerged in humans by utilizing the dynamic concepts of salience and pregnance, derived from Thom's catastrophe theory. I clarified the terms "information", "cognition", and "computation", highlighting the fact that their dynamic nature can only be characterized by taking into account their conceptual entanglement from both a cognitive and an epistemological standpoint. To deal with the emergence of computation, I provided a new interpretation of Turing's original theories regarding the genesis of information, cognition, and computation in biological, inorganic, and artifactual agents, seeing physical computing in the context of the ecology of cognition. I placed a lot of emphasis on Turing's novel theories on the development of computing in digital physical things and how they are comparable to the mimicking of human education. Overall, I contended that in present times with the aim of reaching an acceptable theory of the meaning of the concept of information, it is necessary to make an integrated reference to the role of semiosis, cognition, and computation. The meaning of information is strongly entangled with the other three aspects: it is only by taking advantage of a simultaneous consideration of those four components that characterize human behavior that we can philosophically characterize their interactive meanings. Of course, this means that I do not consider the meaning of information as something stable and firm: it depends on the vicissitudes of other aspects, both from the theoretical perspective of the disciplines that study them and from the practical effects that current statuses of semiosis, cognition, and computation generate in our natural and technological world.

Acknowledgments

Parts of this chapter are excerpted from chapter one of L. Magnani, *Eco-Cognitive Computationalism. Cognitive Domestication of Ignorant Entities*, Springer, Cham, 2022.

References

Chalmers, A. F. (1999). *What is this Thing Called Science*. Cambridge, Indianapolis: Hackett.
Chiribella, G., D'Ariano, G. M., and Perinotti, P. (2012). Quantum theory, namely the pure and reversible theory of information. *Entropy*, 14, 1877–1893.
Clark, A. (2008). *Supersizing the Mind. Embodiment, Action, and Cognitive Extension*. Oxford, New York: Oxford University Press.
Dummett, M. (1993). *The Origins of Analytical Philosophy*. London: Duckworth.
Fodor, J. (1984). Observation reconsidered. *Philosophy of Science*, 51, 23–43. [Reprinted in Goldman (1993)].
Fresco, N. (2013). *Physical Computation and Cognitive Science*. Cham, Switzerland: Springer.
Gibson, J. J. (1979). *The Ecological Approach to Visual Perception*. Boston, MA: Houghton Mifflin.
Goldman, A. I. (ed.). (1993). *Readings in Philosophy and Cognitive Science*. Cambridge, MA: Cambridge University Press.
Goyal, P. (2011). Information physics — Towards a new conception of physical reality. *Information*, 3, 567–594.

Lloyd, S. (2006). *Programming the Universe: A Quantum Computer Scientist Takes on the Cosmos*. New York, NY: Knopf.

Longo, G. (2009). Critique of computational reason in the natural sciences. In E. Gelenbe and J.-P. Kahane (eds.), *Fundamental Concepts in Computer Science*. London: Imperial College Press/World Scientific.

Longo, G. (2009). Turing and the "imitation game" impossible geometry. Randomness, determinism and programs in Turing's test. In R. Epstein, G. Roberts, and G. Beber (eds.), *Parsing the Turing Test. Philosophical and Methodological Issues in the Quest for the Thinking Computer*. Dordrecht: Springer, pp. 377–411.

Magnani, L. (2001). *Abduction, Reason, and Science. Processes of Discovery and Explanation*. New York: Kluwer Academic/Plenum Publishers.

Magnani, L. (2006). Mimetic minds. Meaning formation through epistemic mediators and external representations. In A. Loula, R. Gudwin, and J. Queiroz (eds.), *Artificial Cognition Systems*. Hershey, PA: Idea Group Publishers, pp. 327–357.

Magnani, L. (2009). *Abductive Cognition. The Epistemological and Eco-Cognitive Dimensions of Hypothetical Reasoning*. Heidelberg/Berlin: Springer.

Magnani, L. (2017). *The Abductive Structure of Scientific Creativity. An Essay on the Ecology of Cognition*. Cham, Switzerland: Springer.

Magnani, L. (2021). Computational domestication of ignorant entities. Unconventional cognitive embodiments. *Synthese* 198, 7503–7532. Special Issue on "Knowing the Unknown" (guest editors L. Magnani and S. Arfini).

Magnani, L. (2022a). *Discoverability. The Urgent Need of an Ecology of Human Creativity*. Cham, Switzerland: Springer.

Magnani, L. (2022b). *Eco-Cognitive Computationalism. Cognitive Domestication of Ignorant Entities*. Cham, Switzerland: Springer.

Mithen, S. (1999). Handaxes and ice age carvings: Hard evidence for the evolution of consciousness. In A. R. Hameroff, A. W. Kaszniak, and D. J. Chalmers (eds.), *Toward a Science of Consciousness III. The Third Tucson Discussions and Debates*. Cambridge: MIT Press, pp. 281–296.

Raftopoulos, A. (2001). Is perception informationally encapsulated? The issue of theory-ladenness of perception. *Cognitive Science* 25, 423–451.

Raftopoulos, A. (2001). Reentrant pathways and the theory-ladenness of perception. *Philosophy of Science* 68, S187–S189. [*Proceedings of PSA 2000 Biennal Meeting.*]

Raja, V. (2018). A theory of resonance: Towards an ecological cognitive architecture. *Minds and Machines* 28(1), 29–51.

Thom, R. (1972). *Stabilité Structurelle et Morphogénèse. Essai d'une théorie générale des modèles.* Paris: InterEditions. [Translated by D. H. Fowler, *Structural Stability and Morphogenesis: An Outline of a General Theory of Models*, W. A. Benjamin, Reading, MA, 1975.]

Thom, R. (1980). *Modèles mathématiques de la morphogenèse.* Paris: Christian Bourgois. [Translated by W. M. Brookes and D. Rand, *Mathematical Models of Morphogenesis*, Ellis Horwood, Chichester, 1983.]

Thom, R. (1988). *Esquisse d'une sémiophysique.* Paris: InterEditions. [Translated by V. Meyer, *Semio Physics: A Sketch*, Addison Wesley, Redwood City, CA, 1990.]

Turing, A. M. (1969). Intelligent machinery [1948]. In B. Meltzer and D. Michie (eds.), *Machine Intelligence*, vol. 5. Edinburgh: Edinburgh University Press, pp. 3–23.

Wolfram, S. (2002). *A New Kind of Science.* Champaign: Wolfram Media.

Chapter 6

Theoretical Unification of the Fractured Aspects of Information*

Marcin J. Schroeder

Akita International University
Faculty of International Liberal Arts
Yuwa, Tsubakigawa, Akita 010-1211, Japan
mjs@gl.aiu.ac.jp

This chapter has as its main objective the identification of fundamental epistemological obstacles in the study of information related to unnecessary methodological assumptions and the demystification of popular beliefs in the fundamental divisions of the aspects of information that can be understood as Bachelardian rupture of epistemological obstacles. These general considerations are preceded by an overview of the motivations for the study of information and the role of the concept of information in the conceptualization of intelligence, complexity, and consciousness justifying the need for a sufficiently general perspective in the study of information, and are followed at the end of this chapter by a brief exposition of an example of a possible application in the development of the unified theory of information free from unnecessary divisions and claims of superiority of the existing preferences in methodology. The reference to Gaston Bachelard and his ideas of epistemological obstacles

---

*This work is dedicated to the memory of Mark Burgin who contributed to the study of information and in particular to its theoretical development not only through the writing and publishing of many works of fundamental importance for the subject but also by his work on editing books, organizing conferences, and mentoring younger researchers.

and epistemological ruptures seems highly appropriate for the reflection on the development of information study, in particular in the context of obstacles such as the absence of semantics of information, negligence of its structural analysis, separation of its digital and analog forms, and misguided use of mathematics.

1 Introduction

The myth of possible or even necessary separation of the syntactic and semantic aspects of thought and language persists and spills over to the study of information where it is frequently conflated with the belief that the quantitative analysis of information eliminates the need for a seemingly inferior inquiry of its qualitative characteristics, such as structure or modes of existence.

The exclusive focus on the quantitative methodology in the study of information is a product of the much wider tendency in the scientific methodology of many domains of inquiry produced by the illusion of precision and easy understanding of results expressed with the use of numbers. The focus on numbers brings another confusion into the study of information regarding the distinction between analog and digital types of information and information processing.

Yet another result of the belief in the distinction and superiority of the quantitative methodology is the conclusion drawn from the fact that probability theory and statistics have so broad and successful applications in a very large variety of disciplines that the probabilistic description of the concept of information explains its omnipresence in the inquiry of reality. Surprisingly, the calls for reversing the roles of information and probability and the use of the concept of information as a foundation for the study of probability made by several distinguished mathematicians of the 20th century did not receive a sufficient response.

This chapter has as its main objectives the identification of such fundamental epistemological obstacles in the study of information related to unnecessary, hidden methodological assumptions and the demystification of popular beliefs in the fundamental divisions of the aspects of information in the hope that they bring Bachelardian epistemological unifying rupture. These general considerations are followed by a brief exposition of an example of the unified theory of information free from unnecessary divisions and claims of superiority of the existing preferences in methodology.

The reference to Gaston Bachelard and his ideas of epistemological obstacles and epistemological ruptures seems highly appropriate for the reflection on the development of information study. Bachelard was aware of the unavoidable obstacles created by our intuitive, common-sense conceptual and methodological frameworks which have to be identified and finally eliminated to achieve scientific progress (Bachelard, 1986; Tiles, 1984). Bachelardian rupture is desired particularly in the context of obstacles, such as the absence of unrestricted to particular contexts semantics of information, negligence of its structural analysis, and separation of its digital and analog forms.

The study of the epistemological obstacles in the study of information is preceded in this chapter by an exposition of the reasons why their elimination is of great importance not just to satisfy researchers' curiosity. The main argument for revisiting the methodology of information studies and for maintaining their high level of generality is the relation of information to several other fundamental but insufficiently conceptualized ideas such as intelligence, complexity, and consciousness which makes information a suitable defining concept.

Recent hot discussions of the danger of lost control over information technologies bring the subject to the attention of the global audience. All these three ideas of intelligence, complexity, and consciousness are at the center of the discussions, despite their vague, common-sense understanding. The study of information is of its own interest, but its role as a foundation for the studies of these even less understood ideas makes it a prerequisite for solving one of the greatest challenges for humanity. It is argued in the following text that the only way to prevent the loss of control over information technology is not by blind, uninformed preventive legal regulation but by raising the understanding of the central concept of information.

2 Urgent Motivation for the Study of Information

2.1 *Understanding generative artificial intelligence*

The concept of information, or rather in the absence of a commonly accepted definition its phantom invoked in discourses to create mutual understanding, is haunting the intellectual discussions on virtually every subject. Information is in the ghostly company of other formidable phantoms of concepts, such as consciousness, complexity, intelligence, volition, computation, and life. All of them, or rather their elusiveness, generate myths and anxiety about hidden dangers while the real danger, ignorance of their meaning and role, is in the open view.

Most recently, the biggest splash has been made by the scepter of artificial intelligence whose name in the tradition of all taboos is fearfully expressed with the omnipresent acronym AI ("you know who", or in this case, "you know what"). Yet another source of anxiety expressed in media and on the Internet is the possibility of encounters with extraterrestrial intelligence (ETI). AI seems dangerous as it can escape human rational control as if humans were in rational control of any large-scale phenomena. Of course, there are several well-known existential threats to humanity, such as climate change, non-sustainable use and management of natural resources, pandemics, nuclear weapons, and misuse of nuclear technology. However, each of them is dangerous not because of its inevitability or the lack of knowledge or understanding of prevention but because of the rather irrational actions of humans or the lack of a coordinated effort. There is a common belief that if only humanity decides to be rational, the threats would be overcome, so we do not have to worry about the danger.

The escape of AI systems from human rational control and encounters with ETI seem more threatening because AI can become more intelligent than humanity or humans and ETI would have been more intelligent if it could visit us from far away. It does not matter, at least for many of those who advocate the moratorium on AI research, that it is not clear what it means to be intelligent or why something intelligent out of human control is more dangerous than something that lacks or is deficient in intelligence (for instance, a deranged, immoral, and power-greedy human who managed to take control of nuclear weapons through the corruption of political mechanisms).

The dangers of abuse, misuse, or loss of control of information technologies are real, but this applies to every type of technology (e.g., nuclear technologies). However, instead of calling for a moratorium on AI research before there is a sufficient legal regulation system, it is necessary to stimulate and support research on the concepts and ideas involved in the development of AI technology. Can anyone develop an effective legal regulation system for the developers of AI technologies to prevent all possible harm to humanity when so little is known or understood about the fundamental concepts of information studies?

The only way to prevent the harmful unexpected impact of technologies (all technologies, not just AI) is to create a legal requirement of sufficient investment in independent research on all aspects of their development, creation, and use. The requirement should apply not only to commercial developers of technologies but also to governmental agencies interested in their development. The less predictable the consequences of a given technology, the more fundamental and more intensive research should be mandated. In the case of AI technologies, it is almost impossible to make any predictions in the absence of sufficiently developed studies of consciousness, complexity, intelligence, volition, computation, and life. Thus, any organization engaging in the development of AI technologies should demonstrate investment in these studies. The lack of such investment could and should be used in the future as evidence for possible future liability. There should be no excuse for the insufficient knowledge of the possible consequences of technological innovation in the absence of documented sufficient investment in independent research not just on the subject of engineering, economic, or social aspects of its products but also on the more general task of understanding all phenomena involved in technological processes.

The first step in the direction of understanding AI technologies is the prevention of confusion proliferated by the everyday language of the news and commentaries on the subject. The expression "AI can do..." or "AI becomes more intelligent than humans" suggests that AI is an entity or agent. There are important issues involved in understanding the words "artificial" and "intelligence", but even if we ignore them, the expressions suggesting the uniform and independent ontological status of AI are misleading. Fortunately, this category error has been identified and criticized in the recent editorial in *Nature Reviews Physics* (Shevlin and Halina, 2019; Editorial, 2023).

> "Anthropomorphic language is widespread in physics: masses 'feel' the gravitational potential, photons 'know' the state of their entangled partner and spins generally 'want' to align. [...] First, we will try to avoid at all costs the use of 'the AI/an AI' due to its unfortunate suggestion of agency. Instead, we will either change to 'the AI system/an AI system' or be very clear what we are talking about." (Editorial, 2023)

There are many reasons why this category error may have detrimental consequences for understanding the real reasons for concerns about AI. In the context of our paper, it makes an impression that AI is an existing and independent entity that can be considered in separation from the more fundamental study of information. There is no way to acquire knowledge and understanding of AI without a prior deep and extensive understanding of information and its involvement in phenomena related to consciousness, complexity, intelligence, volition, computation, and life. Of course, there is nothing wrong with using an informal abbreviation AI for the name of the entire complex of information technologies, but statements that AI can or cannot do something are an abuse of language.

AI systems are simply instances of devices designed by humans with the possible help of technological tools in which information dynamics is used to perform some actions. Thus far, the operation of such systems is controlled by human agents, but the control is declining. The devices of generative AI systems are designed to minimize the involvement of slow human agents. This follows the intentional and commonly accepted direction of technological progress which started two hundred years ago from the mechanization of work (elimination of the work of human or animal muscles) and was followed by automation (elimination of human control of machines). The difference in the design of the generative AI systems is in the elimination of meta-control. The automata used in manufacturing follow the process designed and controlled by human programmers. The generative AI systems increasingly act as "black boxes" whose operation is in principle known, but whose actual functioning is more and more autonomous based on the input not from particular human agents but from the data obtained from the Internet. Thus, the control by humans is relinquished to the huge data reservoir on the Internet that is "teaching" the system how to operate.

The question of whether generative AI systems are intelligent can be reformulated as to whether the Internet (or any other dynamical data set) is intelligent. Another possibility is to look for the intelligence of generative AI systems in the ability to use the Internet as a non-intelligent information resource consisting of discarded byproducts of human activities. This may be reassuring because these resources are human products and it seems that the exceeding of human capacities based on rather random human individual contributions is unlikely, but such optimism is unwarranted as information technology may detect patterns in the collective human activities that are beyond individual human comprehension. It seems a bizarre idea that any new great development in science (comparable to relativity theory or quantum mechanics in physics) could arise without human engagement from the patterns in the data stored on the Internet when the generative AI system is trained on social media, but training on the archive of the entire scientific and philosophical heritage makes it more likely. Is the detection of patterns in collective knowledge sufficient for intelligence capable of creativity significantly exceeding human capacities? There is no answer to such questions about the intelligence of AI systems as long as we do not have a clear understanding of intelligence artificial or natural.

The next step is to consider the possibility that some AI systems may acquire the ability to build other AI systems (both as informational and natural/physical entities) and become autonomous natural devices. Here is the essence of the escape from human control. The danger is that some AI systems equipped with physical instruments that make them agents may acquire the ability to proliferate and act independently from human control and understanding. We already have examples of simpler systems that proliferate themselves (e.g., computer malware), but they still rely on devices created by humans and on human (usually unintentional) actions. But we know that self-reproducing physical automata are possible.

This possibility of escape is scary, but its threat is not new as the development of technologies generated concerns about autonomous devices several times in the last two centuries. However, is it so different from the issues related to invasive species in ecological systems or pandemics? In all cases, we need a better understanding of the dynamics of information in multiple contexts. The escape of a virus from a laboratory may be equally dangerous. Does it help that the virus has intelligence incomparable to human intelligence? The key point is that to prevent escapes of natural or artificial agents from human control, we have first to acquire this control in the form of the knowledge of information dynamics.

2.2 Can artificial intelligence be conscious?

We could continue the search for multiple confusing and concerning aspects of AI and each time we will arrive at the same obstacle of the lack of understanding of the fundamental concepts related to information.

Can AI systems be conscious? The answer depends on how we understand consciousness, and whatever consciousness is, its understanding requires an explanation of how consciousness is related to information and its dynamics. The danger is in making assumptions based on popular common-sense metaphors such as that the mind or brain is a type of computer or vice versa that computers are artificial minds or brains. Even more dangerous is the lack of distinctions between mind and brain. The popular analogy of software and hardware appeals to common sense, but it is based on the lack of knowledge of both, information mechanisms in the computer and in the brain.

The question about the possibility of conscious artificial intelligence systems is crucial for the consideration of their agency. When we talk about artificial intelligence and the possibility of systems independent from human control or comprehension, it is a legitimate question about not only their cognition but also conation. As was observed before, there are many systems (ecological, social, economic, cultural, etc.) that are independent of individual or collective human control. However, they are definitely devoid of purpose and their dynamics are governed by natural or social laws typically well known. They do not set or modify their goals, and their mechanisms are driven by well-understood external forces. This gives hope to humanity for gaining control over them. The possibility of the intentional or unintentional construction of conscious artificial intelligence systems brings into consideration the danger of actual lost control. This requires some additional elements such as the capacity for self-consciousness, volition, and independence from human control normativity (the ability to set own values and goals). Only systems equipped with consciousness having these additional characteristics can compete with humanity. Otherwise, the main threat to humanity is humanity itself. However, is artificial consciousness possible?

There is no doubt that without substantial progress in the study of information, there will be no answers to the questions about the prevention of harm caused by present and future technologies. After all, the control of any artifacts comes not from watching and directing their work (technological progress was always generated by the interest in the elimination of both) but from the knowledge and understanding of the phenomena involved in their mechanisms allowing for the prediction of the outcomes of this work and prevention of their deviation from human goals and values. The expectation that the official moratorium on the research and development of AI can prevent future disasters is naive. Instead, there should be more support for the study of information going way beyond its technological aspects. The actions of the external support for the study of information should be informed and guided by the research done by the community dedicated to this study. This chapter is intended as a contribution to this goal.

3 The Role of Information in Understanding Intelligence, Complexity, and Consciousness

3.1 *Information and intelligence*

Without claiming the achievement of the ultimate resolution of the still open issue of what intelligence is, I use in this chapter the concept of intelligence as a capacity to eliminate or decrease the complexity of information (Schroeder, 2020a).

The use of the concept of complexity in defining intelligence may generate an objection that I listed above both as poorly understood and insufficiently conceptualized. However, the study of complexity has acquired quite an advanced level in more specific contexts of inquiry (e.g., computing and dynamical systems). Moreover, complexity, no matter how defined, is a more general concept that can be applied at any level of abstraction to qualify arbitrary subjects of inquiries and actually is applied in a wide variety of contexts, while intelligence can characterize only systems capable of action, the action itself, or its outcome. It is easy to find examples of something complex that cannot be considered intelligent. On the other hand, everybody would agree that human intelligence is associated with the extreme complexity of the brain. This of course does not demonstrate that intelligence can be defined by complexity but only that complexity cannot be defined by intelligence. Finally, complexity is frequently studied in the context of information. The following section is devoted to their relationship.

The difficulties in defining intelligence are well known. Even in the case of human intelligence, there is no consensus on the feasibility of establishing its unique and uniform conceptualization (de Silveira and Lopes, 2023). The concept of human intelligence is so difficult to define for the reason that almost everyone believes in their good understanding of it and whatever seems obvious is almost always highly non-trivial. It is amplified by the Dunning–Kruger effect (actually recognized already by René Descartes who famously and sarcastically prized God for giving everyone a sufficient amount of reason to make them happy). David Dunning and Justin Kruger empirically confirmed the correlation between the lack of competence or intelligence and the conviction of their possession (Kruger and Dunning, 1999). Due to this effect, people develop their views of intelligence consistent with their self-image, with a prominent example of a certain "stable genius".

However, the bias of idiosyncratic views on intelligence is not the main obstacle to its conceptualization. Even more confusing is the projection of common-sense criteria used to assess intelligence derived from the practice (or malpractice) of assessing candidates' suitability for some tasks (including the assessment of children or youngsters applying to schools). An example of a criterion in such an assessment is frequently proficiency in "problem-solving". Not only this expression is meaningless for describing intelligence and the evaluation of such a skill is highly problematic, especially in a diachronic perspective, but it is also dangerous for carrying hidden ethnocultural bias.

It is meaningless because it is based on the assumption that the words "problem" and "solution" have a clear and objective meaning. Once again, the issue is in the illusion of obviousness. What does it mean "problem"? Here too, the answer is highly non-trivial if we want to have it comprehensive. We can try to formulate it as a call for action which can be answering a question, performing some action leading to the desired state of affairs, or some desired behavior including inaction. These are only a few possibilities out of many. To answer the question of what constitutes a solution or correct solution is an even more difficult task. The history of science gives an extensive gallery of solutions ignored or attacked by all contemporaries. Even in mathematics, the most important moments in its development were associated with changes in understanding what constitutes a correct solution to a problem (usually correct proof of a theorem). For instance, one of the main motivations for the development of set theory was justification for proofs using mathematical induction and non-constructive methods. Moreover, very often in hindsight, the solutions that violate the standards of evaluation are later or in different contexts highly prized as "thinking outside of the box" and considered as indications of exceptionally high intelligence.

There is a good example of ambivalence in the evaluation of human intelligence in the story (possibly legendary) of the solution to the long-standing problem of untying the Gordian knot that on the order of King Midas of Lycia was tied to hold the oxcart of his father attached to the column in the temple. The most popular modern story presents the highly intelligent achievement of the solution by Alexander the Great by cutting the knot with his sword. The more reliable ancient sources (e.g., writings of Plutarch and Arrian of Nicomedia) report that Alexander did not cut the knot (questionably intelligent act) but that he pulled the lynchpin from the column, uncovered the ends of the rope, and easily untied the cart (Fredricksmeyer, 1961). The story in both versions (especially in the second version) identifies the high intelligence of Alexander with the elimination of complexity and the latter is clearly about the reduction of complexity of information. However, more importantly, the difference between the versions shows the difference between the ancient and modern views of intelligence.

Not only the normative idea of a good or correct solution is vague, but the development of any scale of assessment is purely conventional. With this conventionality of the assessment of proficiency in problem-solving comes the danger of cultural bias. Different cultures develop different norms and values which naturally influence the way people identify, formulate, solve problems, and evaluate these solutions. What is a legitimate problem or proper solution in one culture that is prized for being evidence of wisdom or intelligence may be considered an expression of stupidity in another.

Since all standard evaluations of human intelligence can be considered variations of problem-solving (e.g., social intelligence can be understood as the ability to solve problems in human interactions, and emotional intelligence can be understood as solving problems in managing own emotions and the emotions of others), they are of little value for the general definition of human intelligence.

The same type of issue is with pragmatic views of intelligence which associate intelligence with the capacity for effectiveness in more general actions that are not necessarily in the context of problem-solving. Not only the assessment of this type can only be *a posteriori*, or based on the circular reasoning "capacity for effective action means having been effective in action", but also the meaning of effectiveness requires conventional, normative, and culturally loaded criteria as described above. Moreover, when we try to generalize the effectiveness of action to eliminate human aspects, we may have to accept the intelligence of objects that are unlikely intelligent. For instance, the motion of mechanical objects is effective in the sense of the principle of the least action. It is difficult to accept the intelligence of falling stones because they optimize their trajectories. Mark Levi (2009) in his book *The Mathematical Mechanic: Using Physical Reasoning To Solve Problems* presents an extensive exposition of many examples of physical phenomena whose measurements can be used to find solutions to mathematical problems and of course, this does not mean that the mechanical systems involved in them solve any problems or manifest any intelligence.

Naturally, the generalization of intelligence beyond human beings is even more difficult. Alan Turing gave up this task in his inquiries of the possibility of artificial intelligence (in 1950, long before this expression was introduced, he called it machine intelligence) and proposed his imitation game (today called the Turing test) as a functional method to judge the intelligence of artificial systems. It was based on human judgment of success in performing some tasks in which human cognitive abilities are involved, more exactly the failure of human ability to distinguish the performance of humans and machines in such tasks. In time, the test became of mainly historical interest, but the efforts to identify the common characteristics of human and machine intelligence continue the same methodological framework of comparing human and machine performance in tasks involving human cognitive functions (de Silveira and Lopes, 2023). This performance is always related to information processing, communication, and its role in expressing behavior (execution of action). Moreover, in the case of human intelligence, the frequently invoked criterion of adaptability serves the purpose of establishing a normative characteristic associated with it. However, adaptability can be considered yet another form of the result of information communication (in the form of feedback and feed-forward loops).

Inquiries of intelligence go beyond humans or machines to include different levels of collective intelligence in individual cells, unicellular and multicellular organisms, and their populations. There is no big difference between organismal, human, and collective intelligence except for the problem of agency. In human beings, agency is associated with a conscious (rational) choice of goal/purpose and the ability to make choices between the direction of complex actions. In the absence of consciousness or the internal, centralized mechanisms of making choices, the concept of agency loses its meaning, unless we consider a mechanism at the collective level of an evolutionary feedback control reducing the multiplicity of behavioral choices of the members of the collective.

Not everyone seeks an evolutionary explanation. Michael Levin, who claims that all instances of intelligence are collective (in humans, it is a collection of neurons responsible for cognition), invokes teleonomy "[...] not the final step on a continuum of agency — it is a primary capacity" (Levin, 2023). Teleonomic explanation eliminates the normative aspect of intelligence, but it does not influence the informational character of the mechanisms involved in the intelligent behavior of the collectives which can be identified in Levin's empirical study of its mechanisms. Brian Ford considers the intelligence of cells as a driver of the evolutionary process shaping the entire organisms (Ford, 2009).

Whatever the explanation of the superorganismal characteristics of collectives engaged in intelligent behavior is (using the century-old term of superorganism introduced by William Morton Wheeler (1910)), the association with an organism brings us back to human intelligence, although we don't have any more conscious or rational purpose and we don't have human physiological mechanisms. What remains are informational processes involving interaction with the environment with reduced but effective choices from a much wider range of possible states of the system.

At this point, we can enter complexity, or rather information complexity used in our definition of intelligence. The environment (in Levin's terminology problem space (Levin, 2023)) may be characterized by a high level of complex information, yet an intelligent system (individual or collective) has the capacity to reduce this complexity either in its internal modeling (e.g., in the case of human intelligence involving consciousness) or in the structures governing the behavior of a system.

3.2 *Information and complexity*

The complexity of information, in turn, can be understood as a qualification of information in terms of its quantitative characteristics (the number or measure of the components of information) and qualitative characteristics (the structure of these components and the degree to which these components are bound together) (Schroeder, 2013, 2017).

Complexity must be evaluated not only in terms of the number of components but also in terms of their mutual relations. For instance, the system of preferences of an individual customer (or information about it) is much more complex than the system of preferences of a crowd of customers (or information about it), even if the latter has a much larger number of degrees of freedom and has individual customers as their components. This leads to an easy predictability of the crowd's actions in contrast to practically unpredictable individual behavior. The entire discipline of statistics is based on this distinction.

The assessment of intelligence (here, the ability to reduce the complexity of information) can be made through the observation and analysis of the overt or covert behavior of an agent. Thus, it can and should be predicated only on entities capable of transforming information. The assessment of the degree of intelligence (qualitative or quantitative) is usually possible for the agents, i.e., entities capable of goal-oriented actions, making choices, and acting based on these choices. This predication in informal contexts can be extended from the information-transforming agents to their actions or the products of these actions but only as an abbreviated expression. So, we can say that the response to the question was very intelligent or that someone's behavior was intelligent. However, the absence of an agent disqualifies the object from being intelligent. This is why it is meaningless to say that AI is intelligent, but only that a given AI system capable of some actions executed with the use of informational interaction with its environment can be considered intelligent.

This particular choice of the understanding of intelligence and complexity will not lower the generality of my arguments as long as the reduction of information complexity in turn associated with the quantitative and structural characteristics remains relevant to intelligence. Someone who prefers a more elaborate description of intelligence may consider my definition just a terminological abbreviation. It should be noted, however, that any further specification in the description of intelligence may reduce its generality. For instance, the loss of generality by any reference to human or organismal physiology will exclude the application of this concept to collective systems of human or organismal populations, and at the same time, the intelligence of artificial systems will lose its meaning. For our purposes, it is only important to recognize that intelligence which is the central concept of AI can be identified or at least closely associated with the transformation of information and resolving its complexity.

3.3 *Information and consciousness*

The relevance of information and intelligence to the study of consciousness may be obvious, but it turns out, it is highly non-trivial. Since it is a very broad topic, I will refer the reader to my detailed exposition elsewhere which here is reduced to a brief remark followed by a report of the most recent developments (Schroeder, 2011a).

The tradition of the study of consciousness as a distinctive subject of inquiry goes back at least to William James (1890). James (1947) noted one distinctive characteristic of the phenomenal experience of consciousness — its unity. Since it was such a distinctive feature, from the very beginning, it remained the focus of all inquiries. It stimulated a direction of study seeking the mechanisms responsible for consciousness in quantum-mechanical phenomena where the superposition of states and entanglement exemplified physical processes leading to absolute unity. The vague ideas of holographic analogy became in time more specific in the search for quantum-mechanical mechanisms in the brain (Beck and Eccles, 1992).

The problems of inapplicability of the quantum formalism to the description of a large and warm brain are amplified by the fact that information processing in the brain contributing to cognitive processes is distributed in its many regions stimulated attempts to modify the processes in the brain or append quantum-mechanical description culminated in works of Stuart Hameroff and Roger Penrose from the end of the 20th century (1996). The attempts were not successful and although this direction of study was never completely abandoned, it is rather dormant at present.

Another direction of the study seeking an explanation of consciousness in terms of the integration of information was initiated and widely promoted by Giulio Tononi. Tononi proposed a measure Φ of consciousness understood as integrated information not only in the human brain but also in any system including individual elementary particles. In the earliest papers, he did not refer explicitly in presenting his measure to integration of information but rather to functional integration of the brain or brain complexity (Tononi et al., 1994). However, soon later he made his measure a quantitative description of consciousness, and the measure Φ became the central concept of what he called the Integration of Information Theory of Consciousness (IIT) (Tononi and Edelman, 1998; Tononi, 2007). The function Φ was derived from purely statistical analysis of the simultaneous firing of neurons without any attempt to provide a structural analysis of consciousness or information integration. The mysterious non-zero value of Φ for objects such as elementary particles instead of being used as the evidence for the error in interpreting Φ as a measure of information integration became the argument for the attribution of consciousness to everything.

The panpsychism of IIT is its least problematic feature. After all, in the history of science, there were many instances of contributions that contradicted common sense but later became commonly accepted. Much more serious deficiencies were in repetitions of old methodological errors and misinterpretation of the mathematical concepts used for justification of the claims. The belief that it is enough to define a measure of something to give it an ontological status is quite an extreme instance of philosophical poverty. The claim that this measure can be associated with the integration of anything and in particular with information integration, without any explanation of the meaning of the concepts of information or its integration, is another example of methodological poverty. The belief in the applicability of all mathematical concepts that involve in their name the word information to inquiries of information in all possible contexts is surprisingly naive.

More specific issues in IIT are in the complete negligence of the relationship between the external observations of simultaneity of nerve firings with the phenomenal experience of unity. The inquiries of consciousness were haunted by the homunculus fallacy for centuries. In my critical appraisal of IIT (Schroeder, 2011a), I objected to the fallacy that I called "homunculus' watch" involved in the claim that what is simultaneous in the brain for the external observer becomes phenomenal, spatiotemporal unity.

Even these clear deficiencies of IIT are not the main disqualifying features of Tononi's approach. His derivation of the measure is based on the consideration of bi-partitions of the brain for which he calculates mutual entropy which he calls mutual information. The formula proceeds to the consideration of all bi-partitions (partitions into two complementary subsets). This is a curious framework as if integration was a result of only binary interactions between the regions of the brain, while it is well known that multiple regions of the brain are involved together in every cognitive process. The binary framework contradicts the very idea of integration and makes the entire IIT irrelevant to the study of consciousness.

Someone could respond that this can be avoided by considering not bi-partitions but multiple-component partitions representing known functionally distinct regions of the brain. This may seem like a good resolution of the issue (never considered in IIT), but it turns out that this can produce the negative values of multiple-regional mutual information. This is an elementary information theory theorem that the binary partition is the exceptional case of non-negative mutual information as for all higher-level partitions, even into three regions, their mutual information can be negative (Reza, 1994). Panpsychism is a bizarre but still conceivable view of reality, but probably nobody would accept negative consciousness.

The IIT was promoted with so noisy fanfare that its proponents and supporters did not hear the criticism "of the leading theory of consciousness" growing over the years until the very recent burst of the bubble when 124 researchers signed the open letter calling IIT a pseudoscience (Fleming *et al.*, 2023). This is an extremely embarrassing and unprecedented situation in science. In response to this letter, several leading researchers of consciousness who in the last quarter of the century never voiced any objections to IIT suddenly worry that the use of the word pseudoscience is "unfair" for what they admit now actually is "bonkers" and that it may slow down research in this important subject (Lenharo, 2023).

This type of defense is as bizarre as IIT. The elimination of erroneous methodology will not slow down but rather accelerate progress. The criticism is of the errors made in a particular approach of particular individuals and of the uncritical acceptance of its products, not of the inquiry of information integration. The harm was done not by the critique of the errors but by the lack of it in the quarter of a century time when IIT was promoted as a "leading theory of consciousness". The more than a century of inquiries of consciousness as a process in which information is integrated will not be wasted when the alternative already existing and future approaches correct old errors.

The key lesson from this dramatic development is that the study of consciousness as integrated information requires a good understanding of what information is, what its integration is, expressed in the proper theoretical description, and proper formalization (Schroeder, 2009).

4 Conceptual Obstacles in the Study of Information

4.1 *Defining information*

The association of intelligence and information is mutual. Not only we can study intelligence by making inquiries of the way information is transformed and used, but the other way around, we can consider an intelligent way to perform inquiries of information. The attempts to define information are simply instances of more or less intelligent actions to organize a very broad range of phenomena with similar characteristics into a unified conceptual system with lower complexity.

These attempts have to satisfy two conditions. The first condition is that the concept of information has to be sufficiently general and at the same time sufficiently specific. It has to include all unquestionable instances of the use of the term information. Any definition of information that does not apply to language and other forms of communication, semiotics, processes of genetic inheritance, processing of information in living organisms and their populations, mechanisms of control and governance, computing devices, and other contexts of the use of the term information cannot be considered adequate in the general study of information. On the other hand, it is necessary to distinguish between the concept of information and concepts such as knowledge, belief, opinion, and wisdom. Thus, we have to avoid over-generalization.

The preceding sections of this chapter provided arguments for the study of information as a fundamental tool for inquiries of intelligence, complexity, and consciousness. Such inquiries require a very high level of generality exceeding specific interests in intelligence in the context of life not only because we want to consider intelligent information systems designed and implemented by humans as a part of the technological progress that we cannot predict but also because we have to have intellectual tools for the search for ETI. There is no reason to claim that intelligence requires the forms of life that we know from our direct environment. We have to consider the possibility that life can have very different forms, based, for instance, on an alternative configuration of chemical elements. Moreover, we even cannot claim that intelligence requires its implementation in life forms.

There is an additional condition for the definition of information, much less obvious and frequently challenged that it should follow the rules of logic. The form of inquiry of information is a matter of choice, i.e., convention, and nobody is prevented from other forms of inquiry, including artistic expression of subjective, intuitive perceptions of informational phenomena. The choice of logical rules for defining concepts is simply a matter of the style of inquiry. This may be considered unnecessarily restrictive. After all, the famous Bateson's "definition" of information as "difference that makes a difference" by the use of an idiomatic expression "makes a difference" is metaphoric and far from being logically and ontologically correct (once again, its strictly logical interpretation leads to a category error).

Bateson intentionally made it open-ended to extend its generality and without a doubt achieved great success in finding followers who accept it as a definition while presenting very different and frequently contradictory interpretations. Batson himself presented at least half a dozen different interpretations with increased logical precision, but each of them significantly restricted the generality of the concept of information. However, it is possible to choose some reformulations of Bateson's information to make it a well-defined concept (Schroeder, 2019). Therefore, with admittedly lower precision, Bateson's information is sufficiently close to meeting logical requirements.

A much bigger problem is with other attempts that seemingly follow logical rules of defining concepts but which use as defining concepts equally unclear undefined concepts or concepts that in the language of the discourse are species of the genus information (e.g., "data" which means "given" in Latin and for which it would be difficult to avoid invoking full expression "given information"). The requirement to avoid the use of undefined concepts in definitions is obvious but not sufficiently obvious to prevent the import of apparently obvious common-sense words or expressions that everyone can interpret freely (such as "difference").

If we follow the directives of the formulation of logically correct definitions avoiding under- and over-generalization of the concept of information, the large variety of attempts is reduced to only a few that did not gain much popularity. It should not surprise us that the discussion of the concept of information has never been finalized by those engaged in this inquiry and the majority of people using the term information are not even aware of the absence of a commonly accepted definition.

Finally, it has to be stressed that there is no need or justification for the claim that there is only one "correct" definition of information. Definitions can be logically correct or incorrect, but only theories that are built over those definitions can be tested and evaluated. There is no non-trivial concept in the history of science or philosophy with an uncontested, single definition. Thus, the choice of definition matters only when it is a part of developing a theory of the concept.

4.2 *Misunderstanding of the concept of information*

Claude Shannon did not define information or even make a distinction between information and uncertainty and could develop such a successful theory of communication that his followers convinced him to rename it "the" theory of information, so why anyone should care (Shannon, 1948; Shannon and Weaver, 1949/1998)? After all, Shannon provided three principles for the quantities H measuring how much choice is involved in the production of discrete information in the form of a sequence of events with some probabilities. The principles determine the functional form of H in terms of the probability that "play a central role in information theory as measures of information, choice and uncertainty" (Shannon and Weaver, 1949/1998) and which are similar to entropy in statistical mechanics (Shannon and Weaver, 1949/1998). What Shannon called "entropy of the set of probabilities" (Shannon and Weaver, 1949/1998) became a powerful tool in the study of communication with a myriad of applications. Since it gives a quantitative description of something, why not dispose of the choice and uncertainty and settle on this something being information?

Not everyone agreed with this idea, and very soon the price for the hidden assumption in Shannon's principles that the order of events in the process of production of information is irrelevant became the source of criticism (Bar-Hillel and Carnap, 1952). Is the measure of information in the words "dog" and "god" and in a meaningless sequence of letters "ogd" really equal? Shannon prevented such criticism by declaring that the semantic aspects of communication are irrelevant to the engineering problem and that therefore the words dog and god differ only in their meaning. It is easy to agree that the semantic aspects of communication are irrelevant to the engineering problem of the speed of transmission of information in communication. However, it is clearly false when other engineering problems are considered. Can we consider an engineering problem of efficient transmission of information solved by sending a report about the number of occurrences in the message for all characters in the alphabet or their relative frequencies? The entropy will be the same as for the original message, but the message, if long, would be completely lost in transmission.

Shannon was aware of the issue even if he did not write explicitly about it in a critical way. He wrote in his famous paper entire two sections, Section 2 "The Discrete Source of Information" and Section 3 "The Series of Approximations to English", about the analysis of the sequences forming messages using conditional probabilities of the choice of a letter or word based on preceding letters or words (Shannon and Weaver, 1949/1998). In 1948, the task was too difficult to have any practical application for the structural analysis of information. It may be surprising that in some sense the generative AI systems can be considered a form of realization of Shannon's idea with the training of neural networks replacing the calculation of conditional probabilities.

The calculation of conditional probabilities even now would be unrealistic. Instead, generative AI systems such as ChatGPT are trained with the methods of deep learning on the large data set from the Internet to acquire the ability to choose which character or word should be selected next in the generation of text. This is a much better solution to an engineering problem to generate meaningful responses to inquiries, but we are not closer to the methodology of structural information. The process of training does not involve any structural analysis of information or its semantics as it functions as a "black box" without accessible memory. It is just reproducing typical (i.e., highly probable) structures of the instances of information in the training reservoir. The fact that the choice is highly probable does not mean that the sentences in the generated text are true or that they make any sense. These hallucinations of generative AI systems demonstrate that Shannon's idea of replacing the structural analysis of information with probabilistic methods was faulty.

Shannon was aware of the importance of the order of characters in the message (i.e., the structure of a message) which he believed could be described by conditional probabilities. However, he did not provide or even try to find any tools for the structural analysis of information. One of his principles for entropy was that the order of characters does not matter. Moreover, he believed that his main achievement was going beyond what Ralph Hartley did ten years earlier by assuming that the characters have equal probabilities (Shannon and Weaver, 1949/1998). Did Shannon read Hartley's article? He cited it, but the word "probability" does not appear in Hartley's article (1928) and this concept does not play any role in it which is in clear contradiction to Shannon's interpretation that Hartley assumed equal probabilities.

Contrary to what Shannon wrote in the introduction to his paper, Hartley (1928) did not use the assumption of equal probability of characters (or probability of anything else) to derive the logarithmic measure of information. He simply observed that the encoding of information can be optimized, i.e., changed without any essential information change. According to Shannon's conceptual framework, this optimum is achieved for the uniform probability distribution of characters. However, Hartley referred not to probability but to the experience of operators encoding messages who certainly knew about Morse's optimization of encoding based on the frequency of characters in the language.

Hartley (1928) derived his formula corresponding in probabilistic interpretation to the special case of Shannon's for uniform distribution from the assumption that the measure of information should be invariant with respect to the change of its encoding. Hartley programmatically avoided any use of psychological considerations such as the way human beings achieve an understanding of the meaning, but his view that the same information can be encoded in one or another language, or using encoding systems with different numbers of characters, indicates that information is invariant with respect to such changes, although he did not refer explicitly to the meaning.

Yehoshua Bar-Hillel and Rudolf Carnap (1952) were probably the first to reject Shannon's theory as a theory of information because of its disregard for semantics. They considered Shannon's work (quite rightly) a study of signal transmission and attempted to develop a theory of information based on the logic of language. Their proposal of the logical theory of information equipped with semantics was not very successful in directing further research.

My diagnostic (2012) of the limited resonance of their approach in the literature on the subject of information was that despite the promises to deliver a semantic theory of information as semantics is understood in the logic of language, their approach was still syntactic. This fact was hidden in the substitution described in one sentence declaring that "for technical reasons" they replaced the states of the world addressed by information with their descriptions. Surprisingly, this shift from semantic to syntactic analysis of information was usually overlooked.

As nobody ever noticed that despite the declaration that their theory of information was semantic, it was actually syntactic, probably a more direct reason for the lack of popularity of the approach proposed by Bar-Hillel and Rudolf Carnap was in their attempt to direct the development of the theory to arrive at results comparable to those of Shannon. As a result, it was not clear how their approach was better. At least, it did not resolve the issue of the meaning of information.

The logical tools of their theory did not help much as logical semantics had more questions than answers at that time. Since they substituted syntax for semantics in their considerations without excluding intentionality as a basis for meaning, they did not identify the actual source of difficulties and did not overcome these difficulties in the search for the meaning of information. At least, they were the first who, using Shannon's expression, did not "jump the bandwagon" and openly criticized his programmatic disinterest in the meaning of information.

The prolonged discussion of the conceptualization of information and the lack of consensus on its definition is not a problem. It is just evidence of its importance and relevance. The much bigger problem is that the diverse attempts to formulate a definition are rarely followed by developments of comprehensive theories of information. Competing definitions of information can be evaluated only through comparisons of the theories of information based on these definitions. Regretfully, the definitions of non-trivial concepts are rarely compared using the criterion of the explanatory power of the theories in which they were used.

We could see that the main unresolved (or not satisfactorily resolved) problems in the study of information were related to two somewhat related obstacles: the lack of tools for the structural analysis of information and insufficiently general semantics of information that usually mimicked linguistic semantics. The minimal criterion for an adequate conceptualization of information is to allow the development of a theory of information that helps overcome these obstacles.

5 Semantics of Information

5.1 *Attempts to develop semantics of information*

After the initial attempt by Bar-Hillel and Carnap to develop a semantic theory of information which already was limited to information in its linguistic form in which logic could be employed, there were not many other attempts to develop a general semantics of information not limited to particular contexts. This should not be a surprise considering the formidable task of answering the question about the meaning of meaning in the limited context of language and human comprehension that remained not achieved in several centuries despite multiple attempts.

One of the reasons for the difficulties in understanding meaning came with its association with intention, a mysterious capacity of the mind to cross the Cartesian precipice separating *res cogitans* and *res extensa*, the mental and material realms, reaching from the thought in the former realm to point at the denotation residing in the latter realm. Cartesian duality was the main obstacle to understanding consciousness which resided in the former realm but was influenced by the objects from the latter realm and in turn could affect these objects. In the case of intention, the situation was even more complicated. Signs belong to the realm of *res extensa*, but they acquire their symbolic characteristic only after they are interpreted by a conscious subject in the realm of *res cogitans* pointing at their meaning in the objects among *res extensa*. The divide between the two realms had to be crossed twice. This led to the most typical tripartite models of semiosis with the *sign, interpretant,* and *denotation* (the last one acquiring the status of the meaning of the sign) in the terminology of Charles Sanders Peirce's *semiotics*. For Peirce (2015), it was a tertiary relation expressing a cooperative action not reducible to its binary components. In his explanations, interpretant was called sometimes interpretant sign as it was rather the effect of the sign on some agent (quasi-mind), not necessarily mind and itself can serve as a sign.

Peirce (1977) explicitly wanted to maintain a higher level of generality by not limiting the interpretant to a conscious person, using this simplified interpretation of the interpretant to make his explanation easier. However, the generalization could require a sequence of tripartite relations compounding interpretant signs over interpretant signs considered as signs. The tripartite relation is more of a fundamental framework underlying the process of semiosis. On the other hand, the consistent use of the expression of the semiotic process places semiosis within time and space limiting its generality.

This attempt to maintain a level of generality exceeding the psychological explanation of the intention as a mental phenomenon was unusual at the time. Franz Brentano (1874/1995) who followed the

Cartesian cut made a clear distinction between what he called, using the Scholastic terminology, the intentional or mental inexistence of an object directing toward an object which was an exclusive characteristic of mental phenomena always including something as an object in themselves, as opposed to "physical phenomena". Obviously, the reference to mental phenomena reduces generality to the purely psychological level.

The subject of "The Meaning of Meaning" (1923) was brought to the attention of a broad intellectual audience by the book of this title published by Charles Kay Ogden and Ivor Armstrong Richards. Their approach was again based on the triangular scheme involving symbol, thought, and referent engaged in the binary relations thought — symbol qualified by correctness, thought — referent qualified by adequacy, and symbol — reference qualified by truth. The triangular scheme described the instrument of both human communication and thought in a culturally determined context. Although their approach introduced yet another aspect of culture to the study of meaning, it did not lift the generality of the perspective above the level of the use of language by human beings.

The first substantial generalization came with the very rich direction of the study in which *bio* joined *semio* (using Kalevi Kell's expression for the early biosemiotics (1999)) initiated by Friedrich Salomon Rothschild (1962). Rothschild (1962) did not even mention information in his three laws of biosemiotics investigating "the communication processes of life that convey meaning in analogy to language", but his work was already preceded by Erwin Schrödinger's epoch-making small book "What is Life?" (1944) that stimulated the revolution of genetics so that it was just a matter of time for biosemiotics to become information theory for living organisms in their complex multilevel architecture from molecular to organismal and beyond to their populations.

In the immensely extensive literature on biosemiotics, there is a very frequent reference to the meaning of information but always within the context of its function in life at some level of its organization. The meaning of information becomes a secondary concept explained by its function in either a causal, deterministic, or teleonomic way. Although the tripartite semantics of the earlier authors is not necessarily invoked directly, it is hidden in the ecological framework. Life at any level of organization cannot be considered without its environment and the constraints imposed by it. An additional limitation of the biosemiotic information at the organismal level comes with agency characterizing living objects. Without ecology and agency, the meaning of bio-information is losing its meaning. This of course does not make biosemiotics inferior or incomplete, but it shows that in looking for context-independent semantics free from the triangular relation engaging pragmatic aspects, we cannot simply import the bio-semiotic framework.

There is nothing wrong with crossing the border separating semantics with pragmatics. The commonly invoked borders between syntactic, semantics, and pragmatics in the study of language popularized by Charles Morris in his useful classification of the subjects in the study of language may lose their application in the case of information (Posner, 1992; Schroeder, 2011b). Even in the study of language, the use of language (i.e., the subject of pragmatics) dominates the inquiry of the meaning (semantics). A prominent example is in the "language games" of Ludwig Wittgenstein (1953).

Thus, the problem of using pragmatic explanations of meaning in the context of information (which includes language as only one of the possible information systems) is not in blurring the divisions between the domains of linguistic studies but in the relativization of the meaning to the user of information. While someone could insist that every language requires a user or rather a community of users, in the general study of information, this assumption brings an unacceptable restriction of the concept of information. If we want to consider information understood in its sufficient generality explained above in this chapter, applicable to information systems in the early stages of the universe before the existence of any forms of life became possible or in the regions of the universe where even now this existence is not possible, we have to eliminate the tripartite framework of semantics engaging concepts of a user, interpreter or thought.

Some attempts to present the semantics of information or the semantical concept of information referred to the idea of "the true information". If we want to include in the theory of information language as a special instance of information, the attempts to define the truth are doomed by Tarski's theorem on undefinability of the truth (Tarski, 1983). The statements about the truth can belong to the metalanguage but cannot be expressed within the language of theory. Thus, all attempts to define information with the use of the qualification of being true, for instance, "true data", do not make much sense unless again we restrict our inquiry to a particular context in which the linguistic form of information is excluded giving us freedom from the constraints of logic.

This is conflated with another issue arising in using data in the definiendum for information as if information could be considered a qualified type of data against the etymology of this word and virtually all its applications. Can we consider data that are not information? If not, then data and information become synonymous. If yes, what are they, and in what sense are they given? It can be claimed that data that are not information are false. Then we are trapped in the undefinability of truth.

In the common-sense view, "true" can be interpreted, for instance, as "effective" in achieving some goals. However, in this situation, we can simply eliminate the former as it just creates confusion by the suggestion of unjustified generality without adding any explanatory power. We can find an analogy with the infamous explanation by Herbert Spencer of the Darwinian concept of natural selection as "the survival of the fittest". Which species are the fittest? Those are that survived. Which instances of the information are effective? Those that are true. Which are true? Those that are effective.

Thus, there is no sound and sufficiently general semantical theory of information. I presented the possibility of overcoming the difficulties in my earlier publications (2011b). The proposal was based on the assumption that the obstacles in the development of the theory of intention (aboutness) were created by the faulty view that elements of the language (e.g., nouns) are about objects that are entities of a very different ontological status. The correspondence between symbols and their denotations requires a theory overarching the study of both types of entities which seems impossible.

There are two possible solutions to this obstacle. We could consider uninterpreted signs and their denotations as entities of the same type (physically created with the ink inscription "dog" on a paper and an instance of an animal in our environment that is a dog) and then the correspondence is between entities of the same status. This could be one of the possible interpretations of the substitution made by Bar-Hillel and Carnap when instead of considering the states of the universe they involved the descriptions of such states.

In this case, we have to consider the description as an object of the same ontological status as the described object. If we claim that the object and its description are both devoid of informational content and that information has a relational character of secondary ontological status, then we encounter the old obstacle of mysterious correspondence between a sign and its meaning. This correspondence is purely arbitrary (there is nothing in the reality of these two types of entities that connect them) and it is not one-to-one. Since we can have multiple sign systems, addressing the same entities and signs may have some level of abstraction addressing, for instance, many animals that we consider dogs, the theory of intention within the reality of denotation breaks up.

There is another way to overcome the difficulties of crossing the precipice between entities of different ontological statuses. This solution is possible at the level of the study of information but not at the level of language. Intention can be considered a relationship between two informational systems. One of them is symbolic, the other possibly not. Thus, the word "dog" is not about an entity of a different ontological status "dog in itself" (paraphrasing Kant's "thing in itself") but about the information associated with this "dog in itself". We can only claim that the object of our inquiry is a dog (we give meaning to the object) through information carried by this dog (for instance, how it looks, how it smells, what sounds it makes, etc.) which is transmitted to us and perceived by us. These are our cognitive mechanisms that link the perceptions of the information about the writing "dog" on the paper and the perceptions about the animal. On both sides, we have only informational entities. Their correspondence is established based on their structural characteristics. We cannot apply this to the linguistic logical theory because we cannot claim that an animal that we call a dog is a linguistic phenomenon. After all, dogs existed before languages developed. However, nothing prevents us from saying that the inscription "dog" stands for the information carried by some type of animal or an instance of such an animal.

5.2 *Reverse semantics and encoding of information*

The second solution of interpreting meaning as the relationship between informational entities can be understood as "reverse semantics". The adjective "reverse" refers to the change of the paradigm in linguistic semantics which starts from the pre-existing language (of any type) with fixed rules and vocabulary and proceeds to establish a way (intention) in which entities and their relations are represented within language.

This is in my view placing a carriage ahead of the horse. In the approach proposed here, the starting point is in the inquiry of the informational structures of entities, and the role of semantics is to inquire a variety of ways in which this information can be encoded. Some of these encodings may have the form of a language, but encoding should not be considered a human or intentional action. The meaning of information in its linguistic form consists not of the entities of reality devoid of information content but of informational structures which are independent of the language.

The study of these informational structures may involve physics but is not limited to the objects of study in this domain. Reality has a hierarchical architecture of multiple levels of complexity irreducible to simpler ones. The study of these levels (for instance, in living organisms) can be supported by physics but cannot be reduced to it. Yet, in one respect, physics can be a resource for the study of other levels of complexity due to a long experience in the inquiry of symmetry. After all, the invariance of information with respect to the transformation (change) of encoding is a form of symmetry and symmetry is one of the most important tools of science with a long tradition of its use.

This idea of intention as an informational relationship is not completely detached from the earlier ideas expressed in terms of linguistic logic, but, of course, these other older ideas were expressed without the use of the term information or concept of information.

First, let's notice that the trick employed by Bar-Hillel and Carnap in their attempt to formulate a semantic information theory was to some extent similar. They replaced "the state of the world" with "the description of the state of the world" (Bar-Hillel and Carnap, 1952). This way, in their approach, although they considered this a marginal, technical procedure, information was not about the state of the world but about the description of the state of the world, obviously an informational entity. The problem here is that in their approach the description of the state of the world requires already existing semantic correspondence which they did not elaborate on as just a secondary technical issue. However, this way, when they assume the existence of the description, their reasoning becomes circular.

Much closer to my approach presented above was John Stuart Mill's view of the meaning (1843). Mill involved in his description of the meaning of a term two concepts, of its denotation (basically similar to what commonly is understood by meaning, i.e., the set of entities to which the term can be predicated) and connotation (the organized system of predicates or properties which can be predicated on all entities from the denotation).

We can identify a similar idea in the works of Peirce (1867), although his view was expressed in a metaphorical way of multiplication of numbers, that information is equal to the product of extension and intention ("breadth × depth of the concept"). Although Peirce considered information in a rather narrow context of the characteristics of concepts, it is interesting probably the first occurrence of this term in the relevant literature.

These views are surprisingly intuitive and at the same time surprisingly absent in the contemporary discourse of meaning. In particular, Mill's view that it is the connotation that determines the meaning is of special importance. We cannot comprehend the denotation which may consist of a large or even infinite number of entities and the only way we can understand the meaning of the term is by the connotation that gives us characteristics of the objects in the denotation. We know what the meaning of the word "dog" is not because we know all dogs but because we know what properties of dogs make them dogs. The connotation can be easily identified with the informational content of the term.

Moreover, the Aristotelean genus-species (classical) concept of definition is based on a similar but slightly modified approach. Aristotle refers to the connotation of the defined concept (genus) and then seeks differentia, a description of differences that distinguish the defined concept from all other species of the genus. This approach refers to the same order structure but involves an additional instrument — differentia analogous to Bateson's difference. His "that makes a difference" can be interpreted as the distinction of differences of species within the genus. Which differences do not make a difference? Those which involve species not included in the genus.

We have here the first Bachelardian epistemological rupture tearing down the separation of semantics and syntactic of information. The next step in removing epistemological obstacles is the reversal of the relationships. Orthodox semantics starts from the pre-existing language and assigns intention to its components (terms) that carry meaning. This is against the historical order of affairs. At first, there were objects of our (human) comprehension whom we gave informational nature as carriers of information. Then gradually humans associated with them (also informational) symbolic entities. This we can call encoding information associated with the process of assigning symbols (thus, reverse semantics), but the original information was already encoded (without our awareness) in the objects of our comprehension. These objects are our construction in the sense that our comprehension selects the information characterizing them, i.e., information encoded in them. Thus, instead of focusing on intention as action directed to the entities of external reality, we should focus on encodings of information in different information systems (consciousness, entities, etc.) and the correspondence between structures of these encodings. This approach emphasizes the structure of information which can be analyzed through its manifestation in encoding. The isomorphism of the structures defines meaning.

In this perspective, Hartley's observation of the invariance of information with respect to the change of encoding acquires special importance (Hartley, 1928). Although he considered this invariance as obvious and wrote about it mainly for the derivation of his measure of information, the recognition of the variety of encodings of the same information makes his study the actual initiation of the study of information. Of course, it would be an anachronism to claim that Hartley was aware of the fundamental issues of the semantics of information and that he appreciated the role of invariance of information in its different encodings for the development of the study of information. For him, it was an obvious characteristic of information derived from the practical experience of telegraphy useful for the measuring of information. However, the way he approached the matters of encoding information in the contexts not only of texts but also of images shows that his approach based on an intuitive insight formulated twenty years before the publication of Shannon's work in some aspects was superior.

6 Qualitative and Quantitative Methodologies

One of the most common myths about science is the belief that numbers are the ultimate tools of scientific inquiry and that quantitative methodologies of inquiry are always better than qualitative ones. I already published a critique of this unfortunately very popular view of science and will not repeat its arguments here (2020b). I focus here on the methodologies of the study of information. Certainly, the popularity of this view among the members of the general audience is perpetuated by the false opinion that mathematics is a discipline studying numbers. Media, with the best intention, promote the idea of "numeracy" as an important part of education or lament its decline in contemporary societies. Teachers are warned about the disability called "dyscalculia" as a source of difficulties in learning mathematics.

Of course, there is a very important subdiscipline of mathematics called number theory with multiple applications across many other subdisciplines. However, there are many other subdisciplines where number theory is absent or does not play any important role. Moreover, number theory studies not numbers but structures defined on sets of numbers. It is powerless to answer the questions about particular numbers although "numerologists" (not "number theorists") will tell you a lot about terrible numbers, such as 4, 13, and 666 (the last is the worst for sure). I am sure that all promoters of numeracy would be terrified if they learned that the arithmetic of natural numbers or the theory of real numbers cannot answer the question about the result of $2+2$ without a prior clarification of the convention in writing names of numbers. The result can be 4 (if we use the decimal convention) or 11 (if we use the convention with three digits inherited from the decimal system). Number theory does not deal with the questions about the result of $2+2$ but about what we can say about numbers when we eliminate conventions.

The enthusiasts of quantitative methodologies frequently refer to Galileo Galilei's view on the use of mathematics in the study of nature: "In *The Assayer*, (Galileo) wrote 'Philosophy is written in this grand book, the universe ... It is written in the language of mathematics, and its characters are triangles, circles, and other geometric figures;...'" (Drake, 1957). There is nothing about numbers here. Of course, Galileo was a pioneer in experimental methodology and his work involved measurements and therefore numbers. But his understanding of numbers was still in the tradition making them derivative from the geometric intuition. It took an additional three centuries to make real numbers independent from geometry, mainly through the work of Dedekind who provided the description of their construction now known as "Dedekind cuts" in the second half of the 19th century.

It was the time when the infatuation of scientists and philosophers with numbers became common. A prominent example of this infatuation can be found in William Thomson's (Lord Kelvin's) *Popular Lectures and Addresses (1891–1894)*, where he wrote, "When you can measure what you are speaking about, and express it in numbers, you know something about it; but when you cannot measure it, when you cannot express it in numbers, your knowledge is of a meager and unsatisfactory kind (...)" (Thomson, 1981).

Although the fallacy of the belief that mathematics is about numbers and not much else is so common, it is not more dangerous than any other expression of ignorance. Much more dangerous are false beliefs about some concepts in mathematics held by people with a relatively high level of familiarity with mathematics, frequently even teaching mathematics at some levels of education.

A typical example of such false belief continuously promoted in the majority of introductory college textbooks for algebra is that the set of real numbers has an easy explanation by its one-to-one correspondence to the set of points on a straight line. We have to choose one point representing 0 and another different point representing 1. The increase in numbers will be in the direction of 1 from 0. Then we relate with a non-negative number x the point on the side of 1 which is in distance equal to x. If x is negative, we relate to it the point in the distance equal to the absolute value of x on the left side from 0. Conversely, if we have a point on the side of 1, we relate to it the real number equal to its distance from 0. For points on the other side, we relate to them numbers opposite to their distances. So, it is concluded that we showed that points of the straight line and real numbers are in a one-to-one correspondence and the order of numbers corresponds to the order of points.

In more than 40 years of teaching math to undergraduate students (some of them very intelligent), I never received a correct answer to the question of why this reasoning is completely invalid and without any hope of making it valid by revisions. It was not the students' fault. In addition to being confused by the deceptively intuitive terminology that can fool even more experienced learners, they were already brainwashed in their high schools. It did not help that before the class about this so-called "real line", I always warned the students about the error of circular reasoning.

Students are equally surprised when I demonstrate that the geometric constructions in the style of Descartes cannot help in the determination of the position of points corresponding to all but a small, countable subset of real numbers. It does not help consider all computable real numbers or numbers that can be identified using any logically formulated description (again both are countable subsets).

Any reasoning about all real numbers involving one of the premises "If we know what x is, then ..." has to be invalid because the majority of real numbers cannot be identified. The theory of real numbers is about the structure of real numbers (formally called the field of real numbers) concerning algebraic operations, not about individual numbers. Thus, the "real line" does not prove anything and does not explain much about real numbers. It only explains how we can create a consistent but rather arbitrary model of the geometric one-dimensional space using an algebraic conceptual framework when we assume that there is some measure of distance defined on it. Of course, without assuming a specific measure there is no correspondence between real numbers and points of the line.

This demonstrates that the belief in the superiority of the quantitative methodologies involves the fallacy that the association of the elements of reality with numbers has an explanatory power because we can comprehend numbers easily, although this comprehension may be illusionary. However, there is another fallacy involved in the opposition between the quantitative and qualitative methodologies. It is based on the conviction of their essential difference. In reality, the mathematical concepts associated with numbers such as the concept of a measure or a distance are simply tools for modeling or generating structures within the field of scientific inquiry. Specific numbers do not have any meaning, only their mutual relations invariant with respect to some transformations.

Numbers obtained in measurements involve conventions of the choice of the system of units. The measurements serve the purpose of determining the structures of investigated objects and phenomena in which these objects are engaged. The structures are not numbers, magnitudes, or quantities, and therefore structural analysis belongs to qualitative methodologies. Quantitative methodologies are just components of more general qualitative methodologies. The confusion about the distinctive forms of inquiry arose from the fact that some forms of inquiry, especially in the contexts of the lower level of abstraction, may involve very simple methods of the collection and analysis of information that do not require advanced mathematical formalisms. Typically, the low level of abstraction is associated with structures that do not require numerical tools. However, there are also many forms of structural analysis engaging sophisticated mathematical theories in which concepts of numbers are absent. There is no reason to claim the essential difference between quantitative and qualitative methodologies and no reason to claim the superiority of either of them.

7 Digital and Analog Information

The most popular distinction between digital and analog computing introduced by von Neumann (1963) is based on the difference between the symbolic, digital, and discrete representation of numbers and their apparently continuous representation as magnitudes characterizing the states of physical objects. In reality, although this distinction is highly intuitive and seems to reflect objective differences between the forms of information, it is purely conventional. The functioning of all computing devices involves the manipulation of the physical states of their operating systems, and at the same time, the digital representation of numbers is achieved by a conventional discretization of continuous magnitudes.

In this chapter, I am using the distinction between analog and digital information and computing that I introduced in my earlier work in which the difference between analog and digital information is similar to the difference between the concepts of physics characterizing physical systems by physical states (analog) and observables (digital). This distinction in physics acquired fundamental importance with the rise of quantum mechanics but was already present earlier. I wrote "similar" because essentially identical distinctions can be identified elsewhere. For instance, the foundations of probability theory can be built starting from the concept of a family of events understood as measurable subsets of an outcome space and proceeding to random variables, alternatively, starting from an appropriate algebraic structure of random variables and proceeding to special class of random variables that can be interpreted as characteristic functions for subsets of an outcome set corresponding to events.

The depth of the distinctions between the fundamental concepts that we can interpret as a state of the system and that of observable in both quantum theory and probability theory is rarely recognized, at least not in an open way. In quantum theory, the issues are hidden, for instance, by the use of *ad hoc* terminology of "hidden variables" (that sounds better than the oxymoron "unobservable observables"). More recently, to avoid the name "observable" suggesting engagement of human inquirer, the name "beable" was introduced by John Bell. In probability theory, the standard trick to avoid complications is to focus on just two special cases of discrete and continuous random variables while excluding anything else that causes trouble. Probably the closest to an honest denunciation of the forgotten problems was the series of the 1998 Turin Lectures by Gian-Carlo Rota (2001): *Twelve Problems in Probability No One Likes to Bring Up*. A more extensive and detailed exposition of some issues addressed by Rota is in the book created in collaboration with others (Kung et al., 2009).

Rota's lectures addressed not only issues within probability theory but also the study of information. In particular, Rota addressed the issue of the formulation of the orthodox information theory derived from probability theory while it should precede probability. This concern was not new as it was already voiced by Andrey Kolmogorov (1983) a long time ago when he proposed his solution in the form of a description of algorithmic complexity. Kolmogorov's approach did not bring a sufficiently general solution and Rota directed the future inquiry toward a new logic of information in terms of the lattice of partitions.

A similar issue is in the relationship between quantum physics and quantum information theory. Quantum computing became the hottest topic of this century, but it seems that here too the carriage was placed in front of the horse. The usual approach is to study quantum computers considered as a special case of a quantum system, and quantum information is just an engineering concept necessary for the use of such computing devices. With the increasing role of information as the most fundamental physical concept as promoted by Rolf Landauer (1991, 1996, 1999a, 1999b) (*Information is Physical*) and John Archibald Wheeler (1990) (*It From Bit*), quantum theory should be derivable from quantum information theory (Wheeler and Ford, 1998).

The main problem is in the main focus on quantum computing formulated almost exclusively in terms of qubits, i.e., quantum systems that are described in terms of two-dimensional Hilbert spaces. The description is appropriate for quantum computer systems built with processors whose state is described in terms of spin. However, the theory of qudits (quantum information systems that can serve as quantum logical gates described by Hilbert spaces of dimensions higher than two) which was initiated at the end of the 20th century remains in the status of future or early inquiry (Rains, 1999; Gottesman, 1999; Vourdas, 2004).

The source of the problem is a well-known but frequently ignored fact that quantum theory in two-dimensional Hilbert spaces is fundamentally different from all cases when the dimension is higher than two or infinite. For instance, Hilbert spaces of dimension three or higher have to be infinite if we want to have orthocomplementation defined for their subspaces, while there are finite two-dimensional Hilbert spaces with orthocomplementation. Since obviously the restriction of the quantum theory to one special binary case does not make sense, there is no hope that quantum mechanics can be derived from the quantum information theory developed in terms of qubits.

A closer look at the formalisms involved in quantum and probability theories brings into focus another similar type of formalism developed in the semantic inquiries of modal logics based on the idea of possible worlds (initiated by Rudolf Carnap but already considered by Leibniz), in particular in the frames of Kripke semantics. Semantics in logic is introduced with the use of truth/false valuations of sentences in the two-element Boolean algebra. This can be implemented through functions on the set of all sentences with values in a set $\{0,1\}$ with appropriate conditions of consistency distinguishing the sentences with values 1 as (descriptions of) possible worlds. Then the necessarily true sentences are those that have valuations 1 in all possible worlds, and possibly true sentences have valuations 1 in at least one possible world. Thus, a similar semantics for information in the style of Kripke semantics can be developed in terms of valuations. However, this can be done only with the prior development of the logic of information.

8 Brief Outline of an Example of Theory of Information Meeting the Postulates of the Rupture

8.1 *Definition of information*

Thus far, this chapter presented a critical analysis of methodological assumptions inhibiting progress in the study of information as a fundamental concept for inquiries of intelligence, complexity, and consciousness together with postulates to eliminate obstacles. There is a legitimate question about how realistic these postulates are. This section is intended as a confirmation that a theory of information meeting the postulates is possible.

The remaining part of this chapter presents an outline of a theory of information developed in my earlier publications formulated here in a way consistent with the postulates promoted in this work (2011c, 2022). The reason for this short presentation (extensive and detailed presentations are published elsewhere in my multiple articles) is not intended as a closure of the theoretical study of information but rather its opening at a sufficiently high level of generality for inquiries of intelligence, complexity, consciousness, etc. Its direct objective is to demonstrate or rather illustrate practical applications of the methodological tools described in this chapter. This illustration of the use of methodological tools may be helpful for alternative conceptualizations of information.

At this point, it is important to recall that there are some alternative approaches to the study of information that explicitly denounce the shortcomings of the Shanonian tradition and offer ways of their elimination. It would have been too extensive a task to present them all in this chapter. Since none of them resolves or addresses all the issues analyzed here, this analysis may be helpful in their continuation. The approach with the most interesting results in my subjective judgment, an advanced theoretical formalism, and applications in more specific domains was developed by René Thom in his famous but currently rarely and insufficiently revisited book *Structural Stability and Morphogenesis: An Outline of a General Theory of Models* (1975). Thom provided excellent tools for the structural analysis of information. His way of thinking about the relationship between qualitative and quantitative methodologies was similar to that presented in this chapter, for instance, in his criticism of Rutherford's dictum "Qualitative is nothing but poor quantitative".

The content of this section includes an outline of a mathematical formulation of a variety of information systems that refers to several mathematical concepts and their algebraic theory (not explained here, but in referred sources). As such, it can be omitted without any loss of understanding of the general idea presented in the following four paragraphs.

The definition of information used here is very general (Schroeder, 2011c, 2022). It does not refer to any other definable concepts but exclusively to one categorial opposition of one and many. This generality that someone could object to as being excessive is intentional. Since the concept of information appears in virtually all possible contexts and has as a particular instance the main tool of any intellectual inquiry in the form of language, any other more specific and less general concept would lose some important applications.

There is a natural question about why not go further and simply consider information a category, an undefinable (primitive) concept characterized by axioms. The reason is that this would obscure some unquestionable features of information present in all its contexts. Also, the fundamental features of the one-many opposition can be derived from the immense body of philosophical reflection on it in the diverse philosophical traditions of several civilizations. These features direct the methods for differentiation of a variety of different types of information. Moreover, the rich philosophical tradition of the one-many opposition creates a valuable intellectual environment for the study of information that cannot be replaced by even a long list of axioms. Finally, the opposition influenced the development of set theory, and mathematical models of information are formulated in the language of set theory.

Thus, information is understood as an identification of a variety understood as a resolution of the categorial opposition of one and many. The study of information is focused on this resolution as a transition from many to one. This transition can be achieved by a selection of one out of many (selective manifestation of information) or by the identification of a structure binding this multitude into one (structural manifestation of information). The two manifestations are always present together.

This definition, or rather any definition of information, becomes meaningful only when followed by a theory providing theoretical tools for the study of information. For instance, the statements of the theory should have consequences that can be empirically tested. Moreover, they should be consistent with already accumulated results of inquiries of information in the more specific domains of its applications. On the other hand, the concept of information has so large variety of applications that its theory requires a high level of abstraction and therefore a mathematical formalism.

8.2 Outline of a mathematical formalism of the theory of information

Information has multiple contexts and each of them requires a specification of its manifestation in terms of an information system. An information system specifying the type of information in mathematical terms is in this approach a closure space and all mathematical concepts mentioned here for modeling information are expressed in terms of the theory of such spaces (Birkhoff, 1967). In this study, there is no need to restrict this concept by additional conditions unless we proceed to its application in one of its specific domains. Thus, information can be of the geometric, topological, logical, or physical type associated with additional defining conditions for an appropriate closure space describing an information system. However, here we want to consider a general formalism. This is the reason for using the concept of a general closure space which generalizes formalisms of all these (geometric, topological, logical, or physical theories) and many more mathematical theories (Birkhoff, 1967).

The logic of such a general information system is a complete lattice of closed subsets of the information system. At this point, it is important to indicate that the term "logic" as used here has a much more general meaning than usual which only in the case of linguistic or probabilistic information systems can be identified with the familiar Boolean lattice defined by the connectives between propositions of some language. The less conventional application of this term consistent with the approach of this paper can be found in quantum logics defined on the closed subspaces of a Hilbert space (alternatively, on the projectors on these subspaces) or in Rota's logics of information identified with lattices of partitions.

The instances of the information within an information system are filters (sometimes called dual ideals) defined on the lattice of closed subsets (the logic of the information system). Filters are collections of closed subsets selected from the logic of the information system, such that with each closed subset all closed subsets including it belong to the filter too (i.e., filters are hereditary), and which are closed with respect to finite intersections. Filter in the logic of information systems is a direct generalization of Mill's connotation in the traditional Boolean logic. Ultrafilters, principal filters, and prime filters characterize special types of information. Since filters representing instances of the information are defined on logics that are not necessarily Boolean lattices, and the theory of filters is typically studied in this particular context (e.g., Stone theorem), it is important to be aware of the ramifications of the theory of filters when we transcend the Boolean context. For instance, ultrafilters are not necessarily prime filters anymore, which is a fact that frequently confuses physicists in discussing the question of hidden variables.

The formalism based on filters reflects structural characteristics of information and filters represent a state of some universe of inquiry (quite frequently simply called "possible worlds"). Certainly, the universe of inquiry should not be confused with the world as the filters can be defined only after a particular logic of information is chosen, i.e., after the type of information is established.

Information defined or characterized as filters can be identified as analog type as they constitute the connotation of information characterizing the state of the inquired system. However, we have an alternative tool for inquiry of information referring to observed numerical characteristics associated with digital type. Here we have σ-fields of subsets and measures defined on them. The measures can be arbitrary, associated with magnitudes characterizing the objects of inquiry, possibly restricted to probability type, or further restricted to binary logical valuations. In each case, we can construct a corresponding lattice that can be interpreted as a logic of information. Furthermore, we can distinguish filters representing instances of information.

Finally, we can establish the relationships between the analog and digital descriptions of information in different types of information systems classified by their logics, i.e., lattices of closed subsets. These relationships are complex and they heavily depend on the specifics of the systems. In general, the best-known and simplest relationships in the Boolean type of information systems become more complex and ramified when their logics are unconventional, i.e., the lattices of closed subsets differ substantially from Boolean algebras, for instance, in quantum logics. Then, the logic of qubits is different from the logic of qudits.

The explanatory power of the formulation of information theory in terms of closure spaces can be appreciated even more when we recall that the famous construction of real numbers in terms of Dedekind cuts is nothing else but the identification of real numbers with the closed subsets of the set Q of rational numbers with respect to the Galois closure (polarity) defined by the order relation of Q. This makes it possible to interpret magnitudes and measures as structure-preserving functions (morphisms) between the lattices of closed subsets (logics) and to get better insight into the role of real numbers in the study of information.

Further details of this theory of information can be found elsewhere in my already published papers (2019, 2022), while additional details and demonstrations or proofs of the mathematical claims made here will appear in a paper currently in preparation.

9 Conclusions

This chapter demonstrates the urgent need for further intensive studies of information as the only tool for securing human control of rapidly developing information technologies. Legal regulations may restrict or direct human actions, but even if effective in such tasks (which is doubtful) they will not prevent the dangers of unpredictable developments in technology.

The further development of the study of information has to be coordinated with the studies of related important but poorly understood concepts of consciousness, complexity, intelligence, computation, and life (in this chapter, the first three are considered). The wide range of phenomena involving these concepts requires that adequate theories of information have to be at a sufficiently high level of abstraction. On the other hand, they have to be sufficiently specific and precise in their methodologies to have explanatory power for the entire complex of studies not only of information but also of consciousness, complexity, and intelligence.

An outstanding deficiency in the study of information at a high level of generality is the negligence of the semantics of information which if considered at all is typically formulated in restricted contexts, such as the context of life in biosemiotics. All existing semantic theories of information have in their center a triangular relation relativizing the meaning to non-informative elements, such as an interpreter and thought. A binary approach excluding the mediation of a third party is proposed.

The most important epistemological obstacles identified in this chapter have methodological character. They are related to misunderstandings of the traditional divisions into quantitative vs. qualitative methodologies, the role of mathematics, in particular, the role of number theory, measure theory, and probability theory. Another source of confusion identified in this chapter was a more specific misunderstanding of the distinction between analog and digital information. In each case, some proposals were presented for how to eliminate confusion.

The final part of this chapter provided a very brief and general exposition of an example of a theory of information in which the epistemological obstacles are avoided.

Further study in this direction may address a question about generative AI. Can we identify essential differences between generative AI systems and human intelligence? My working hypothesis is that the most important difference is in their information logics (understood in the way described in the preceding section). In the former, the logic of information is built based on Large Language Models (LLMs) derived from the characteristics of the language. Human intelligence has its logic reflecting not relations within the language but in the model of reality or world that we develop in our living experience. Thus, to acquire human intelligence, the AI systems have to be built not based on the patterns of the language but based on the patterns of reality which we could call "Large World Models". The formalism of the theory of information presented above is consistent with this idea.

Acknowledgment

The author would like to express his gratitude to the anonymous reviewer for helpful suggestions regarding the format of this work.

References

Bachelard, G. (1986). *The Formation of the Scientific Mind: A Contribution to a Psychoanalysis of Objective Knowledge.* Boston, USA: Beacon Press.

Bar-Hillel, Y. and Carnap, R. (1952). An outline of a theory of semantic information. Technical Report No. 247, Research Laboratory of Electronics, MIT. Reprinted in Bar-Hillel, Y. (1964). *Language and Information: Selected Essays on Their Theory and Application.* Reading, MA, USA: Addison-Wesley, pp. 221–274.

Beck, F. and Eccles, J. C. (1992). Quantum aspects of consciousness and the role of consciousness, *Proc. Natl. Acad. Sci. USA* 89, 11357–11361.

Birkhoff, G. (1967). *Lattice Theory,* 3rd. edn. American Mathematical Society Colloquium Publications, Vol XXV. Providence, RI, USA: American Mathematical Society.

Brentano, F. (1874/1995). In McAlister, L. L. (ed.), *Psychology from an Empirical Standpoint* Routledge, London, UK.

de Silveira, T. B. N. and H. S. Lopes, H. S. (2023). Intelligence across humans and machines: a joint perspective, *Front. Psychol.* 14, 120761, doi: 10.3389/fpsyg.2023.1209761.

Drake, S. (1957). *Discoveries and Opinions of Galileo.* New York, NY, USA: Doubleday.

Editorial. (2023). How to edit anthropomorphic language about artificial intelligence, *Nat. Rev. Phys.* 5, 263, https://doi.org/10.1038/s42254-023-00584-1.

Fleming, S. *et al.*, (2023). The integrated information theory of consciousness as pseudoscience. Preprint at PsyArXiv https://doi.org/10.31234/osf.io/zsr78.

Ford, B. J. (2009). On Intelligence in cells: The case for whole cell biology. *Interdisciplinary Sci. Rev.* 34(4), 350–365.

Fredricksmeyer, E. A. (1961). Alexander, Midas, and the Oracle of Gordium, *Class. Philol.* 56(3), 160–168.

Gottesman, D. (1999). Fault-tolerant quantum computation with higher-dimensional systems. In Williams, C. P. (ed.), *Quantum Computing and Quantum Communications, QCQC 1998.* Lecture Notes in Computer Science, 1509. Berlin, Heidelberg, Germany: Springer, pp. 302–313.

Hameroff, S. R. and Penrose, R. (1996). Orchestrated reduction of quantum coherence in brain microtubules: A model for consciousness, *J. Conscious. Stud.* 3, 36–53.

Hartley, R. V. L. (1928). Transmission of information. *Bell Syst. Tech. J.* 535–563.

James, W. (1890). *The Principles of Psychology.* New York, NY, USA: Holt.

James, W. (1947). The one and the many. In W. James. *Pragmatism: A New Name for Some Old Ways of Thinking.* New York, NY, USA: Longman's Green and Co.

Kolmogorov, A. N. (1983). Combinatorial foundations of information theory and the calculus of probabilities. *Russian Math. Surv.* 38(4), 29–40.

Kruger, J. and Dunning, D. (1999). Unskilled and unaware of it: How difficulties in recognizing one's own incompetence lead to inflated self-assessments. *J. Personal. Soc. Psychol.* 77(6), 1121–1134.

Kull, K. (1999). On the history of joining *bio* with *semio*: F. S. Rothschild and the biosemiotic rules. *Sign Syst. Stud.* 27, 128–138.

Kung, J. P. S., Rota, G.-C., and Yan, C. H. (2009). *Combinatorics: The Rota Way*. New York, NY, USA: Cambridge University Press.

Landauer, R. (1991). Information is physical. *Phys. Today* 44, 23–29.

Landauer, R. (1996). The physical nature of information. *Phys. Lett. A* 217, 188–193.

Landauer, R. (1999a). Information is a physical entity, *Phys. A* 263, 63–67.

Landauer, R. (1999b). Information is inevitably physical. In Hey, A. J. G. (ed.), *Feynman and Computation: Exploring the Limits of Computers*. Reading, MA, USA: Perseus, pp. 77–92.

Lenharo, M. (2023). Consciousness theory slammed as pseudoscience — Sparking uproar: Researchers publicly call out theory that they say is not well supported by science, but that gets undue attention. *Nature NEWS* 20 September 2023, available online at Consciousness theory slammed as 'pseudoscience' — sparking uproar (nature.com).

Levi, M. (2009). *The Mathematical Mechanic: Using Physical Reasoning To Solve Problems*. Princeton, USA: Princeton Univ. Press.

Levin, M. (2023). Collective intelligence of morphogenesis as a teleonomic process. In Corning, P. A., Kauffman, S. A., Noble, D., Shapiro, J. A., Vane-Wright, R. I., and Pross, A. (eds.), *Evolution "On Purpose": Teleonomy in Living Systems*. Cambridge, MA, USA: MIT Press, pp. 175–197.

Mill, J. S. (1843). *A System of Logic, Ratiocinative and Inductive, Being a Connected View of the Principles of Evidence, and the Methods of Scientific Investigation*. London, UK: John W. Parker.

Ogden, C. K. and I. A. Richards, I. A. (1923/1989). *The Meaning of Meaning: A Study of the Influence of Language upon Thought and the Science of Symbolism*. Jovanovich, San Diego, CA, USA: Harcourt, Brace.

Peirce, C. S. (1867). Upon logical comprehension and extension, *Proc. Am. Acad. Arts Sci.* 416–432.

Peirce, C. S. (1977). A letter to Lady Welby, C. S. Peirce. *Semiotic and Significs: The Correspondence between Charles S. Peirce and Lady Victoria Welby*. Bloomington, ID, USA: Indiana Univ. Press.

Peirce, C. S. (2015). *The New Elements of Mathematics*. Berlin, Germany: Walter de Gruyter.

Posner, R. (1992). Syntactics, semantics, and pragmatics revisited half a century after their introduction. In Charles W. Morris, Deledalle, D., Balat, M., and Deledalle-Rhodes, J. (eds.), *Signs of Humanity: Proceedings of the IVth International Congress. International Association for Semiotic Studies Barcelona/Perpignan, March 30-April 6, 1989*. Berlin, Germany: De Gruyter Mouton, pp. 1349–1354.

Rains, E. M. (1999). Nonbinary quantum codes. *IEEE Trans. Inf. Theory* 45(6), 1827–1832.
Reza, F. M. (1994). *An Introduction to Information Theory*. New York, NY, USA: Dover.
Rota, G.-C. (2001). Twelve problems in probability no one likes to bring up. In Crapo, H. and Senato, D. (eds.), *Algebraic Combinatorics and Computer Science*. Milano, Italy: Springer, pp. 57–93.
Rothschild, F. S. (1962). Laws of symbolic mediation in the dynamics of self and personality. *Ann. NY Acad. Sci.*, 96, 774–784.
Schroeder, M. J. (2009). Quantum coherence without quantum mechanics in modeling the unity of consciousness. In Bruza, P., Sofge, D., Lawless, W., van Rijsbergen, K., and Klusch, M. (eds.), *QI 2009, LNAI Volume 5494*. Heidelberg, Germany: Springer, pp. 97–112.
Schroeder, M. J. (2011a). Concept of information as a bridge between mind and brain. *Information* 2(3), 478–509, http://www.mdpi.com/2078-2489/2/3/478/.
Schroeder, M. J. (2011b). Semantics of information: Meaning and truth as relationships between information carriers. In Ess, C. and Hagengruber, R. (eds.), *The Computational Turn: Past, Presents, Futures? Proceedings IACAP 2011, Aarhus University — July 4–6, 2011*. Munster, Germany: Monsenstein und Vannerdat Wiss., pp. 120–123.
Schroeder, M. J. (2011c). From philosophy to theory of information. *Intl. J. Inform. Theories Appl.* 18(1), 56–68.
Schroeder, M. J. (2012). Search for syllogistic structure of semantic information. *J. Appl. Non-Class. Logic* 22, 101–127.
Schroeder, M. J. (2013). The complexity of complexity: Structural vs. quantitative approach. In *Proceedings of the International Conference on Complexity, Cybernetics, and Informing Science CCISE 2013 in Porto, Portugal*, https://proceedings.informingscience.org/InSITE2013/CCISE13pE005CC195GT0409Schroeder.pdf.
Schroeder, M. J. (2017). Structural and quantitative characteristics of complexity in terms of information. In Burgin, M. and Calude, C. S. (eds.), *Information and Complexity, World Scientific Series in Information Studies*, Vol. 6. New Jersey: World Scientific, pp. 117–175.
Schroeder, M. J. (2019). Theoretical study of the concepts of information defined as difference and as identification of a variety. In Dodig-Crnkovic, G. and Burgin, M. (eds.), *Philosophy and Methodology of Information — The Study of Information in a Transdisciplinary Perspective*. World Scientific Series in Information Studies, Vol. 10. Singapore: World Scientific, pp. 289–314.

Schroeder, M. J. (2020a). Intelligent computing: Oxymoron? *Proceedings* 47(1), 31, https://www.mdpi.com/2504-3900/47/1/31.

Schroeder, M. J. (2020b). Contemporary natural philosophy and contemporary *Idola Mentis*. *Philosophies* 5, 19, https://www.mdpi.com/2409-9287/5/3/19/htm.

Schroeder, M. J. (2022). Symmetry in encoding information: Search for common formalism. *Symmetry: Art and Science, Special Issue: Symmetry: Art and Science 12th SIS-Symmetry Congress* (1-4), pp. 292–299, 588708.pdf (up.pt).

Schrödinger, E. (1944). *What is Life? The Physical Aspect of the Living Cell.* Cambridge, UK: Cambridge Univ. Press.

Shannon, C. E. (1948). A mathematical theory of communication. *Bell Sys. Tech. J.* 27, 323–332 and 379–423, https://doi.org/10.1002/j.1538-7305.1948.tb01338.x.

Shannon, C. E. and Weaver, W. (1949/1998). *The Mathematical Theory of Communication.* Urbana, IL, USA: University of Illinois Press.

Shevlin, H. and Halina, M. (2019). Apply rich psychological terms in AI with care. *Nat. Mach. Intell.* 1, 165–167.

Tarski, A. (1983). The concept of truth in formalized languages. In Woodger, J. H (eds.), Tarski, A. and Corcoran, J., *Logic, Semantics, Metamathematics.* Indianapolis, IN, USA: Hackett pp. 152–278.

Thom, R. (1975). In Fowler, D. H. (Transl.), *Structural Stability and Morphogenesis: An Outline of a General Theory of Models.* Reading, MA, USA: Benjamin.

Thomson, W. (Lord Kelvin) (1891). *Electrical Units of Measurement,* Vol. 1 in *Popular Lectures and Addresses (1891-1894).* London, UK: Macmillan, p. 80.

Tiles, M. (1984). *Bachelard, Science and Objectivity.* Cambridge University Press, Cambridge, UK.

Tononi, G. (2007). The information integration theory of consciousness. In Velmans, M. and Schneider, S. (eds.), *The Blackwell Companion to Consciousness.* Malden, MA, USA: Blackwell, pp. 287–299.

Tononi, G., Sporns, O., and Edelman, G. M. (1994). A measure for brain complexity: Relating functional segregation and integration in the nervous system. *Proc. Natl. Acad. Sci. USA* 91, 5033–5037.

Tononi, G. and Edelman, G. M. (1998). Consciousness and complexity. *Science 282*, 1846–1851.

von Neumann, J. (1963). The general and logical theory of automata. In Taub, A. H. (ed.), *John von Neumann, Collected Works,* Vol. V. Oxford, UK: Pergamon Press, pp. 288–326.

Vourdas, A. (2004). Quantum systems with finite Hilbert space. *Rep. Prog. Phys.* 67, 267–319.

Wheeler, J. A. (1990). Information, physics, quantum: The search for links. In Zurek, W. H. (ed.), *Complexity, Entropy, and the Physics of Information*. Redwood City, CA, USA: Addison-Wesley, pp. 3–28.

Wheeler, J. A. and Ford, K. (1998). *Geons, Black Holes, and Quantum Foam: A Life in Physics*. New York, NY, USA: W. W. Norton.

Wheeler, W. M. (1910). *Ants: Their Structure, Development and Behavior*, Vol. 9. New York, NY, USA.

Wittgenstein, L. (1953). *Philosophical Investigations*. New York, NY, USA: Macmillan.

Chapter 7

A Naturalistic Approach to the Study of Information Philosophy

Zhensong Wang

Philosophy Department, School of Humanities and Social Sciences, Xi'an Jiaotong University, No. 28 Xianning West Road, Xi'an 710049, China
zhensong@xjtu.edu.cn

There are two approaches in the studies on information philosophy: idealism and materialism. After analysis, we find out that the idealistic study on information philosophy actually limits our understanding of information and the materialistic study on information philosophy faces a more complex predicament. Therefore, in order to promote the research of information philosophy, the research approach of naturalistic study on the philosophy of information has been put forward. It has a critical attitude toward idealistic approach and critically inherits some basic propositions of materialism path as well. This approach further distinguishes itself through its reliance on a novel logic, distinct from standard logical frameworks, which serves as the underlying logical foundation for its theoretical constructs. And it advocates a naturalistic monism, to avoid the problems that previous information philosophies brought about, which will further defend the positive significance of unified information science. At last, this chapter explains how this approach could be treated as a new form of panpsychism.

1 Introduction

The objective of information philosophy is to formulate a groundbreaking model for elucidating the essence of "mind". This novelty lies in its departure from treating the mind as an entity or quality separate from matter, instead positioning "information" as a more fundamental philosophical construct than that of the "mind". Moreover, this approach heavily draws on the advancements in information science and technology since the mid-20th century. Regarding the interplay between information and the mind, proponents of information philosophies consistently maintain that the mind is composed of information. However, due to variations in their comprehension of the nature of information and the mechanisms by which the mind forms, different research methodologies have emerged within the field of information philosophy. In the early 21st century, Floridi (2011) and Wu (2005) both established mature systems within the realm of information philosophy. While they both assert that the philosophy of information constitutes a fundamentally new form of philosophical inquiry — termed Philosophia Prima by Floridi (2002) and Meta-philosophy by Wu and Brenner (2017), respectively — their perspectives diverge significantly. They present differing interpretations of the intrinsic nature of information, the relationships it bears with the mind and matter, and their stance toward information science itself. Nonetheless, these two schools of thought share common ground as part of the broader umbrella of information philosophy. A prime example of this consensus is the recognition that information constitutes a foundational domain of existence. Further details are expounded upon in the subsequent analysis.

This chapter aims to highlight the limitations inherent in both prevailing approaches and concurrently propose a third perspective on the study of information philosophy. To summarize, Floridi's stance is characterized by an idealistic interpretation of information philosophy, while Wu advocates for a materialistic approach. I endeavor to introduce a naturalistic framework for understanding information philosophy, which perceives the informational realm as an integral part of the natural world — a domain that, though distinct, coexists with the material world within an epistemological context. Acknowledging the philosophical lineage from naturalism to materialism and physicalism, the Naturalistic Information Philosophy (NIP) fundamentally builds upon certain core concepts of the materialistic philosophy of information. However, it also espouses unique perspectives that diverge from its predecessor, thus offering a novel contribution to the discourse.

2 A Critique of Cartesianism

The Cartesian doctrine of the absolute dichotomy between mind and matter, as well as subject and object, emerged within the historical context of the burgeoning modern scientific era. Its aim was to clear the path for natural sciences by dispelling mystical impediments, concurrently safeguarding the rationality of God's existence. However, this stance carries a limitation of the times; its original intent to subsume the mind under divine authority has significantly impeded the process of scientification or naturalization of the study of consciousness in the nearly 400-years trajectory of philosophical and scientific thought since Descartes. Even after God was explicitly ousted from the philosophical discourse in the 19th century, the inertia fostered by Cartesian dualism's influence on scientific development eventually led us into the quagmire of investigating consciousness. The surging philosophies of consciousness in the 20th century failed to deliver a genuine science of consciousness because they could not extricate themselves from the Cartesian premise of human exceptionalism — the notion that humans are fundamentally distinct from all other material systems, including other primates. This uniqueness is epitomized in the semantic manufacturing capacity of the human mind. While the philosophical inquiry into semantics gained prominence in the 20th century, its significance can be traced back to Descartes' deliberations on the role of human language, highlighting its pivotal position in this discourse.

Descartes' philosophical stance provides a foundational prerequisite for the evolution of epistemology. Spanning from Locke to Berkeley, Hume to Kant, reflections on the essence of knowledge have consistently remained anchored within the vessel of human cognition. Even in the metaphysical systems of thinkers like Spinoza, Leibniz, and Hegel, there is an overarching emphasis on demonstrating human rational faculties. The analytical philosophy movement sought to clear the philosophical landscape by ridding it of what they deemed "metaphysical debris". The Vienna Circle, organized with an anti-metaphysical agenda, endeavored to render epistemology scientific through pure linguistic analysis and logical reasoning — a philosophical endeavor premised on human's inherent thinking abilities. It becomes evident that despite more than three centuries having passed, 20th century philosophers did not stray far from Descartes' footsteps. They believed that the logic embedded in language was their beacon out of the "labyrinth of mind" crafted by Descartes. However, this pursuit inadvertently led them into the *cul-de-sac* of consciousness research, as highlighted by Chalmers (1995). Thus, Floridi, building upon the legacy of analytical philosophy, attempts to advance philosophical inquiry in the 21st century along this path, shifting the thematic focus from "semantics" to "information". Upholding the tradition of anthropocentrism, he posits that "information" constitutes the singular entity of existence and regards information philosophy as the ultimate haven for philosophers. In Floridi's information philosophy, the mind functions as an information processor, where information serves as the raw material to construct a knowledge system, and it is this very knowledge system that underpins the existence of the material world. This Platonic-like position is known as "Informational Structural Realism" (Floridi, 2008).

Wu shares both similarities and differences with Floridi, and they at least concur on two key points: the critique of Cartesian dualism and the acknowledgment of anthropocentrism. However, their divergence lies in the following aspects.

First, the outcomes of their critiques diverge significantly. Floridi's critique of Descartes leads him to advocate a return to Platonic concepts. In his perspective, the conundrum in modern philosophy originates from the dualistic thinking that was constrained by the philosophical climate of that era. Succinctly, he posits that mind and matter are not fundamentally dualistic, rather matter is inherently embedded within the mind. And a mind is an information processor, existing directly. Conversely, Wu's criticism of Descartes does not position him as either an idealist or a typical materialist monist; instead, it prompts him to propose a dual-existence theory involving both matter and information, with matter being more fundamental. Hence, his stance can be characterized as a materialistic approach to information philosophy. This "dual-existence" principle bifurcates existence into two domains: one being direct existence, which is matter itself, and the other indirect existence, represented by information. Furthermore, information can be further divided into subjective and objective aspects (for a concise introduction to this theory, see Brenner, 2011; Wu and Da, 2021). Wu views the material world as more foundational than the informational world yet acknowledges the irreducibility of the informational realm. His understanding of the relationship between matter and information bears resemblance to the connection between matter and form in Aristotelian philosophy. Thus, we may regard Wu's critique of Descartes as a return to Aristotelian principles, a point that will be explored more extensively in Section 5. This also constitutes a point of critique from the NIP. The logical foundation of naturalistic path is rooted in a non-Aristotelian logic, known as the Logic in Reality (LIR), Brenner (2008), whereas Wu's conception of information relies on Aristotelian formal logic for its division of the realm of existence. A detailed discussion of this is presented in Section 6.

Second, the origins of their recognition of anthropocentrism differ. Floridi's stance on anthropocentrism is overt; his philosophy of information pivots around semantic information, a construct created and exchanged exclusively among humans due to our lack of evidence for any other natural species using human-like languages. This perspective stems from his philosophical upbringing background. However, elucidating Wu's anthropocentric position requires a different approach. Wu posits that "information is a philosophical category signifying indirect existence, serving as the self-manifestation of material (direct existence) patterns and states" (Wu, 2005, p. 45). But who is this manifestation aimed at? The act of manifestation necessitates an observer. Given that we can only meaningfully communicate with other human beings, it follows that this manifestation is directed toward humanity. Furthermore, only through an anthropocentric lens can information be categorized as subjective or objective. It becomes evident that Wu's epistemology is underpinned by a materialist theory of reflection. On the other hand, Floridi adopts a transcendental epistemological standpoint, which diverges from Wu's perspective. In summary, while both acknowledge anthropocentrism, they do so from distinct vantage points — Floridi via the creation and exchange of semantic information unique to humans, and Wu through the manifestation of material existence patterns requiring human observation and interpretation.

Third, the incorporation of evolutionary thought constitutes another distinction between Wu and Floridi. Although the theory of evolution has been around for nearly two centuries, its purview typically encompasses the development of the universe, Earth, life, and society. Mind and consciousness (which can be treated as separate entities based on their differing degrees of scientific study difficulty) are often perceived as outcomes of natural evolution. Wu argues that mind and consciousness do not evolve from the material world but rather from the informational world, given that an individual embodies a unit of both material and informational existence. His dual-existence theory of matter and information, bolstered by evolutionary theory, offers a novel perspective to re-evaluate cognitive behavior, suggesting it is fundamentally an act of information transmission. Moreover, this theory posits that such information transmission behavior is not exclusive to humans but inherent in any material system. However, the materialistic underpinnings of Wu's theories somewhat hinder the realization of this idea because materialism tends to eliminate or downplay aspects of existence beyond matter. Conversely, Floridi's philosophy, intentionally or unintentionally, overlooks evolutionary sciences, much like most contemporary philosophies that aim to investigate the nature and foundations of science — a reflection of modern philosophical thought which questions the veracity of science and seeks solid grounding for it. Yet, the advancements in evolutionary sciences have prompted some philosophers to reassess the fundamental premises of philosophical inquiry, culminating in a new 20th century trend known as naturalism. This approach integrates science and philosophy closely, eschewing the traditional separation and treating philosophy as a study of the basis and essence of scientific knowledge. NIP adheres to this principle, and we delve further into its significance in the subsequent part of our discussion.

3 The Significance of Naturalistic Study on Philosophy of Information

At this juncture, we require the perspective of naturalism to effectively embrace the concept of the "co-evolution and interdependence of matter and information". In reality, nature itself constitutes the sole evolutionary domain. The scientific study of information and the scientific examination of matter represent two distinct pathways to understanding nature; each field's subject is an existing but not exhaustive depiction of nature, rather different models that elucidate its essence. From a naturalistic standpoint, cognitive behavior is not solely the preserve of humans or even animals in a broader sense. Moreover, the extent to which it can be generalized depends on our definition of cognition. If we conceive cognitive behavior merely as the process of information transmission, consider the mind an informational system, and view conscious activity as the decision-making mechanism within that system, then such a conception could potentially be extended to every facet of nature.

Wu's critique of Floridi is rooted in the general evolutionary theory. The semantic suspension introduced by Floridi, regarding the death of "the great programmer of the game of existence" (Floridi, 2011, p. 20), appears shortsighted from an evolutionary standpoint because semantics is not solely a human domain. This realization has gradually dawned upon some late 20th century analytical philosophers. Wu shares a significant degree of agreement with Dretske (Wang, 2019), who employed information theory to explore the naturalization of semantics (Dretske, 1981). The process of semantic creation (meaning emergence) is merely a ubiquitous form of information marking in nature. And this process depends on the recursiveness in nature, according to Brenner's conception (Brenner, 2008). During the course of natural evolution, this information marking activity has become increasingly complex as evolution progresses, culminating in highly sophisticated semantic engines found within advanced informational systems like those possessed by humans, capable of conducting intricate information transmission and marking activities within the realm of abstract self-consciousness. Thus, the rudimentary forms of semantic activities are relatively simple information transmission and marking processes occurring between basic material systems. As we delve into the realm of basic consciousness, which is typically associated with certain stages of biological development, we can observe that at least within the studies of biological macromolecules (West and Stock, 2001), prokaryotic cells (Marijuán et al., 2010), and lower life forms (Thompson, 2004; Lyon, 2006; Margulis, 2010), there are fundamental conditions and processes that lay the groundwork for the emergence of consciousness or its primary attributes. In these instances, the exchange of information between these entities that exhibit rudimentary forms of consciousness utilizes more foundational material and energetic interactions as the means of conveyance. The increasing complexity of material structures leads to a concomitant sophistication in the methods of information transmission, inevitably evolving into more complex decision-making mechanisms within informational systems. These decision-making mechanisms dictate how and at what efficiency information is exchanged between the material system or the informational system itself (viewed as a unified whole) and the external world. If such complexification continues, it will inevitably give rise to symbiotic systems composed of multiple interconnected material structures, exemplified by animals which have at least physical, chemical, and biological structural movements. Such complex organisms not only possess decision-making mechanisms for information exchange with the outside world but also require internal decision-making mechanisms to ensure the effective functioning of the entire system. The control mechanism governing the internal information exchange process constitutes a prerequisite for the formation of consciousness.

Human self-consciousness represents the most intricate decision-making mechanism for information exchange within an organism, enabling us to engage in clear reflective thought. This includes introspection on our own actions, contemplation of priori knowledge formed in evolution process, and even reflection upon our own reflections, thereby giving rise to thinking activities with distinct formal logical attributes. This elucidates why our logical constructs can be effectively applied to the empirical world — not because the experiential realm shares the Platonic world of ideas but because human logical ability is a product of natural evolution. Although the process of acquiring logical capability through evolution is imperceptible to individual humans, its traces can be discerned not only across human and primate populations and throughout the animal kingdom but even also across the entire spectrum of life. Thus, our limited perspective often leads to the presupposition of logic as an *a priori* faculty. As Quine astutely realized, epistemology grounded in absolutely clear logical structures is philosophers' idealized and beautiful vision (Quine, 1951). Instead, knowledge and even logic itself should be understood within the context of empirical psychology (Quine, 1968). Sellars similarly posits that logical abilities are acquired rather than innate, originating from experience (Sellars, 1997). Therefore, it is evolution that accounts for the isomorphism between logic and nature. Ascribing the power of reason to an abstract objective world is nothing more than a rationalization of human theological aspirations. This explains why the Pythagorean school, a pioneer in rationalism, was inherently religious, and why Russell contended that philosophy and theology jointly pursue ultimate existence, albeit through differing methods — philosophy via reason and theology through authority. The theory of evolution pulls us back to "earth" from "heaven".

If the explanation of mind and consciousness from the evolution of matter seems somewhat far-fetched, we might reconsider it through the lens of information's evolution. The rapid advancements in information technology over recent decades have profoundly revolutionized human society at an exponential pace, compelling us to deeply contemplate the essence of information. This introspection drives us for the first time to introduce scientific concepts into philosophy — a reversal of the traditional approach that seeks to render philosophical concepts scientific. At the turn of the last century, forward-thinking philosophers and scientists alike embarked upon this endeavor. Scientists often pondered the novel worldviews engendered by information science, such as Wheeler's "it from bit" principle (Wheeler, 1989) and Wiener's assertion that "Information is information, not matter or energy" (Wiener, 1961). Meanwhile, philosophers endeavored to construct comprehensive frameworks to explore the nature of information, exemplified by Floridi's information structural realism and Wu's dual-existence theory of matter and information. However, as analyzed earlier, despite both Floridi and Wu's commitment to critiquing Cartesian dualism, they haven't fully broken free from the constraints of Cartesian epistemology, particularly anthropocentrism. Under its influence, Floridi gravitates toward absolute idealism while Wu finds himself in a more intricate quandary. If he adheres strictly to materialism, he confronts the issue of dissolving information; if he leans toward informational epiphenomenalism, his theory will relinquish the autonomy of the informational realm that he proposes; and if he advocates a new form of dualism, his materialist stance becomes untenable, and he must elucidate the interaction mechanism between matter and information, presenting a new iteration of the mind–body problem. To circumvent these dilemmas, I put forward a fresh research approach — NIP, rooted in Wu's foundational propositions. This could be characterized as a naturalistic monism, offering a potentially viable alternative to overcome the existing philosophical impasses.

In the subsequent sections, I initially (in Section 4) delve into the ontology and epistemology underpinning Wu's materialism information philosophy, critically examining its significance, pinpointing the incompleteness of its critique, and presenting a naturalistic revision strategy. Subsequently (in Section 5), through an argumentative discourse, I demonstrate that Wu's philosophy of information, despite being akin to a return to Aristotelian principles, similarly fails to fulfill its intended meta-philosophical purpose, as has been the case with Floridi's *Philosophia Prima*. In the subsequent part (in Section 6), I argue that a non-Aristotelian logic, specifically Lupasco–Brenner logic, can serve to strengthen and consolidate the study of NIP. Then (in Section 7), from the standpoint of Unified Information Science (UIS), I discuss the theoretical advantages inherent in NIP. Lastly (in Section 8), I expound upon why a naturalistic approach to the philosophy of information can be construed as a novel form of panpsychism.

4 The Ontology and Epistemology of Materialistic Study on Philosophy of Information

4.1 *The dual-existence theory of matter and information*

Cartesian dualism inherently presents a conundrum in cognitive research. When the cognitive process is reduced to operations within the central nervous system and consciousness is studied based on computational and neuroscientific models, it inadvertently sidelines the complex issue of subjective experience — a fundamental component that underpins all intellectual activities. This constitutes an internal inconsistency within Cartesian dualism (Chalmers, 1995). Thus, the dual-existence theory of matter and information serves as a critique of mind–body dualism due to the following reasoning.

First and foremost, matter and information are isomorphic and coexistent entities. While the essence of matter is indeed an extensive property, the nature of information should not be misconstrued as a thinking property, rather it is characterized by its transmissive or processual quality. For instance, consider the high-energy particle collisions in a particle collider: when a particle undergoes structural change due to bombardment, this transformation can be likened to a record left behind — a trace that reveals the occurrence of the collision. The new configuration of the impacted particle may potentially yield insights into undiscovered particles, encapsulating information about the impacting particle (such as energy levels, spin, velocity, and so forth). In parallel, the impacting particle retains information regarding the particle it struck. Consequently, every instance of material movement and interaction inherently involves the presence and exchange of information. Thus, wherever there is dynamic engagement between matter, there exists a corresponding flow of information.

Second, the principle of general evolution can be employed to elucidate the relationship between information and the mind. As matter evolves and its movements become more intricate, so too does the transmission of information due to their isomorphic connection. This escalating complexity eventually gives rise to the formation of a decision-making mechanism within the system, which manifests as an informational system and serves as the wellspring of consciousness. This evolutionary perspective aligns with Laszlo's general theory of evolution and systems philosophy (Laszlo and Salk, 1987). Wu's endorsement of "information thinking" (Brenner et al., 2013a, 2013b) extends the notion of "systems thinking" initially presented by Bertalanffy (1969). However, a common drawback in systemic and evolutionary philosophies is that they regard consciousness or the mind as a higher-level product of material evolution. This evolutionary paradigm fails to escape the criticism inherent in mind–object dualism and cannot satisfactorily explain the emergence of the mind through a physical substrate, particularly when it comes to the question of "meaning". To bridge this divide effectively, the evolutionary model proposed by the dual-existence theory of matter and information presents a more viable alternative. In this framework, "meaning" can be naturalized into the fundamental fabric of the informational realm, thus addressing the issue of how the mental and physical realms interact and coexist.

Third, the mind is conceptualized as a high-level informational system. General evolution should not be misconstrued as a hierarchical progression from physics to chemistry, biology, and then to mind and consciousness but rather as a co-evolution of the dual yet isomorphic domains of matter and information (this may appear to suggest a form of dualism; however, I would argue that these "domains" are not composed of separate entities when considering scientific models about nature. Instead, the science of information and the science of matter each construct distinct models of the same reality — nature itself. The "quandary" alluded to at the end of Section 3, which could create a misleading or ambiguous impression, can be circumvented through a naturalistic study of information philosophy). In evolutionary terms, material systems and informational systems evolve in tandem, growing in complexity. Cognitive behavior is the manifestation of information control and processing within various levels of the informational system, with consciousness being the defining attribute of informational systems capable of decision-making at different degrees of complexity. In exercising control, a complex matter system must manage not only its behavioral interactions with the external environment but also the intricate dynamics of internal environmental behaviors. While simple interactions between internal and external environments might be facilitated by stimulus-response mechanisms, complex interplay necessitates the presence of consciousness endowed with selection and judgment capabilities. Therefore, inspired by advancements in information technology and buttressed by the general theory of evolution and the scientific principles governing complex systems, the philosophical insight treating the subject of consciousness as an informational system (also termed as "informosome" (Wang, 2018) by Wu, a terminology borrowed from biology. I alternately employ this concept alongside the term "informational system") emerged in Wu's thought. This led to the development of the dual-existence theory of matter and information, where both domains are seen as integral parts of a single evolving process.

4.2 *A naturalistic revision of ontology*

However, this theory, despite its intentions, inadvertently maintains a contradiction with its initial aim. While it purports to challenge Cartesian dualism, it does not fully escape the dichotomous paradigm. Due to its materialistic inclinations, it subtly adheres to the Cartesian philosophical doctrine that segregates mind from matter and subject from object. Historically, matter has symbolized all entities of scientific inquiry. Yet, as science progresses, its research domain has expanded beyond mere matter. The advent of information science brings to light the existence of an independent informational realm. Consequently, the scope of scientific investigation must now encompass both matter and information. This shift necessitates that the position of materialism becomes less tenable, giving way to naturalism as a more fitting perspective.

Naturalism perceives matter and information as dual, interdependent facets (in epistemological terms) of a single natural entity, which is the only genuine ontological existence. The essence of nature is thus articulated and elucidated through the combined understanding of both its material and informational properties. Evolution, therefore, fundamentally represents the unfolding development of these natural entities. This naturalistic stance illuminates two critical insights: first, it acknowledges the coexistence of an informational world that parallels the material world in an isomorphic relationship — a coexistence characterized by mutual reliance. Second, this "isomorphism" suggests that the informational world and the material world are, in essence, complementary epistemological aspects of the same natural entities. They can be likened to the two sides of a coin, with one side embodying the material dimension and the other representing the informational realm.

Here, I believe it is necessary to clarify my understanding of the commonly used philosophical concepts in a general sense — materialism, idealism, physicalism, and naturalism — and their interconnections:

- **Materialism:** Materialism posits that an objective world exists which can be scientifically studied through rational cognition, with its roots stemming from the discovery of "mass" in Newtonian mechanics. This perspective asserts that empirical methods can discern the nature of reality.
- **Idealism:** Idealism, on the other hand, questions the solidity of matter and instead emphasizes the primacy of the mind or consciousness. During the nascent stages of modern science, when evolutionary theory was not yet widely embraced, philosophers often

defended the autonomy and singularity of the mind as a fundamental entity.
- **Physicalism:** The advent of Einstein's energy–matter equivalence principle led to the rise of physicalism — a more refined variant of materialism, where matter is essentially reduced to its energetic essence.
- **Naturalism:** With the gradual popularization of evolutionary theories following Darwin, the philosophical stance of naturalism became increasingly prominent. Unlike materialism and idealism, which are seen as opposing viewpoints, naturalism seeks to reconcile the seeming conflict between matter and mind by incorporating both within the framework of evolution. From my standpoint, naturalism transcends the confines of general material sciences, suggesting that mind does not abruptly emerge from matter as some contemporary philosophies of mind propose. Instead, I envision the future development of information philosophy endeavoring to elevate the status of information science from a formal discipline to a full-fledged natural science, though this requires further substantiation. The aim is to explore the evolutionary mechanisms governing the mind within the domain of information science or related fields. Ultimately, the science of information and the science of matter will together reveal different facets of the same underlying reality: the natural entity.

As illustrated in Fig. 1, when observing a cylinder from two perpendicular perspectives, we perceive two distinct forms: a rectangle and a circle. If this cylinder symbolizes a natural entity, the rectangle can be interpreted as the material aspect of our understanding of the entity, while the circle represents its informational aspect. This universal isomorphic coexistence between matter and information pervades throughout the evolutionary cosmos. In the context of a famous Daoist saying, "The Dao gives birth to the One, and the One produces the Two...", matter and information embody the "Two" aspects, with the natural entity representing the "One". The Dao itself corresponds to the Big Bang or the process of evolution itself. Brenner also sees the link between the Daoism and philosophy

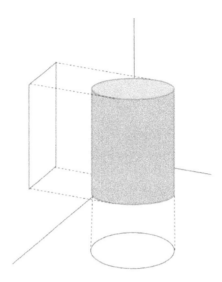

Figure 1. The cylinder of nature.

of information (Brenner, 2017). Although we have not yet conclusively determined whether matter and information are indeed the sole dual facets of the "One" — the natural entity — the natural entity can be regarded as a harmonious integration of matter and information.

4.3 *The epistemology of the information medium*

Drawing from the dual-existence theory of matter and information, Wu constructs an epistemological framework in which the mind is conceptualized as an informosome deeply embedded within the informational realm. In this model, due to the inherent interdependence between information and matter, the informational system can glean knowledge about the material world by extracting it from the vast informational ocean. This particular approach is termed as Informational Medium Epistemology (IME), characterized by a naturalistic leaning. However, despite its apparent naturalistic feature, this epistemological theory does not fully achieve naturalization. It remains confined within the boundaries of anthropocentrism because of the limitations exposed in Wu's previously discussed theoretical predicament. A truly naturalized information epistemology would extend beyond these confines, treating all informational systems as potential cognitive agents, not just those associated with human minds.

The manner in which informational systems acquire and process information fundamentally shapes the structure of knowledge. While this kind of knowledge may not necessarily depict an exact representation of the substance's true essence, it represents the most optimal understanding that can be achieved given the system's capabilities. As a classic philosophical analogy, Nagel (1974) points out that our human perception of the world is vastly different from how bats perceive theirs due to divergent information-gathering methods. Human beings' unique aptitude for acquiring, manipulating, and organizing information sets us apart from other species. Up until now, we have yet to identify any non-human natural systems capable of reflecting on their environment or themselves. Most biological entities aside from humans rely on information exchange solely for the sustenance and survival of their own existence. Contemporary neuroscience research has provided valuable insights into how nature might have evolved into such a distinctively self-reflective entity as humanity (Dehaene and Naccache, 2001; Frith *et al.*, 1996; Leon-Dominguez *et al.*, 2014; Merker, 2007). However, to make substantial progress in understanding this phenomenon, it is crucial to delve deeper into the exploration of the informational world itself. This involves investigating not only how information is perceived but also how it is structured, processed, and utilized to generate complex cognitive phenomena like self-awareness and reflective thought.

4.4 *A naturalistic revision of epistemology*

To begin this section, it is essential to highlight the distinction between naturalistic epistemology and modern epistemology. While the latter primarily focuses on questioning, critiquing, and examining the validity of scientific knowledge, the former fundamentally employs scientific understanding as its foundational theoretical framework to delve into the mechanisms underlying cognition.

From Wu's materialistic perspective, the informational world permeates both the objective and subjective realms, with the objective realm being considered more fundamental than the subjective. However, I find it difficult to concur with this viewpoint. In actuality, due to the pervasive nature of consciousness and meaning when viewed from an informational standpoint, the traditional epistemological distinction between subject and object becomes inherently paradoxical. In a natural and informational context, there should not exist a clearcut division between subject and object. The informational field is indeed the domain where conscious activities unfold, and consciousness is not confined to human consciousness alone. Rather, human consciousness represents the higher-level control and decision-making capabilities within a sophisticated informational system in the informational world — a development that evolved from lower-level informational system functions. This implies that consciousness exists as a continuum across various levels of complexity within the informational world.

Even though Wu's materialistic information epistemology endeavors to challenge the Cartesian cognitive model, it still adheres to the fundamental structure of Cartesian cognitive model. In contrast, a Naturalistic Information Epistemology (NIE) would decisively break away from this cognitive pattern. Similarly, Floridi's theory also employs the Cartesian cognitive model framework. Moreover, NIE critiques the classical computational representationalist stance in cognitive science, offering instead a more expansive and nuanced understanding of cognition. This generalized cognition is dynamic, embodied, embedded, extended, and enacted — thereby aligning naturally with the principles of the second generation of cognitive science. Several studies have proposed that phenomenology can serve as a philosophical underpinning for the second wave of cognitive science (Dreyfus, 2002; Dreyfus and Dreyfus, 1999; Varela et al., 1991). However, I argue that NIE provides a more compelling philosophical foundation compared to phenomenology. Unlike both analytic philosophy and phenomenology, which maintain anthropocentric positions by attributing meaning-making capabilities solely to humans — analytic philosophy through logical thinking and phenomenology via consciousness itself — NIE transcends the constraints of materialism and abandons idealistic biases. It advocates that all natural systems possess the capacity to generate and transmit meaning. Regarding the concept of "meaning", I concur with the following statement: "... and meaning emerges at the threshold between entities that are the equivalent of the LIR subject and object" (Brenner and Igamberdiev, 2021, p. 191). Here, "LIR" stands for the "Logic in Reality", which is a logic of evolution. This perspective suggests that meaning is not confined to human cognition but emerges from information world. Brenner believes that the capacity to generate meaning is the recursiveness in nature, and we see more in Section 6.

Furthermore, I concur with the notion that the idea of meaning being inherent in nature aligns, to a certain extent, with the stance held by the second generation of cognitive science. Moreover, NIE does not merely advocate for a metaphorical interpretation of cognitive processes and principles; instead, it proposes a more integrated and sophisticated research methodology within the field of information science to uncover these cognitive processes and principles. The superiority of information epistemology over phenomenology alone is evident in the following citation. In light of the above analysis, however, I contend that Wu's IME is not sufficiently naturalized. Despite this, these two authors recognize its potential merit reflected in the following quote (Brenner and Igamberdiev, 2021, p. 263), acknowledging its value while also pointing out areas where it could be further developed to align better with a comprehensive and naturalistic understanding of cognition:

> However, the difference between Wu's theory and that of Husserl is obvious: the purpose of Wu's original Philosophy of Information is to clarify the nature of the dual existence and dual evolution of material and information in the objective world, starting from the logic of the existence and dynamics of the natural human self. The phenomenology of Wu, unlike that of Husserl, does not have to be "naturalized", that is, brought into the domain of natural science. It is already there. The naturalization of Husserlian phenomenology was the subject of a major study. Wu's approach eliminates the arduous task of finding natural equivalents for Husserl's transcendental intuitions.

Additionally, the NIE that eschews the dichotomy between subject and object can provide a robust epistemological foundation for the advancement of a UIS. This approach posits that cognitive behaviors across different natural systems at all levels are fundamentally consistent, viewing the study of cognitive behavior as essentially an examination of the information control system's operations. It also delves into the mechanisms underlying judgment and decision-making capabilities, which are central to the functioning of consciousness. Thus, by examining the cognitive processes and consciousness attributes in lower-level natural systems, we can analogously infer and analyze those in higher-level systems. By investigating the laws governing the evolution of conscious entities as informational judgment and decision-making systems, we can deduce the extent or level of consciousness capabilities within our own consciousness. The quantitative analysis and measurement of informational systems at various evolutionary stages would then become a primary research goal and method in a UIS. In summary, from a naturalistic perspective, all material systems — spanning physical, chemical, biological, social, and multicomposite domains — are considered to possess cognitive abilities in a broad sense, given that they too are informational systems.

5 Return to Aristotelianism

In light of the critique of Floridi's theory, Wu's information philosophy inherently carries a natural opposition to Platonism, aligning Wu more closely with the Aristotelian tradition. If the discourse revolves deeply around existence, such an alignment is indeed inevitable. However, if this stance is genuine, it poses a significant challenge to Wu's declared mission of pursuing meta-philosophy, as this positioning reveals that its essence has yet to fully transcend the philosophical constraints imposed by dualism, oscillating between matter and mind, object and subject. While emphasizing the ontological importance of information, this position inadvertently situates itself in an awkward space between a new form of dualism and materialistic monism. It should be noted that ancient Greek philosophy did not assign "mind" the same unique status as later philosophies; instead, the concept of "soul" primarily served the role akin to what "mind" does in subsequent philosophical discussions. Moreover, the notion of soul inherently contains elements of the subject–object dichotomy, which can also influence how we perceive the relationship between information and these fundamental philosophical constructs.

Critiques of the traditional subject–object dichotomy lead us toward naturalism from materialism. A naturalistic approach to the philosophy of information integrates the study of mind into the domain of information science, aiming to demystify its supernatural qualities. While materialism also aspires to use material sciences to dissolve these supernatural attributes, it appears that this endeavor may be less than optimistic. Wu's information philosophy, by virtue of its emphasis on the autonomy of the informational world, cannot be strictly categorized as a philosophy of materialism, rather it is more aptly characterized as having a materialistic inclination. It is precisely this attribute that aligns Wu's philosophy with Aristotelian vital materialism, evident in several aspects:

(1) In Wu's philosophy of information, the relationship between matter and information mirrors that between material and form in Aristotelian thought. Etymologically, the connection between "information" and Plato's concept of "form" can be traced back to its linguistic roots. The Latin term *informatio* has historically been used to denote essence, idea, type, form, and representation within ancient Greek philosophical discourse. Here, "idea" and "form" are what Plato refers to as the entities seen by the "eye of the mind". However, Aristotle's critique of Platonic philosophy involved a reorientation of ideas and forms from an otherworldly realm to an earthly one. In his doctrine of hylomorphism, Aristotle posited material and form as two fundamental causes for material movement. This critical shift is highly resonant with Wu's advocacy for the interdependence between information and matter. Just as form (information) serves as the purpose, driving force, and rationale behind the movement of material (matter), Wu asserts that material (matter) acts as the vessel for form

(information). While material (matter) is considered more foundational, form (information) also holds a certain necessity and singularity for existence. Thus, both theories underscore the symbiotic nature of these elements, where neither can exist independently without the other.

(2) Wu's advocacy for the relationship between philosophy and science aligns closely with Aristotle's perspective. Grounded in the unity of matter and information, Wu's philosophy of information strives to transcend the study of mind and consciousness that has dominated modern philosophy since Descartes. He posits that the historical philosophical inquiry that aimed to construct a solid foundation for science has reached its end. Instead, philosophy should be seen as the scientific philosophy, while science is inherently the philosophical science; they are mutually intertwined and interdependent. Aristotle's categorization of knowledge laid the groundwork for modern scientific disciplines, where his metaphysics, which delves into the first cause, represents a highly abstract, comprehensive, and generalized synthesis of empirical sciences. Thus, in Aristotle's view, science and philosophy represent distinct yet interconnected levels and domains of research — ranging from concrete empirical investigations to abstract theoretical explorations. Wu repeatedly and explicitly underscores this relationship (Wu, 2013, 2014; Wu and Liu, 2014), emphasizing the seamless integration and reciprocal dependence between these two intellectual pursuits.

(3) The evolutionary perspective that Wu's philosophy of information offers on mind and consciousness naturally steers toward a new stance akin to panpsychism or panprotopsychism, which is rooted in the pervasive presence of informational systems, with the differences among these systems stemming from an evolutionary process. Although Aristotle did not explicitly articulate the concept or theory of evolution, his classification of soul does exhibit a gradational progression from lower to higher forms, reflecting a primitive form of panpsychism where consciousness or some semblance of it permeates through various levels of existence. Wu's philosophy posits that the informational essence of

entities leads to an understanding where all natural systems are imbued with some level of cognition, and the complexity of these cognitive abilities increases as the evolutionary trajectory unfolds. This view resonates with Aristotle's hierarchical arrangement of souls while extending it into a contemporary framework grounded in information theory.

Through meticulous analysis, it becomes evident that Wu's philosophy of information is retracing the path back to Aristotelian philosophy. However, this return is not a regression but rather a dialectical and spiraling progression. It incorporates the concept of general evolution, the research methods derived from information science and technology, as well as the holistic perspective inherent in complex system science. Thus, this reversion can be seen as an elevation of philosophy through the incorporation of new scientific advancements. Nonetheless, this upgrade is not without its challenges; it has yet to achieve the smooth integration anticipated. The limitations of its critique of traditional philosophies are still apparent, which is why Wu's materialistic inclination falls short in resolving the quandaries encountered in the study of mind and consciousness. To make such breakthroughs, there is a pressing need for a comprehensive understanding of the significance of information science, coupled with an innovative philosophical approach that involves deeper introspection and reflection. This would allow for a more profound re-evaluation and synthesis of ancient wisdom (including those from the Eastern world) with contemporary scientific insights.

At the opposite end of the spectrum is Floridi's Information Structural Realism (ISR), which, as he posits, "is a version of structure realism (SR). As a form of realism, ISR is (0-order) committed to the existence of a mind-independent reality addressed by and constraining our knowledge" (Floridi, 2011, p. 360). In essence, Floridi's information philosophy can be encapsulated by several distinguishing features:

(1) **Cartesian Epistemology:** In this framework, knowledge about objects is perceived as inherently subjective, with data serving

as the foundation for knowledge, and semantic information being the direct raw material that constitutes understanding. The semantics derived from logical analysis assumes a pre-existing or innate meaning structure. This reliance on congenital semantics implies that cognition based on such structures is exclusive to human cognition, which in turn fosters cognitive anthropocentrism. Consequently, it supports the classical cognitive science approach of modeling intelligent behavior through computational representation, where intelligence is understood as a process of manipulating symbols and representations akin to human thought processes.

(2) **Platonic Ontology:** Floridi's Information Structure Realism (ISR) suspends the primacy of the material or physical world, suggesting that "as far as we can tell, the ultimate nature of reality is informational" (Floridi, 2011, p. 361). This nature, akin to Plato's "forms", is the ubiquitous "dedonema" suggested by Floridi. Its ubiquitous characteristic is explained as follows: "This clarifies why a datum is ultimately reducible to a lack of uniformity" (Floridi, 2010, p. 24). Cognitive subjects organize data to form semantic information, thereby constructing knowledge about the realm of data. In this view, data, much like Plato's world of ideas, serve as the source of both knowledge and information. This ontological aspect, which aligns with objective idealism, is inextricably linked to Floridi's background in analytical philosophy. While Frege constructed a typical Platonic philosophy centered on the world of "thought" (Chen, 2012), Floridi replaces this world with the world of "information". Thus, Floridi's ISR reimagines the traditional Platonic ontology by asserting that the fundamental nature of existence is not in material substance but rather in the informational structures that pervade and underpin all reality.

Thus, from Floridi's perspective, philosophy and science remain fundamentally distinct domains. Conversely, Wu endeavors to critique and reverse the historical trajectory that has led to the separation of philosophy and science — a schism that began with Descartes but finds its philosophical roots in Plato. Therefore, it is a logical progression for Wu to advocate a return to Aristotelian thought through critiquing Floridi's stance. This approach aims to reconcile these disciplines by emphasizing their inherent interconnectedness and mutual dependence, thereby transcending the dichotomy that has characterized much of Western philosophical history since Plato's time.

From this analysis, it becomes apparent that the foundational objectives of Wu and Floridi's information philosophies have yet to be fully realized. Their ideas can be viewed as novel incarnations of two persistently opposing philosophical perspectives in the era of information. Neither holds an unambiguous advantage over the other, each perpetuating its own blend of merits and flaws within its respective traditions. However, the naturalistic approach to information philosophy discussed here has the potential to dissolve the long-standing confrontation between idealism and materialism. Moreover, it goes a step further by providing a direction for the advancement of new scientific disciplines, thereby bridging the gap between philosophy and science and offering a more cohesive understanding of knowledge and reality in the context of the modern information age.

6 The Logical Foundation for NIP

This section integrates the study of NIP with Brenner's LIR (Brenner, 2008) system, thereby establishing a logical foundation for NIP. This integration presents a logical model that effectively delineates the evolution of natural informational systems. This model not only uncovers the principles underlying the natural emergence of meaning but also elucidates the natural development mechanisms of mind, consciousness, and rationality.

6.1 *The scientific basis and core principle of LIR*

Brenner's system of LIR is built upon the non-Aristotelian logic developed by Lupasco (1986, 1987). In the mid-20th century, the French-Romanian philosopher Stéphane Lupasco posited that a new logic could be constructed based on the particle-field worldview presented by quantum mechanics — a logic designed to describe the workings of the real world. The axioms and rules within this logical framework provide a means for explaining the evolutionary processes in the real world. Building upon Lupasco's logical system, Brenner further develops his own system of LIR, which he applies to describing the dynamical mechanisms inherent in the natural information world, thereby establishing a connection with information philosophy. At its heart lies the Principle of Dynamical Opposites (PDO), a core principle suggesting that complex processes in reality evolve logically and functionally in conjunction with their opposites or contradictory aspects. This novel philosophical approach rooted in this logical principle represents a form of natural philosophy, distinct from those philosophies that uphold idealist positions and rely on *a priori* elements.

Brenner points out that during the early development of quantum mechanics, it was discovered that the law of non-contradiction and the law of excluded middle (Aristotle's second and third axioms) do not apply straightforwardly within the world of quantum mechanics. Lupasco reinterpreted these axioms by introducing a logic of energy to describe the dynamic patterns in the thermodynamic world, which he then extended to encompass macroscopic phenomena and their evolutionary mechanisms in the real world. Lupasco's logic differs significantly from classical logical concepts, which are built on propositions aimed at deducing truth or falsity. Prior to the advent of quantum mechanics and the evolution sciences exemplified by systems science and complexity theory, propositional logic and its mathematical formulations served as an apt logical foundation for the majority of scientific and philosophical inquiries.

However, it is important to note that these two logical systems are not inherently contradictory. Rather, they deal with different levels and even qualitatively distinct objects. One could say that standard logic represents the abstract principles and methods for information processing that necessarily emerge in conscious entities when higher-level evolutionary subjects, as illuminated by the LIR, confront complex environments. In essence, while standard logic may be a natural progression in the way information is handled at more advanced stages of evolution, it does not negate the applicability of the broader and deeper insights provided by the LIR.

6.2 The naturalization of information

There exist two distinct understandings of information. The first views information as data, where information is seen as embedded within language, and since language is an abstraction and representation capability unique to humans, meaning is thereby derived from the human subject. This constitutes an anthropocentric approach to the study of information. The latter understanding, embraced by naturalists, offers a naturalized conception of information — where information is conceived as processes. More specifically, it regards information as complex processes involving the generation, transmission, reception, and interpretation of meaning.

Our understanding of meaning is often situated within the context of metaphysics and logic, referring to the significance of propositions or sentences and being closely tied to the comprehension of truth. However, this constitutes a limited view of meaning. In light of LIR, akin to how information exists in nature, meaning too is embedded within reality, just as information. The perspective on meaning within the framework of LIR challenges and critiques the more narrow semantic conception typically found in traditional philosophy.

Naturalism rejects the meaning-theoretical models provided by a binary formal logical system. From the perspective of LIR, meaning is often acquired through "felt", that is, as experiential meaning and the experiencing "self" in relation to the world. It's crucial to note that here, "experiencing" does not solely refer to human perception nor just animal or biological sensation, rather any natural system (or "informosome") possesses the capability to perceive and be perceived. Moreover, the term "self" in this context goes beyond just humans. In progression from physical systems, through chemical systems, biological systems, societal systems, and complex megasystems, this "experiencing" capacity becomes increasingly sophisticated with the evolution of systems and their growing complexity, culminating in cognitive patterns that necessitate abstract language symbols for meaningful communication. The emergence of abstraction in human thought, seen from the lens of complex systems, is due to the overwhelming amount of information needing to be processed; only through abstraction can the efficiency of information exchange within and between systems be increased, ensuring these complex systems operate effectively and accurately. Essentially, the more complex a system, the greater the need for abstract modes of information exchange; abstraction serves as a process to filter out irrelevant information, which is vital for the survival of the system itself. However, this doesn't mean that only complex systems with linguistic abilities have the capacity to generate meaning. Therefore, NIP aims to critique the Platonic interpretation of meaning, placing it instead within the natural realm. This enables us to metaphorically express the realness and universality of meaning: if the world has meaning to us, then we also have meaning to the world.

Thus, we argue that the semantic understanding of meaning obscures the real dynamic and interactive relationship between the world and ourselves. Naturalists do not adhere strictly to classical logical methods and conceptions of truth in either logic or philosophy. This constitutes a naturalistic critique of standard logic and semantic information theory, embodying a non-semantic form of scientific realism stance.

6.3 *The relationship between meaning and information*

According to the semantic understanding of information, which states that "Information = Data + Meaning", data can be viewed as potential information, characterized by its ability to be encoded in a binary logical format. When human subjects impose structure on data and imbue it with meaning, these data then constitute information. However, under the framework of LIR, this relationship between meaning and information is inverted. Meaning exists inherently in nature; thus, the definition of information based on the relationship between meaning and data loses its validity because the latter's meaning is only constructed by the unique semantic engine of humans. From a naturalistic standpoint, the understanding of meaning becomes generalized, and information is seen as a more expansive existence than mere data. It could be said to be ubiquitous, akin to what Floridi refers to as "dedomena" (which also is called as "data in the wild" by Floridi).

Floridi views "dedomena" as an ontological entity, the very source of data and "the fractures of fabric of being". Implicit in this is that "dedomena" contain elements which can be encoded as digital data and others that cannot. Thus, for us "information beings", "dedomena" represent a primordial database akin to sediment in a riverbed mixed with sand and gold, where the gold (metaphorically standing for "data") is what we seek to extract. In contrast, from LIR perspective, it is precisely these dedomena inferred from experience, demanded by experience, and ubiquitous that constitute information itself, because so-called "information beings" are abstract processors of information, while actual informosomes are pervasive processors that do not necessarily rely on, or may even be independent of, abstracting abilities to process information. Floridi's rejection of the argument that "information requires material implementation" underscores the ontological status of dedomena. However, from our viewpoint, while it might seem that information does not require material implementation *per se*, it is nonetheless impossible without materiality; for, fundamentally, it is the processes relying on material existence itself. Information embodies the complex process of generation, transmission, reception, and interpretation of meaning, which involves a transition from potentiality to actuality and back again. Processes and material entities together constitute nature itself. Our sciences have traditionally focused on the material aspect of natural entities while overlooking processes (informational aspect). Alfred N. Whitehead once urged us to pay attention to processes, but his call ultimately resonated more with theologians and postmodernists than with mainstream philosophers and scientists.

Thus, from a naturalistic standpoint, meaning arises from information rather than the reverse. Brenner further asserts that meaning emerges from initially meaningless physical and chemical signals, with the recursiveness found in nature providing a scientific basis for the naturalization of meaning. It is in the world of natural information that meaning can be generated through informosomes capable of recursively processing information. As repeatedly emphasized, such informosomes are by no means exclusive to humans but abound in the natural world. The naturalization of meaning depends on the naturalization of information, which in turn relies on the naturalization of logic itself within the context of LIR. That is, the understanding of semantic information is predicated on binary formal logical systems, whereas the understanding of natural information is grounded in Brenner's LIR system. However, it is crucial to reiterate that the LIR system does not contradict binary formal logic systems but rather situates them within the broader evolutionary context of nature. In essence, binary formal logic represents an abstraction that emerges at higher levels of evolutionary development, akin to how Einstein's theory of relativity confines Newtonian mechanics to describing low-speed motion and Prigogine's work confines the second law of thermodynamics to the evolution of closed systems.

6.4 *Recursiveness in nature and the emergence of meaning*

The study of recursiveness initially gained concentrated attention in mathematics and linguistics, where it is considered a core characteristic of rationality and language. Recursiveness is also recognized as a crucial mechanism for the generation and understanding of meaning. In semantics, recursiveness provides natural languages with infinite expressive power and structural complexity, which in turn influence our interpretation and comprehension of meaning.

With the development of 20th century natural sciences, the trait of recursiveness has been identified across various disciplines, gradually revealing its presence in nature. At different levels within the natural hierarchy — from atoms and molecules to cells, organisms, ecosystems, and even entire Earth systems — each level can be viewed as an organization that reflects structures at lower levels to some extent, exemplifying recursiveness. Natural phenomena like branching patterns in plants, coral reef formations, and snowflake shapes are visual representations of recursiveness in nature, often resulting from simple rules iterated into complex structures. Even in the quantum world, where particle behavior is typically random, certain experimental setups exhibit periodic or quasi-periodic behaviors akin to those found in classical physics, reflecting recursion to some degree. On a grand cosmic scale, the distribution and structure of galaxies and nebulae demonstrate a form of self-similarity, suggesting the existence of recursiveness on an astronomical level. Fractals, geometric shapes characterized by self-similarity, embody this recursive phenomenon and abound in nature. Furthermore, chaotic systems, which are inherently unpredictable, often exhibit recursive traits; their phase space trajectories may repeatedly cycle through certain regions, forming so-called "loops". Therefore, recursiveness does not confine itself to mathematics, language, and computer science alone but pervades throughout nature. This widespread occurrence implies that the meaning that relies on recursiveness also possesses a foundational basis for its existence in the natural world.

Brenner argues that meaning arises within a dynamic process, and this very process is the recursiveness found in nature. According to the principle of recursiveness, there exist "transfinite" amounts of meaning in the natural world. He specifically emphasizes that meaning cannot be considered "finite" or "infinite" in a conventional sense, as either condition would render it absolute and meaning-ceased.

6.5 The types of meanings

Brenner categorizes meaning into two types: the first type is embedded within natural information, while the second type exists within language. See the following quotes (Brenner and Igamberdiev, 2021, p. 269):

> "Meaning I inheresin all existent entities. It hasipso facto value for conscious entities. Changes in it can be characterized as ontological flows of information (energy). Its logic is the non-propositional logic of energy. Its units are complex dynamic structures—'ontolons'.
>
> Meaning II is generated in the codified, interpreted communications between conscious entities. Its dynamics are epistemological,

without energy change. Its logic is propositional, bi- or multivalent, paraconsistent and/or intuitionist, modal, etc. Its units are linguistic structures—'epistemons', or in LL's term 'kenes', which refers to units or blocks of knowledge."

When we consider meaning as something that is "felt", the essence of this "felt" lies in the exchange of information between informosomes. The information being exchanged itself constitutes both an "ontolon" and an "epistemon"; these terms require an explanation.

Brenner employs "ontolon" as a fundamental term within the framework of LIR to describe the essential nature of entities "existing in reality" and also to delineate forms of existence. This term presupposes a diversity of dynamic processes within finite domains, which includes cognition itself. Hence, the "epistemon" represents the semiotic structure inherent within the "ontolon", or its cognitive representation. LIR provides a framework for describing both the substantial processes and the cognitive processes and their interrelations, integrating them through the recognition of logical operations that transcend potential and actual states of interaction. Brenner argues that the "ontolon" can replace concepts like Leibniz's "Monad" and Heidegger's "Dasein", thereby minimizing the reliance on non-naturalistic concepts to the greatest extent possible.

Based on the categorization of meaning, Brenner draws the following two corollaries (Brenner and Igamberdiev, 2021, p. 269):

Corollary A: For any real system including conscious entities, both forms of meaning are present and influence one another dialectically.

Corollary B: Both forms of meaning are causally efficient for conscious entities, Meaning I directly and Meaning II via Meaning I.

It is worth noting that the two types of meaning developed by Brenner correspond to Grice's distinction of meaning (Grice, 1957). Grice posits that there are two forms of meaning: natural meaning, which aligns with Meaning I, and non-natural meaning, corresponding to Meaning II. Natural meaning can also be referred to as indicator meaning, while non-natural meaning is sometimes termed communicative meaning. Natural meaning is connected to nature itself, whereas non-natural meaning pertains specifically to communication among humans. Natural meaning relies on certain regular relationships, which can be situated within the recursive nature emphasized by Brenner. On the other hand, non-natural meaning is further divided into conventional meaning and speaker meaning.

However, Grice's distinction is made with the intention of excluding a discussion on natural meaning and instead focuses on developing a theory centered around non-natural meaning. This non-natural meaning arises from intentionality and constitutes a central dimension in the semantic study of information. In the realm of non-natural meaning, he argues that speaker meaning is more fundamental than conventional meaning. Consequently, he constructs a semantics theory based on speaker meaning. Brenner's work, on the other hand, aims to integrate and connect these two types of meanings — natural and non-natural — that have been artificially separated by analytic philosophers within the dynamic pattern provided by LIR. He endeavors to break down this illusory metaphysical dichotomy and emphasizes the foundational importance of natural meaning.

Based on this intention, I propose a naturalistic interpretation for the concept of data as follows, which will be seamlessly integrated into the hierarchical progression of informational world:

A is **data**, if and only if,

- **A** originates from natural information, carrying natural meaning,
- **A** is currently being or has been collected by a conscious agent through a perceptual process and stored in memory by encoding and forming or has formed a structure that adheres to grammatical norms,
- **A** acquires semantic content (non-natural meaning) by the conscious agent during the collection process,
- **A** can convey instructional information and facilitate factual judgments after being imbued with semantic content (non-natural meaning),
- **A** becomes a direct source(semantic information) for knowledge upon verification of its truth in a factual judgment.

7 The Unity of Information Sciences

Having delved into the logical underpinnings of NIP, I will proceed to investigate its integration with information science. As previously outlined, NIP supports a cohesive and unified study of information science that is designed to elucidate the mechanisms governing informosomes across various stages of evolution.

Floridi argues that different disciplines within the field of information science cannot be inherently unified, rather each area's research should be grounded in its own philosophical analysis of semantic information (Floridi, 2004). To clarify the relationships between various concepts of information, one might explore "... whether some concepts of information are more central or fundamental than others, and should therefore be privileged" (Floridi, 2011, p. 33). This privileged concept is semantic information, which he posits as holding the essence in the philosophy of information due to its factual or cognition-oriented nature. Even under the strict definition of semantic information, it raises the question of whether all disciplines claiming to be part of information science truly meet the criteria. This stance runs counter to common intuition and the historical development trajectory of science.

Consequently, in response to this discourse, a group of information scientists and philosophers convened to deliberate the feasibility of unifying the various branches of information science, culminating in the establishment of the International Society for the Study of Information (IS4SI) in 2011. As a member of this academic society, Wu's research furnishes a philosophical foundation for the vision of a UIS. However, as per the preceding analysis, the Cartesian vestiges within Wu's information philosophy undermine the solidity of his critique. Thus, it is proposed that adopting a critical naturalistic approach to information philosophy can offer a more robust philosophical underpinning for the advancement of a UIS. Concurrently, it presents a logical possibility when combined with Brenner's advocacy of LIR introduced above, which aims to uncover the essence of mind and consciousness through the lens of unified information science. I now reiterate and summarize the core tenets of the NIP approach by examining its relationship with UIS as previously discussed:

(1) The NIP argues that matter and information represent two epistemological facets of natural entities and two coexisting, co-evolving, and isomorphic worlds. It critiques the philosophical stance that posits information as inherently dependent on matter. This perspective can lead to either the dissolution or reduction of information due to physical determinism or the tendency to regard information as an ancillary phenomenon tied to matter. However, maintaining a degree of autonomy and uniqueness for the informational realm risks degenerating into a new form of dualism, which contradicts its materialist foundation. Yet, in practice, the decision-making function inherent in informosomes facilitates more efficient information transmission within their corresponding material counterparts. Therefore, to uphold both the autonomy and uniqueness of mind while avoiding a slide into new dualism, the NIP presents a promising research avenue. Thus, it's more reasonable to examine the real world of natural entities from the dual perspectives of matter science and information science. This position provides a philosophical underpinning for Marijuán's theory of the four domains of science (Marijuán, 1998, 2014) and serves as a philosophical core for Hofkirchner's advocacy of a UIS program (Hofkirchner, 2011, 2013).
(2) The NIP espouses a monistic view of natural entities, critiquing both anthropocentrism and the epistemological division between subject and object. Wu conceptualizes information as an indirect form of existence distinct from matter (direct existence), distinguishing between subjective and objective information, and he regards objective information as the more foundational informational domain. This perspective inadvertently aligns with anthropocentric inclinations. As previously mentioned, the separation of subject and object is an outdated philosophical perspective that naturalism seeks to transcend. For a UIS, the goal extends beyond unifying the various understandings of information across disciplines; it necessitates treating this unified

concept of information as neither purely subjective nor objective but rather as inherently natural. Material phenomena and informational phenomena are isomorphic, representing dual aspects of the same natural entity. Matter embodies extensive properties, while information embodies transmissive properties (process). The extensive properties give rise to the possibilities for transmissive properties, while the transmissive property is the transmission of extensive properties. Hence, it can be said that these two interrelated phenomenal realms coexist within and are underpinned by the evolving world of real and natural entities.

In essence, the evolutionary laws and characteristics of the informational realm require elucidation through a more mature and cohesive UIS. By examining the informational world, we can reveal another facet of the natural world, much as we have done with the material world. For such a UIS, it is crucial to establish not only a shared research subject — the informosome — but also a common research objective — uncovering the operational mechanisms inherent in ubiquitous informosome. The stance of naturalistic monism underscores not only the isomorphic relationship, coexistence, and coevolution of matter and information but also the unity between information science and information philosophy, along with the internal cohesion among various branches of information sciences. Similar to the internal consistency found within the material sciences, where different theories often share theoretical connections, forming a knowledge network for mutual reference, the UIS must integrate philosophical thought. This encompasses not just contemplation on the basis, prospect, structure, goal, object, and other topics of information science but also reflection upon its interconnection with the science of matter. Moreover, the UIS and a Unified Material Science form two integral knowledge frameworks that collaboratively offer a holistic understanding of the natural world.

8 A New Form of Panpsychism in Informational World

At the end of this chapter, I argue that NIP represents a novel interpretation of panpsychism. Panpsychism is an old-fashioned philosophical doctrine that manifested in its rudimentary form during ancient Greek times. Post-Cartesian era, it re-emerged under the banner of anti-Cartesian metaphysics, most notably exemplified by Spinoza's monism of substance and Leibniz's monadology. So, what exactly constitutes the panpsychism theory? Let's see the quote from *Stanford Encyclopedia of Philosophy* Philip, et al., 2022):

> Panpsychism is the view that mentality is fundamental and ubiquitous in the natural world. The view has a long and venerable history in philosophical traditions of both East and West, and has recently enjoyed a revival in analytic philosophy. For its proponents panpsychism offers an attractive middle way between physicalism on the one hand and dualism on the other. The worry with dualism—the view that mind and matter are fundamentally different things—is that it leaves us with a radically disunified picture of nature, and the deep difficulty of understanding how mind and brain interact. And whilst physicalism offers a simple and unified vision of the world, this is arguably at the cost of being unable to give a satisfactory account of the emergence of human and animal consciousness. Panpsychism, strange as it may sound on first hearing, promises a satisfying account of the human mind within a unified conception of nature.

It can be discerned that panpsychism theory embodies a quintessential anti-Cartesian philosophical stance, serving as "an attractive middle road between physicalism and dualism", which aligns with the objectives of NIP. Despite these similarities, NIP offers distinct perspectives:

- Primarily, NIP diverges from most proponents of panpsychism by eschewing the enigmatic qualities often attributed to spirit or maintaining an affinity for divine nostalgia — this very attribute has historically distanced the theory from mainstream metaphysics during the era of modern science's robust development.
- Second, NIP does not consider mind, psychology, or consciousness as the fundamental entities within nature.
- Third, NIP is underpinned by novel scientific and philosophical theories: the principle of general evolution, the contemporary methods in information science and technology, the holistic worldview fostered by complex systems science, the natural logic described by LIR, and the new metaphysical insights emanating from information philosophy itself.

In contrast to classical panpsychism, rooted predominantly in classical mechanics and mechanical philosophy, NIP presents a fresh opportunity for the panpsychism theory to break through its age-old conundrums.

This novel panpsychism theory shares common characteristics with its predecessors, which warrant the use of a similar nomenclature, suggesting that mental faculties or the functions thereof are not exclusive to humans but are indeed universal in nature. From the vantage point of NIP, psychological activities and conscious states of the mind become functions of an informosome, with their functional complexity varying according to the evolutionary advancement of the informosome. Here, the term "informosome" replaces "mind", representing a philosophical concept's substitution with a scientific one. Thus, the aim of NIP is more explicit; it not only seeks to evade criticisms leveled at dualism and physicalism but also constructs theoretical groundwork for the scientific inquiry into the mind. However, by asserting that information is more fundamental than the mind, it can be mistakenly likened to a "pan-informationalism" stance on a literal level. In addition to famous proponents such as Wheeler and Chalmers, (Stonier, 1990, 1992, 1996, 1997) also advocated for a pan-informationalism perspective akin to Wheeler's views. Floridi, though difficult to categorize under pan-informationalism due to his highly constrained definition of information, could still be included in this category in a broader sense. It should be emphasized that pan-informationalism represents a philosophical standpoint that regards information as the sole existence and ultimate reality. Clearly, NIP does not align with this contemporary variant of idealism.

In reality, the NIP is more inclined to endorse a related but distinct stance — panprotopsychism. This position posits that mind or consciousness is not an ultimate fundamental entity but arises from certain more rudimentary attributes through specific structural arrangements. This perspective is often conflated with "emergentist panpsychism", and it does afford space for materialistic interpretations. However, the NIP diverges here, as it does not propose that mind or consciousness emerges from a more basic, physical attribute; instead, they originate in the realm of information, the process in nature. The material world and the informational world are isomorphic, coexisting harmoniously, and both serve as phenomenal facets of the natural world. These two worlds can give rise to two separate yet complementary scientific knowledge systems.

Panpsychism theory continues to garner support in contemporary times, exemplified by James' panpsychism based on "neutral monism" (James, 1909). However, James' formulation of panpsychism is not as clear-cut or robust as he asserts, with the neutral monism he presupposes being less neutral due to the incorporation of spiritual components. This idealistic inclination becomes more pronounced in other panpsychisms from the 19th century, tracing its roots back to Leibniz's metaphysical concept of spiritual monads. In the early 20th century, panpsychism theories reached their zenith in Whitehead's process philosophy and Russell's logical monism, yet they too failed to fully shake off Cartesian dualism's lingering influence. Whitehead's philosophical perspective eventually veered toward theological underpinnings.

Chalmers' naturalistic dualism (Chalmers, 1996, 2013, 2016) has been an influential panpsychist perspective in recent years. He tenaciously employs Shannon's information theory to construct an ontological framework for consciousness, thereby fostering the advancement of consciousness science. However, as previously mentioned, Chalmers subscribes to a form of pan-informationalism. His version of dualism pertains to a duality within the realm of information itself, where he distinguishes between physical information and phenomenal information. This contrasts with naturalistic monism, which perceives information and matter as two facets of the same natural entity. Nonetheless, Chalmers' conceptualization of information surpasses that of many panpsychism theorists because he demonstrates a commitment to leveraging information science for scientific inquiry into consciousness.

The NIP shares a similar objective. The panpsychism it upholds aims to definitively remove the Cartesian label attached to the mind and posits that the informational world characterized by its transmissive properties is fundamentally equivalent to the material world in an isomorphic sense. In the course of evolution, these transmissive properties (presented as a process) within the informational realm progressively develop into cognitive and conscious functions. By treating the informosome as tantamount to a mind, this approach seeks to achieve the goal of scientifically investigating consciousness. However, further elucidation of the intricate details and laws governing this transmissive property, along with a more comprehensive revelation of the scientific attributes of consciousness, hinges on the integration of UIS and NIP.

It is noteworthy to mention that the Orchestrated Objective Reduction (Orch-OR) theory, as advocated by Hameroff and Penrose (Hameroff and Penrose, 2014), endeavors to provide a scientific account of consciousness. This hypothesis endeavors to unify quantum mechanics, consciousness studies, and neurobiology in an effort to elucidate the fundamental nature of human consciousness. According to Orch-OR, consciousness arises as a macroscopic outcome of micro-scale quantum processes within the brain, which ultimately give rise to our subjective experiences and the continuous flow of consciousness. Despite its intriguing insights, this theory has ignited significant debate within the scientific community and remains in need of more robust and extensive experimental corroboration.

From the perspective of panpsychism, there exists a theoretical possibility of extending the principles of Orch-OR to other levels of organisms, given that all living entities are composed of proteins and various molecular structures that could potentially exhibit analogous quantum behaviors. However, applying such a complex framework to non-human organisms raises several critical questions that warrant careful examination. Notwithstanding these challenges, the Orch-OR theory does offer a novel scientific vision for a quantum information-based interpretation of panpsychism, which may provide scientific support to NIP.

References

Bertalanffy, L. V. (1969). *General System Theory*. N.Y.: George Braziller Inc.

Brenner, J. E. (2008). *Logic in Reality*. Dordrecht: Springer.

Brenner, J. E. (2011). Wu Kun and the Metaphilosophy of Information. In K. Markov, V. Gladun (eds.), *International Journal. Information Theories and Applications*, Vol. 18. ITHEA, pp. 103–128.

Brenner, J. E. (2017). Linking the Tao, biomathics and information through the logic of energy. *Pro. in Bio. and Mole. Bio.* 131, 15–33.

Brenner, J. E., and Igamberdiev, A. U. (2021). *Philosophy in Reality*. Springer.

Brenner, J. E., Wu, K., and Wang, J. (2013). A comparision on information thinking and system thinking (part 1). *J. Fo. Univ. (So. Sci. Ed.)* 31(2), 1–7.

Brenner, J. E., Wu, K., and Wang, J. (2013). A comparision on information thinking and system thinking (part 2). *J. Fo. Univ. (So. Sci. Ed.)* 31(3), 1–10.

Chalmers, D. (1995). Facing up to the problem of consciousness. *J. Con. Stud.* 2(3), 200–219.

Chalmers, D. J. (1995). Facing up to the hard problem of consciousness. *Dis. Filo.* 12(19), 223–235.

Chalmers, D. J. (1996). *The Conscious Mind: Towards a Fundamental Theory*. New York: Oxford University Press.

Chalmers, D. J. (2013). *Panpsychism and Panprotopsychism*, Amherst Lecture in Philosophy, Vol. 8.

Chalmers, D. J. (2016). The combination problem for panpsychism. In Godehard Bruntrup, and Ludwig Jaskolla (eds.), *Panpsychism: Contemporary Perspectives, Philosophy of Mind Series*. Oxford: Oxford University Press.

Chen, B. (2012). Reformualtion of Frege's Theory of Thoughts, *Phil. Ana.* 3(05), 18–31 + 197.

Dehaene, S., and Naccache, L. (2001). Towards a cognitive neuroscience of consciousness: Basic evidence and a workspace framework. *Cog.* 79(1–2), 1–37.

Dretske, F. I. (1981). *Knowledge and the Flow of Information*. Massachusetts: The MIT Press.

Dreyfus, H. L. (2002). Intelligence without representation-merleau-ponty's critique of mental representation the relevance of phenomenology to scientific explanation. *Phen. and the Cog. Sci.* 1, 260–383.

Dreyfus, H. L., and Dreyfus, S. E. (1999). *The Challenge of Merleau-Ponty's Phenomenology of Embodiment for Cognitive Science*. London: Routledge.

Floridi, L. (2002). What is the philosophy of information? In J. Moor and T. W. Bynum (eds.), *Metaphilosophy*. Blackwell, pp. 123–145.

Floridi, L. (2004). Information. In L. Floridi (ed). *The Blackwell Guide to the Philosophy of Computing and Information*. Oxford: Blackwell Publishing Ltd, pp. 40–61.

Floridi, L. (2008). A defence of informational structural realism, *Syn.* 161(2), 219–253.
Floridi, L. (2010). *Information: A Very Short Introduction.* London: Oxford University Press.
Floridi, L. (2011). *The Philosophy of Information.* Oxford: Oxford University Press.
Frith, C., Lawrence, A., and Weinberger, D. (1996). The role of the prefrontal cortex in self-consciousness: The case of auditory hallucinations. *Phil. Trans. Roy. Soc. Lond.* 351(1346), 1505–1512.
Grice, H. P. (1957). Meaning, *The Phil. Re.* 66(3), 377–388.
Hameroff, S. and Penrose, R. (2014). Consciousness in the universe: A review of the 'Orch OR' theory. *Phy. of Li. Re.* 11(1), 39–78.
Hofkirchner, W. (2011). Toward a New Science of Information. *Info.* 2, 372–382.
Hofkirchner, W. (2013). *Emergent Information: A Unified Theory of Information Framework.* Singapore: World Scientific.
James, W. A. (1909). *Pluralistic Universe: Hibbert Lectures at Manchester College on the Present Situation in Philosophy.* New York: Longmans, Green and Co.
Laszlo, E., and Salk, J. (1987). *Evolution: The Grand Synthesis.* M.S: Shambhala.
Leon-Dominguez, U., Izzetoglu, M. Leon-Carrion, J. *et al.* (2014). Molecular concentration of deoxyHb in human prefrontal cortex predicts the emergence and suppression of consciousness. *Neu.* 85(2), 616–625.
Lupasco, S. (1986). *L'énergie et la matière vivante, Éditions du Rocher.* Monaco (originally published in Paris: Julliard, 1962).
Lupasco, S. (1987). *L'énergie et la matière psychique, Éditions du Rocher.* Monaco (originally published in Paris: Julliard, 1974).
Lyon, P. (2006). The biogenic approach to cognition. *Cog. Pro.* 7(1), 11.
Margulis, L. (2010). The conscious cell. *Ann. NY Acad. Sci.* 929(1), 55–70.
Marijuán, P. C. (1998). Second conference on foundations of information science, Vienna, 1996: The quest for a unified theory of information. *Bio.* 46(1–2), 1–7.
Marijuán, P. C. (2014). The historical review and prospective forcast of unified information theory and information philosophy. *J. Xi'zn Ji.Uni.(So. Sci.)* 34(1), 1–2.
Marijuán, P. C., Navarro, J., and Moral, R. D. (2010). On prokaryotic intelligence: Strategies for sensing the environment. *Bio.* 99(2), 94–103.

Merker, B. (2007). Consciousness without a cerebral cortex: A challenge for neuroscience and medicine. *Be. & Br. Sci.* 30(1), 63–134.

Nagel, T. (1974). What is it like to be a bat. *Phil. Re.* 83(4), 435–450.

Philip, G., Seager, W., and Allen-Hermanson, S. Panpsychism. In Edward N. Zalta (ed.), *The Stanford Encyclopedia of Philosophy*, Summer 2022 Edition. https://plato.stanford.edu/archives/sum2022/entries/panpsychism/.

Quine, W. (1951). Two dogmas of empiricism. *Phil. Re.* 60(1), 41–64.

Quine, W. (1968). Epistemology naturalized, In W. Quine (ed.), *Ontological Relativity & Other Essays*. New York: Columbia University Press.

Sellars, W. (1997). *Empiricism and The Philosophy of Mind*, 2nd edn. Cambridge: Harvard University Press.

Stonier, T. (1990). *Information and the Internal Structure of the Universe*. Berlin: Springer-Verlag.

Stonier, T. (1992). *Beyond Information The Natural History of Intelligence*. Berlin: Springer-Verlag.

Stonier, T. (1996). Information is a basic structure of the universe. *Biosystems* 38, 135–134.

Stonier, T. (1997). *Information and Meaning: An Evolutionary Perspective*. Berlin: Springer-Verlag.

Thompson, E. (2004). Life and mind: From autopoiesis to neurophenomenology. A tribute to Francisco Varela. *Phen. & the Cog. Sci.* 3(4), 381–398.

Varela, F., Thompson, E., and Rosch, R. (1991). *The Embodied Mind: Cognitive Science and Human Experience*. Cambridge: MIT Press.

Wang, Z. (2018). About the nature and unification relationship of information science and information philosophy—An academic conversation between Wu Kun And Pedro Marijuan, *J. Int.* 37(1), 114–121.

Wang, Z. (2019). A research approach for naturalistic philosophy of information based on objective information, *J. Int.* 38(4), 161–167.

West, A. H., and Stock, A. M. (2001). Histidine kinases and response regulator proteins in two-component signaling systems, *Trends in Bio. Sci.* 26(6), 369–376.

Wheeler, J. A. (1989). Information, physics, quantum: The search for links. In *Proceedings III International Symposium on Foundations of Quantum Mechanics*. Tokyo, pp. 354–358.

Wiener, N. (1961). *Cybernetics: or Control and Communication in the Animal and the Machine*. The Technilogy Press.

Wu, K. (2005). *The Philosophy of Information-Theories, Systems and Methods*. Peiking: Commercial Press.

Wu, K. (2013). The development of philosophy and its fundamental informational turn. *Information* 6(4), 693–703.

Wu, K. (2014). *Philosophy and Philosophical Turns.* Peking: People Publishing House.

Wu, K., and Brenner, J. (2017). Philosophy of information: Revolution in philosophy. Towards an informational metaphilosophy of science. *Phil.* 2(4), 22.

Wu, K., and Liu, X. (2014). The internal unity of science and philosophy embodied by modern physics. *Stu. Phil. Sci. Tech.* 31(2), 9–14.

Wu, T., and Da, K. (2021). The Chinese philosophy of information by Kun Wu. *J. Doc.* 77, 871–886.

© 2025 World Scientific Publishing Company
https://doi.org/10.1142/9789811294921_0008

Chapter 8

Discrimination and Analysis of Concepts and Relationships Related to Information and Intelligence

Kun Wu[*,‡], **Qiong Nan**[†,§], **and Li Luo**[†,¶]

[*]*Institute for Advanced Studies in Philosophy and Social Sciences, SNNU, Xi'an 710119, China*

[†]*Philosophy Department, School of Humanities and Social Sciences, Xi'an Jiaotong University, Xi'an 710049, China*

[‡]*wukun@mail.xjtu.edu.cn*

[§]*nanqiong0418@126.com*

[¶]*chinalaura@163.com*

At present, there are many opinions and confusion in the discussion of the concepts of information, knowledge, data, digits, practice, and intelligence, and even the regulations and appellations and their interrelations in the information society and intelligent society, which directly proves that it is necessary to find a reasonable explanation dimension to explain such concepts, appellations, and their relations clearly and uniformly. This will certainly help promote the progress of research in this field. The creative proposal and development of Chinese information philosophy can meet the above requirements. From the perspective of information philosophy, knowledge is a systematic subjective information formed by human beings through the grasp, processing, and creation of information through subjective spiritual activities. Data are the state information of the object obtained by people through perception (observation and experiment), and it exists in the way of for-itself information. Practice is a process in which the purposeful information created by human beings is realized in the object through the implementation of the planned information created by human beings. Intelligence is the active way and method for the subject to grasp, process, create, develop, utilize, and realize information in the process of cognition and practice.

1 Division of the Field of Existence and the Essence of the Information World

Contemporary science has regarded material (mass), power (energy), and information as the three basic elements of general things. Moreover, at the level of general philosophy, mass and energy are the existing forms of matter, which shows that material (mass) and power (energy) are the matter at the ontological level of philosophy, and so matter and information are the two basic fields that make up the world.

In general, "the division of the field of existence" is the highest paradigm of philosophy because only by determining the field of existence in the world can we explore the relationship between different fields and then make a hierarchical division and interpretation of different things and phenomena.

In the tradition of Western philosophy, existence belongs to three major fields: God (objective idea, absolute spirit), substance world, higher animal, and human spiritual world. With the development of science, God has been gradually dispelled in the field of general science and philosophy, so what remains can only be the latter two fields. This constitutes a basic model of the highest paradigm of traditional philosophy: existence = matter + spirit.

With the rise of modern information science, an information world (indirect existence, which is different from the substance world (direct existence, has been gradually accepted by the field of science and philosophy, so it is necessary to change the basic model of the highest paradigm of traditional philosophy. Chinese information philosophy has put forward a set of systematic theories of the dual existence of substance and information through more than 40 years of development. According to this theory, the expression of the new highest paradigm of philosophy should be existence = matter + information. Information also includes two fields: objective information and subjective information, and human spirit is the subjective information, which is the advanced form of the information world.

As early as the 1980s, Chinese information philosophy defined the essence of information from the ontological level of philosophy: "information is a philosophical category indicating indirect existence. It is the self-manifestation of the existing mode and state of matter (direct existence)" (Wu, 1984, pp. 33–35). Chinese information philosophy divides information into three basic forms: in-itself information (objective information), for-itself information (information grasped by subject's perception and memory), regenerated information (information created by subject's thinking), and a comprehensive form: social information (cultural world created by human beings). Based on the classification of information forms, Chinese information philosophy also puts forward an extended definition that includes all information forms: "Information is a philosophical category indicating indirect being. It is the self-manifestation and re-manifestation of the existing mode and state of matter (direct being), as well as the subjective grasp and creation of information by the cognitive and practical subject, including the cultural world that has been created by human beings" (Wu *et al.*, 2020).

Since matter and information are the two basic fields of existence that make up the world, all other things and phenomena should belong to these two fields. Knowledge, emotions, consciousness, intelligence, etc. are generally considered to be spiritual phenomena, therefore, they should all be contained in the information world rather than detached from the information world.

2 Knowledge, Digit, and Data

Knowledge is a systematic subjective information formed by human beings through the grasp, processing, and creation of information through subjective spiritual activities. From the perspective of the genesis of knowledge, it originates from the cognitive subject's grasp and creation of information, and from the way of the existence of knowledge, it can only exist in the form of subjective information (including its externalized form — the cultural world created by human beings belonging to social information).

Since the second half of the last century, the wave of digitization called by academic circles has been established in the sense of technical processing of classified knowledge. Digitalization is not another field besides informatization; it is only a kind of technical means adopted by human beings to realize informatization at present because it is possible to deal with it only by transforming the corresponding information into the form of digital code and achieve it in a mechanical way.

Mr. Luciano Floridi, a representative of Western information philosophy, puts forward a definition of the nature of information: "information = data + meaning" (Floridi, 2010). According to his explanation, data are excluded from information, and only those data that are given meaning are information. The parochialism of this view is obvious. In fact, data are the state information of the object obtained by people through perception (observation and experiment), and it exists in the way of for-itself information. People can only carry out the process of their thinking activities through the processing of the relevant data they grasp. It is data information that provides the basis of information elements for people's thinking activities.

3 Practice and Intelligence

The practical activities of human beings are different from the unconscious activities of ordinary natural objects. Purpose and planning are the main characteristics of human practice, and any purpose and planning is first of all a form of regenerated information created by human thinking. Chinese information philosophy regards human practice as a process of multilevel information feedback chain operation: the purposeful and planned information created by the subject ←→ the instruction information for the activation of the subject's behavior ←→ the action of the locomotive organs ←→ operating tools ←→ acts on the object ←→ the production of practical products. It is based on such a complex and intertwined information feedback loop that we have reason to conclude that practice is a process in which the purposeful information created by human beings is realized in the object through the implementation of the planned information created by human beings (Fang, 1984, pp. 57–63) Thus, it can be seen how narrow it is to regard practical activities only as a kind of material activities.

Intelligence is a concept that refers to the superposition of the wisdom and ability of the subject with cognitive and practical ability. On the one hand, human intelligence depends on the physiological and genetic structure, and on the other hand, it depends on the acquired construction of socialization. It's possible behavior and possible degree depend on the unity of the comprehensive construction and holographic interaction among physiological function, psychological activity, and behavior mode.

Although human intelligence must also exist in the form of knowledge and information, intelligence cannot be equated with general knowledge. Heraclitus once believed that wisdom was the ability to master knowledge. (Department of the History of Foreign Philosophy, 1981) Intelligence is the active way and method for the subject to grasp, process, create, develop, utilize, and realize information in the process of cognition and practice. In fact, in the process of human perception, memory, and thinking, in the process of the generation of human emotion and will, and in the process of human practice, there must be corresponding intelligent intermediaries to provide corresponding ways and methods. In this way, intelligence embodies the active aspect of the behavior of the subject, which is different from the general thing. It not only plays a role in human cognition and practice, but also plays a role in controlling, guiding, and dominating cognition and practice.

4 Levels of Human Information Activities and Interactions between these Levels under Intelligent Guidance

People's cognition and practice are advanced activities of grasping, creating, and realizing information. This process includes many levels of complex interactions (the basic level of in-itself information activities, the level of for-itself-information activities of perception and memory, the level of subject creation of information by thinking, and the level of practical activities in which the information created by the subject has been socially realized). In the interior of these many levels of information activities, as well as the complex interaction between levels, they are bound to be under intelligent control, guidance, and domination at the same time (Wu, 1993, 2005).

Figure 1 briefly shows the complex interaction between many levels of human cognition and practice under intelligent control, guidance, and domination.

5 Information Paradigm and Information Civilization

The rise of information science and information philosophy has fundamentally changed the way people understand the world. The first standard theoretical paradigm of human science and philosophy, which was originally put forward in ancient Greek philosophy and modern science, is the entity paradigm. This paradigm holds that the world is made up of particles with mass. The atomic theory in ancient Greece is the philosophical representative of this paradigm, while Newton's particle mechanics and the atomic-molecular theory put forward by modern science are the scientific representatives of this paradigm. The second standard theoretical paradigm of human science and philosophy is the energy paradigm. This paradigm holds that the world is constructed by an energy field that does not have a static mass. The theories of electromagnetic field, quantum mechanics, relativity, and modern cosmology in modern science are the scientific representatives of this paradigm, while contemporary realism based on energeticism is the philosophical representative of this paradigm. The third standard theoretical paradigm of human science and philosophy is the information paradigm. According to this paradigm, the nature of things cannot be simply determined by the amount of mass and energy but must consider the combination of mass and energy, the information content of things, and the evolution procedure and process determined by information. The rise

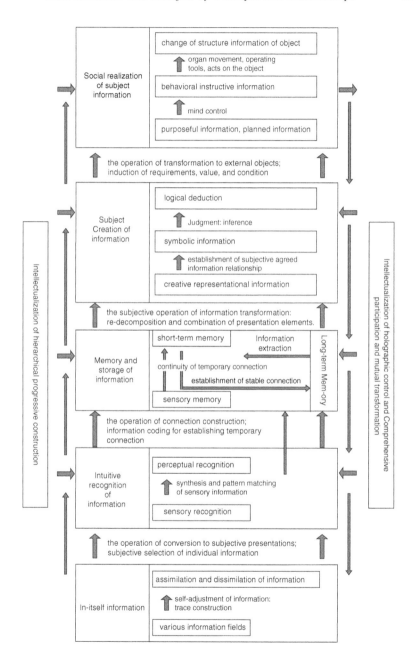

Figure 1. Diagram of Concepts and Relationships

of contemporary complex information system theory is the scientific representative of this paradigm, while the theory of dual existence and dual evolution of matter and information put forward by Chinese information philosophy is the philosophical representative of this paradigm.

As early as the early 1980s, Alvin Toffler, an American futurist, vividly compared the development of human civilization to three waves: agricultural civilization, industrial civilization, and information civilization.

With the advent of information civilization, academia has put forward many theories about the division of times: post-industrial era (post-industrial society), post-modern era, digital era (digital civilization, digital economy), intelligent era (intelligent civilization, intelligent society, and intelligent economy), etc. Although these formulations are more or less meaningful and valuable from the unique point of view of the proposer's thinking, strictly, they all have a certain degree of imprecise parochialism.

The formulation of post-industrial era and post-modernism only makes it clear that the traditional era has come to an end and a new era has been opened up. However, modern and post-modern, new and traditional are relative. With the passage of time, the post-modern will also be transformed into the modern, and the new will also become the traditional. In fact, this formulation not only fails to reveal the specific characteristics and nature of the new era but also leads to the confusion of naming in the following era. If another era of renewal comes, should we call it "post-post-industry" or "post-post-modern"?

Based on the point of view we have discussed before, digitization is only a technical means for us to achieve informatization at present, and using it to mark a new civilization era and economic system obviously cannot properly reveal the essential characteristics of a new civilization and a new economy. Similarly, because intelligence is the active way and method for the subject to grasp, process, create, develop, utilize, and realize information in the process of cognition and practice, it is included in the information world and must exist in the form of information.

Just as human agricultural civilization and industrial civilization have gone through different stages of development, human information civilization will certainly go through several different stages of development. At present, we are in the era of vigorous development of information civilization. As for what is the next civilization after the information civilization, we still do not see any clear signs.

References

Department of the History of Foreign Philosophy, Peking University (1981). *Selected Readings of Original Works of Western Philosophy (Volume 1)*, Beijing: Commercial Press, p. 26.

Fang, Y. (1984) (used pen name of Wu Kun). Discussion on information mediation theory of philosophical epistemology. *Lanzhou Acad. J.* (5), 57–63.

Floridi, L. (ed.) (2010). *The Blackwell Guide to the Philosophy of Computing and Information* (mainly translated by Liu Gang). Commercial Press, pp. 128–150.

Wu, K. (1984). Philosophical classification of information forms. *Beijing: Potential Sci.* (3), 33–35.

Wu, K. (1993). Levels of subject information activities and interactions between these levels. *J. Northwest Univ. (Philosophy and Social Sciences Edition).* (3), 43–49.

Wu, K. (2005). *Philosophy of Information-Theory, System and Method.* Beijing: The Commercial Press, pp. 108–133.

Wu, K. and Luo, L. (2018). This picture was originally published in On the holographic unity of information, knowledge, intelligence, and practice. *J. Intell.* 51, 21–25.

Wu, K. Wang, J., and Wu, T. (2020). *An Introduction to the Philosophy of Information.* Xi'an, China: Xi'an Jiaotong University Press, p. 143.

Chapter 9

Paradigm Change in AI

Yixin Zhong

University of Posts & Telecommunications, AI School
Beijing 100876, China
zyx@bupt.edu.cn

The term "paradigm for a scientific discipline" is re-defined here as "the scientific worldview and its methodology for the scientific discipline", which is thus the uniquely supreme force for leading and regulating the study of the scientific discipline. So, the study of a scientific discipline should follow a paradigm of its own. Artificial intelligence (AI) is the most advanced subdiscipline of information discipline and should thus follow the paradigm for the information discipline. It is found, however, that the paradigm practically followed in AI is not the one of its own but is the one borrowed from the discipline of physical science. The borrowing of the paradigm has caused serious problems to the study of AI, leading to it being confined to the lower stages of its development due to the mismatching between the paradigm and the scientific discipline. Paradigm change, replacing the borrowed paradigm with the paradigm of the information discipline, is the solution to make the study of AI able to walk out from the lower stage and enter into the advanced stage of its development so as to meet the urgent needs of social development. This chapter presents the big problems in current AI research and describes the epoch-making progress achieved due to the paradigm change in AI.

1 Introduction

Paradigm has been defined as worldview and action manner, and pattern, model, mode, etc. (Kuhn, 1962). More precisely, here we re-define it as "the scientific worldview and the associated methodology for a scientific discipline" (Zhong, 2023). Thus, the paradigm for a scientific discipline is the unique supreme force for leading the study of the discipline. Each discipline should follow its own paradigm. This is one of the most fundamental principles in the field of scientific research.

On the other hand, however, an odd phenomenon has been found through our deep and wide investigation: as a discipline of open and complex information systems, research in artificial intelligence (AI) should follow the paradigm for the information discipline (PID), yet, it follows the paradigm for a physical discipline (PPD), resulting in a "mismatch" between the paradigm and the discipline of research.

What is the matter with it? What kinds of consequences will the "mismatch" give to AI research theoretically as well as practically? What kinds of efforts should we make for remedying the situation in AI research if there are negative implications?

To achieve the correct understanding of the problem mentioned above, a thorough investigation of the history of the studies of AI and information is necessary. In the following sections, some significant, interesting, and surprising discoveries from the investigation of the entirety of information discipline, AI in particular, and the related discussions are presented. Before that, some concepts and facts needed for the discussion are explained.

2 Brief Review of Some Facts, Concepts, and Definitions Related

To avoid unnecessary misunderstandings, some basic facts and fundamental concepts related to the studies of information discipline have to be clarified through discussion and redefinition.

2.1 *Human wisdom, human intelligence, and artificial intelligence*

AI is one of the key concepts involved here. There have been a variety of definitions of AI that can be found in the literature. Here the concept of AI is re-clarified through the following definitions.

Definition 1 (Human Wisdom, HW). HW is referring to such a kind of human ability that is able to find the proper problem to solve, to assign the reasonable goal for problem-solving, and to provide the necessary knowledge for problem-solving, so that the human purpose of improving the standard of living and developing can be realized through the problem-solving and the goal-reaching (Zhong, 2021, 2023).

It is important to remember that only human beings possess human wisdom because only humans can understand the human purpose and thus can find the proper problem that must be a new problem and a solvable problem and will be beneficial to the improvement of human living and developing standard when solved. Machines cannot possess the ability of human wisdom and also cannot simulate human wisdom because machines are non-living entities and are never able to understand the human purpose.

Definition 2 (Human Intelligence, HI). HI is re-defined as another kind of human ability that is able to solve the problem and to reach the goal provided that the problem, the goal and the seed of knowledge assigned by HW are solvable (Zhong, 2021, 2023).

It is clear from Definition 2 that the human ability for finding a proper problem (HW) is much more complex and important than the human ability for solving the problem (HI). This is because of the fact that the human ability for finding and defining a proper problem must be based on a deep understanding of whether problem-solving is beneficial to the implementation of human purpose. So, finding a proper problem is much more creative than solving it.

Definition 3 (Artificial Intelligence, AI). AI is the implementation of human intelligence on machine (Zhong, 2021, 2023).

2.2 *Information discipline*

There has been a variety of different understandings of the concept of information discipline. This is due to the fact that there have been a large number of different groups of researchers with different

Table 1. Information technology: primary, intermediate, and advanced stage (Zhong, 2015, 2021).

Stages	Starting point	Members of IT at the stage	Features of the stage
Primary stage	Since 1940s	Sensing, communications, computing controlling	One info-function is performed object dominates
Middle stage	Since 1990s	Internet, Internet of Things (IoT) Primitive AI	More info-functions are performed object dominates
Advanced stage	Since 2020s	Advanced AI/ Standard AI	All info-functions are organized subject–object interaction dominates

backgrounds and various objectives in the studies of information discipline.

However, the following facts about information technology shown in Table 1 would be accepted by the community of information studies.

It is seen from Table 1 that there are three stages in information technology development, namely the primary stage, the middle stage, and the advanced stage. The major features of the different stages are briefly described as follows:

- Only one single kind of information function, i.e., acquiring, or transferring, or processing, or executing the information, can be performed by each member of information technology at the primary stage and only the objective factor is involved in their designing.
- More, but less than all, kinds of information functions, i.e., information processing and transferring for Internet, or information acquiring together with transferring, processing, and executing for IoT, can be performed by each member of information technology at the middle stage and yet only objective factors are involved in the designing.
- All kinds of information functions are harmoniously organized to become the standard/advanced AI at the advanced stage and both

the subjective and objective factors, and the interaction between them are involved in the designing and, even more importantly, the subjective factors play the dominant role at the advanced stage.

It is clear from Table 1 that only the advanced AI, among all members of information technology, meets the requirements for the advanced stage of information technology development due to its full consideration of both the subjective and objective factors and the interaction between them. From the point of view of information function performed and the capability, the advanced AI is the unique representative of the entirety of information technology.

Recalling back to Table 1, we have Definition 4.

Definition 4 (Information Discipline (HI)). Information discipline is referred to as the entirety of all possible research areas, in which the process of information function performing plays the dominant role and then the advanced AI becomes the most suitable representative of the discipline.

In what follows, we concentrate our attention only on the advanced AI and yet the major conclusions for advanced AI will all be meaningful to the entirety of the information discipline.

2.3 The ecological chain of information

Matter, energy, and information are regarded as the three categories of raw resources in reality. Through proper manufacturing and processing, these raw resources can be converted to related kinds of products for human usage, such as materials, power, and AI, respectively.

Information as a raw resource is subject to certain kinds of complex processing by human subjects for achieving certain goals, resulting in a sequence of information products, forming *the ecological chain of information* within the framework of subject–object interaction (Zhong, 2015). More specifically, there are two kinds of information occurring in the ecological chain of information: one is object information and the other is perceived information. The respective definitions for them can be given as follows:

Definition 5 (Object Information (HI)). Object information, which is produced by an object (either living things or non-living things) in the real world, is referring to the object's state and the pattern of change in state, all presented by the object (Zhong, 1988).

Note that object information in the fields of science and technology is known as ontological information in philosophy. It is the raw information originating from the object in the outside world. So, object information pervasively exists everywhere and at every time in the real world, not only on Earth but also in the entire space of the universe.

Note that object information has nothing to do with subjective factors.

Definition 6 (Perceived Information). Perceived information, that a subject perceived from the object information of an object, has three orderly components: (1) the form (syntactic) information provided by the object and sensed by the subject, (2) the utility (pragmatic) information provided by the object and evaluated by the subject with respect to the subject's goal, and (3) the meaning (semantic) information provided by the object and abstracted by the subject via mapping the former two components (1) and (2) into the meaning (semantic) space and then naming the result (Zhong, 1988).

Perceived information is named epistemological information in philosophy. It originates from object information but is modulated by the subject. So, it is not the raw information anymore. It is the integrated outcome produced by both the subject and the object.

Note that the definitions of form (syntactic) information and utility (pragmatic) information are obvious and are easy to understand because the form (syntactic) information can be acquired via sensing and the utility (pragmatic) information can be acquired via evaluation. Both of them are intuitive and specific in nature.

Yet, the definition of the meaning (semantic) information is not intuitive. Instead, it is relatively abstract in nature and cannot be acquired through sensing and evaluation. Hence, the definition of meaning (semantic) information may need a little bit of explanation.

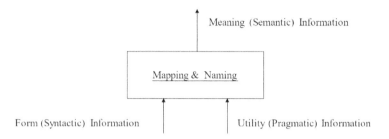

Figure 1. Relationship among form, utility, and meaning.

The principle for abstracting the meaning (semantic) information from the form (syntactic) information and utility (pragmatic) information can specifically be explained in Fig. 1.

Let the symbol X stand for the form (syntactic) information, Y the meaning (semantic) information, Z the utility (pragmatic) information, and λ the logic operator, whose function is to map the pair set (X, Z) into the meaning (semantic) information space and then to give it a name. Then this relationship shown in Fig. 1 can be expressed in the following equation (Zhong, 1988):

$$Y = \lambda(X, Z). \tag{1}$$

It is obvious from Equation (1) that if the meaning (semantic) information Y is obtained, which is the result of the pair set (X, Z), then the form (syntactic) information X and the utility (pragmatic) information Z can also be obtained. This indicates that the meaning (semantic) information Y can serve as the representative of both the form (syntactic) information X and the utility (pragmatic) information Z.

The relationship (1) brings out a big convenience in practice: people can use the name of the event (meaning information) for communication between them and can then understand the corresponding form information and utility information that the name implies.

We would like to point out that the terms syntactic, semantic, and pragmatic have been used respectively for form, utility, and meaning components of the perceived information. They originated from semiotics developed by Morris who declared that semiotics is composed of semantics (defined as the relationship of signs to what they stand for), syntactics (defined as the formal or structural relations between signs), and pragmatics (defined as the relation of signs to interpreters) (Morris, 1938). Yet, there exist serious ambiguities in those definitions that have led to many confusions and controversies. We have a further discussion on this issue in a later section. This is why we use the expressions of form (syntactic) information, utility (pragmatic) information, and meaning (semantic) information to show our preference.

2.4 Methodology of information ecology

According to the definition of paradigm above (see Introduction), the scientific worldview is its main body, whereas the methodology should serve its scientific worldview. In other words, different scientific worldviews should correspondingly require different methodologies. The function that the methodology for a scientific discipline performs is to provide the scientific discipline with instructions on what kinds of requirements the studies of the discipline should observe so that the studies of the discipline can comply with the scientific worldview of the discipline. Therefore, it is reasonable to give the following definitions.

Definition 7 (Methodology of Information Ecology: Holistic Approach). Referencing the principles of general ecology, the methodology of information ecology is a set of instructions for information discipline study, including (1) As the information products are produced by subject–object interaction, the ecological chain of information should be a kind of form-utility-meaning trinity; (2) The ecological chain of information should be complete, both in time and space of the subject-object interaction; (3) The interaction should be a globally optimized process. In view of (1), (2) and (3), it can also be named holistic approach (Zhong, 1988).

It is clear from the above definitions that the advanced AI study, or the nucleus of information discipline study, has a scope not limited to, but much wider and much deeper than, the scope of the traditional Shannon information theory which is the mathematical theory of communication.

3 Paradigm Issue: A Grand Challenge to Information Discipline

Concerning the paradigm issue, the first thing we should do is to clarify the precise meaning that the term paradigm should have. As is well known, in his book (Kuhn, 1962), Kuhn considered the term paradigm to signify not only a "world view" but also a "manner of knowing" or a "mode of scientific practice", etc. Yet, due to the differences existing among these explanations, this concept has in practice led to different understandings and confusions.

In view of the fact that the term paradigm is a major concept in this discussion, it is therefore necessary to give it a clear and precise definition.

Definition 8 (Paradigm for a Scientific Discipline). Paradigm for a scientific discipline is defined as the scientific worldview and the associated methodology for that scientific discipline (Zhong, 2023).

It is well known that the "scientific worldview" of a scientific discipline is the macro view at the highest level of the scientific discipline, answering the question of "what the essence of the scientific discipline is". On the other hand, the "associated methodology" of a scientific discipline is a set of principles and instructions reflecting the requirements of the scientific worldview so that the study of the scientific discipline can appropriately comply with the scientific worldview.

This is the basic understanding of the paradigm of a scientific discipline.

Up to the present time, there have been two major categories of scientific discipline in natural sciences so far, that is, the discipline of physical sciences (including material science and energy science) and the discipline of information science (including information science and intelligence science). The former is a long-lasting discipline having started as early as the agricultural age, while the latter is a newly borne discipline that started about the 1940s.

As we know, the scientific worldview and its associated methodology are the results abstracted and refined from the activities of the related scientific study, and, therefore, the paradigm is regarded as the social consciousness, whereas the activities of scientific study are regarded as the social existence.

There is a definite rule of social development saying that the time for the successful formulation of a kind of social consciousness has to be greatly delayed to the time for the formation of the related social existence, owing to the extremely high difficulty of abstracting ideas from facts.

Facing the constraints from the rule of social development stated above, the discipline of physical sciences, which has a thousand-year history, has already formulated its paradigm, whereas the discipline of information sciences, which has less than a century's history, has not yet completed the task of the formulation of its paradigm till the present.

Because of the facts mentioned above, the research activities carried on in information discipline have borrowed the paradigm from the discipline of physical sciences, leading to the "dis-matching" between the paradigm and the discipline in information studies.

Noticing that the phenomenon of "dis-matching" happened in information discipline is entirely due to the rule of social development rather than researchers' intentions and thus is a kind of compulsive and unavoidable challenge to the studies of information discipline.

To clarify the borrowing of the paradigm in the information discipline, it is necessary to have a specific look at the comparison of the paradigm for the physical discipline (PPD) and the paradigm that has been practically followed in the information discipline.

The PPD has major features as shown in Table 2.

The paradigm practically followed in information discipline has the following features shown in Table 3.

It is seen from Tables 2 and 3 that the scientific worldview in both disciplines requires ignoring subjective factors. This leads to a big problem as to the information studies: Where and how can the intelligence come from if all the subjective factors involved are expelled?

Further, it is seen from Tables 2 and 3 that the methodology employed in both the disciplines of physical sciences and information

Table 2. Major features of PPD.

Scientific Worldview	Object of the study: *Objective entities, expelling any subjective factors*
	Property of the object: *Deterministic*
	Goal of the study: *To know the structure and function of the object*
Methodology	General approach: *Divide and conquer*
	Means for description/analysis: *Pure formalism*
	Criterion for decision-making: *Form matching*

Table 3. Major features of the paradigm followed in information discipline.

Scientific worldview	Object of the study: *Special matter (brain), expelling subject factors*
	Property of the object: *Non-deterministic*
	Goal of the study: *To know the structure and functions of the brain*
Methodology	General approach: *Divide and conquer*
	Means for description/analysis: *Pure formalism*
	Criterion for decision-making: *Form matching*

sciences is the same, namely the so-called "divide and conquer" and "pure formalism". Hence, one can say that the discipline of information sciences has definitely borrowed the paradigm from the discipline of physical sciences as if it were its own.

As a consequence of borrowing the paradigm from the physical discipline, the studies of information discipline have suffered from many problems. The most serious problems, among others, are as follows:

(1) **Absence of a unified theory for the entirety of Information discipline:** Because of the employment of the approach of "divide and conquer", which is the methodology of the paradigm for the physical discipline, the information discipline has thus been divided into a number of branches like sensing (information acquisition), communication (information transferring), computing (information processing), controlling (information

execution), etc. As for (primary) AI, it has also been divided into three branches, such as artificial neural networks (McCulloch and Pitts, 1943; Rosenblatt, n.d.; Hopfield, 1982; Kohonen, 1990; Rumelhart and McClelland, 1986; Zurada, 1992), expert systems (McCarthy et al., 1955; Newell, 1980; Turing, 1963; Newell and Simon, 1963; Feigenbaum, 1977; Nilsson, 1982; Minsky, 1986), and sensor-motor systems (Brooks, 1989, 1990, 1991). On the whole, the isolation and separations among the branches have led to the absence of a unified theory for the entirety of information discipline. In accordance with the system theory, the sum of all parts is much inferior than the whole. So, the absence of a unified theory for the information discipline is really a big problem caused by the methodology of the physical discipline.

(2) **Very low level of intelligence in primary AI Systems:** Due to the application of the "pure formalism" approach, which is another methodology of the paradigm of the physical discipline, both the factors of meaning (semantic) information and utility (pragmatic) information, which are the real nucleus of the ability to understand, have been completely ignored, leading the primary AI to a very low level of understanding. Although deep learning based on deep neural networks has achieved high performance in some special cases since 2006 (Hinton et al. 2006, 2012; Bengio and LeCun, 2007; Bengio and Delalleau, 2000; Ranzato et al., 2007; Delalleau and Bengio, 2011; Pascann et al., 2014; Montufar et al., 2014), the ability to understand and the trustworthiness of its results are severely questionable (Marcus and Davis, 2019).

After all, the "dis-matching" between the paradigm followed and the research needs in information discipline is a very serious problem!

4 Paradigm Issue: A Grand Challenge to Information Discipline

The big problems mentioned above, the lack of a unified theory for the information discipline and the very low level of intelligence for primary AI, are very serious problems for information discipline study and cannot be overlooked for the rapid development of human society nowadays.

What we should do, and also can do today, is to completely replace the borrowed paradigm in information discipline with the paradigm for information discipline (PID). This is exactly what we mean by the "paradigm change" in information discipline.

But how can we effectively implement the "paradigm change" in the information discipline in practice? How many stages and steps are needed for the implementation of a paradigm change?

In accordance with the definition of paradigm (Definition 8), there should be two successive stages needed for dealing with the paradigm issue in a scientific discipline. The first stage is to formulate the paradigm for the discipline, consisting of two elements: the scientific worldview and the methodology. The second stage is to implement the new paradigm in the discipline, including building up the framework of the discipline, elaborating the framework of the discipline, and finally creating the entire theory for the discipline. Table 4 shows the stages and steps (Zhong, 2021, 2023).

Therefore, we have to carry on the task of "paradigm change" by starting first with the formulation of the paradigm for the information discipline which is step 1 in Table 4.

4.1 Step 1: Formulating the paradigm for the information discipline

As is pointed out, there has been no common understanding of the PID in the world academic community of the study of information discipline till the present time.

Yet in the early 1960s, as a student, the author of this chapter learned from Shannon theory that the concept of information was just embodied in the signal waveform, completely ignoring the semantic and pragmatic factors. He felt very uncomfortable. From that time on, he initiated for himself deep studies on the concept of information. Later, in the 1980s, he found a similar problem in artificial intelligence: the concept of intelligence has also no meaning and no value factors. Through nearly 60 years of studies, he and his

Table 4. The stages and steps needed to fulfil the task for the paradigm issue.

Stages	Tasks	Specific steps	Points of the steps
Stage I: Bottom-up	Paradigm forming	(1) The paradigm for the discipline	(1.1) Scientific view of the discipline
			(1.2) Methodology for the scientific view
Stage II: Top-down	Paradigm execution	(2) The framework of the discipline	(2.1) Global model for the discipline
			(2.2) Approach to the study of the discipline
		(3) The elaboration of the framework	(3.1) Scientific structure of the discipline
			(3.2) Special requirement for the discipline
		(4) The theory of the discipline	(4.1) Basic concepts for the discipline
			(4.2) Basic principles for the discipline

Table 5. Major features of the PID.

Scientific Worldview	Object of the study: *Info process produced by subject–object interaction*
	Property of the Target: *Non-deterministic*
	Goal of the study: *To have double win between subject and object*
Methodology	General approach: *Methodology of information ecology (Definition 7)*
	Means for description/analysis: *Holistic Approach (also see Definition 7)*
	Criterion for decision-making: *Understanding-based*

team have gradually summarized and refined the PID (Kuhn, 1962; Zhong, 1988, 2008, 2012, 2015, 2017a, 2021, Burgin and Zhong, 2019) which is greatly different from PPD.

The major features of the PID, including its scientific worldview and the associated methodology, are briefly listed in Table 5.

Comparing the major features expressed in Table 5 with those in Table 2 as well as in Table 3, the following points are worth emphasizing:

(1) There are evident differences concerning the major features of the scientific worldview between PID and PPD:

- The object of the study in information discipline is both the subject and object and more importantly the information process produced by the subject via the subject–object interaction, whereas that is only the object with no subject in physical discipline; therefore, the scientific worldview of PID is really the holistic view, while the one for PPD is a kind of shallow view — no information.
- The goal for the studies in information discipline is to have double win for both subject and object, whereas that is to learn the structure and function of the object in the physical discipline.

(2) There are also evident differences concerning the major features of methodology between the PID and PPD:

- The general approach in information discipline is the methodology of information ecology without "divide and conquer", whereas that in the physical discipline is reductionism.
- The method for description and analysis for the studies of the information discipline is the form-utility-meaning trinity (holistic), whereas that is the pure formalism in the physical discipline.
- The decision-making in the information discipline is based on the criterion of understanding the problem concerned, yet that is on the formal matching in the physical discipline.

It is clear from the comparisons above that PID is almost inverse to PPD. Consequently, the paradigm borrowed from the physical discipline for the information discipline has led the studies of the information discipline to such a situation — the research on information can only be confined within its lower stage, which is the stage that the PPD could permit, with the features of no unified theory established (due to the "divide and conquer"), very low intelligence (due to the "pure formalism") and unable to enter the higher stage of its development.

The conclusion reached so far is that the time for paradigm change in the studies of information discipline has now been completely matured because of the fact that the PID has already been established (Burgin and Zhong, 2019; Zhong, 2017b).

Next are the steps needed for implementing PID for establishing the advanced AI theory.

4.2 Step 2: Implementing PID for establishing the advanced AI theory

4.2.1 Step 2.1: Creating a global model for AI according to the scientific worldview of PID

The first contribution that the newly established PID can make to the studies of advanced AI is to create the appropriate global model for advanced AI study and to deeply explore what the essence of advanced AI should be.

According to the scientific worldview of PID (see Table 5), the global model for AI can be created like the one in Fig. 2.

The symbols P, G, K, and S in Fig. 2 stand for problem, goal, *a priori* knowledge, and *a priori* strategy respectively possessed in, and provided by, human wisdom.

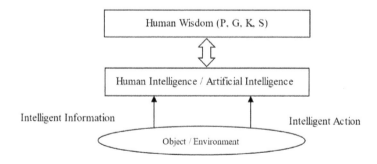

Figure 2. Global model for advanced AI (Zhong, 1992).

Figure 2 shows that the interaction between subject and object is initiated and proceeded through the following successive steps:

(1) The object information acts on the subject, presenting a problem (P) to the subject.
(2) If the subject finds the problem has no relation to the subject's goal (G), it does nothing.
(3) If the subject finds the problem has a relation to the subject's goal, the subject will utilize its *a priori* knowledge (K) to produce the intelligent action (re-)acting on the object for solving the problem.

It is seen from the model that AI is just the artificial agent for human intelligence.

Concerning the global model for advanced AI, one may immediately consider the model of the human brain, which had been, in primary AI, regarded as a special kind of material system able to produce human intelligence, with no subjective factors considered, thus the structure and the related functions of the human brain had been the focuses for the study of primary AI.

However, these considerations are unreasonable. The human brain isolated from environment is surely unable to produce that human intelligence that the subject really needs. This is because of the fact that, on the one hand, there would be no real stimulus from the outside world to the isolated brain and therefore there would be no reason for it to initiate the brain's work. On the other hand, even if the isolated brain could generate some "intelligence", there would be no chance to check whether the brain's strategy is good or not.

4.2.2 Step 2.2: Exploring the scientific approach to advanced AI according to the methodology of PID

The second contribution that PID can make to the study of advanced AI is to create, in accordance with the methodology of information ecology, a unified approach to the research in advanced AI and therefore to instruct how advanced AI should be studied.

Concerning the approach to advanced AI research, people may remember the facts that, due to the "divide and conquer" methodology borrowed from physical discipline, there have been three major approaches to the primary AI research, namely

- the structural approach, leading to artificial neural network research (Brooks, 1989, 1990, 1991; Hinton et al., 2006; Hinton et al., 2012; Bengio and Delalleau, 2000; Ranzato et al., 2007; Delalleau and Bengio, 2011; Pascann et al., 2014; Montufar et al., 2014),
- the functional approach, leading to the physical symbolic system research (McCarthy et al., 1955; Newell, 1980; Turing, 1963; Newell and Simon, 1963; Feigenbaum, 1977; Nilsson, 1982; Minsky, 1986),
- the behavioral approach, leading to sensor-motor system research (Marcus and Davis, 2019).

All of them have been mutually isolated, and even mutually expelled, leading to a serious problem: no unified theory for primary AI.

It is very fortunate and also of great significance that the implementation of the PID will bring about a revolutionary change, never seen in history, of the approach to the study of advanced AI, that is the great change from the three mutually expelled approaches in primary AI to a unified approach, leading to a unified AI, i.e., advanced AI.

The secret for the great change lies in the idea that the approach to advanced AI should observe the methodology of PID. Based on the global model of advanced AI shown in Fig. 2 and following the methodology of information ecology, the approach to the study of advanced AI should be the subject's inner process of information conversions, initiated with the object information and ended with the intelligent action.

The subject's ability for problem-solving is implemented via performing a series of functions. The specific functional processes performed by the subject for producing the intelligent action are shown in Fig. 3 (Zhong, 2007).

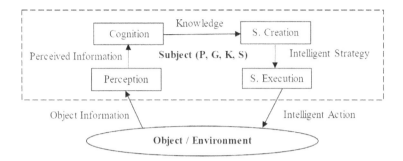

Figure 3. Methodology of information ecology: information conversion and intelligence creation ("S." stands for "Strategy").

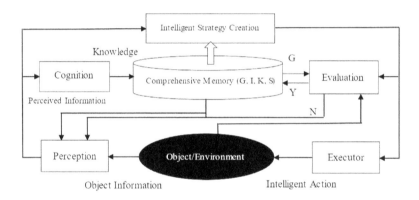

Figure 4. Model of the approach to advanced AI study (Zhong, 2014).

It is seen that Fig. 3 is the same as Fig. 2. The only difference is that the inner process within the "subject", that is, the methodology of information ecology, or more specifically the entire process of "information conversion and intelligence creation", has been explicitly expressed in Fig. 3, whereas it has been completely implicit in Fig. 2. This means that what we are doing is to apply the methodology into the global model of advanced AI.

As can clearly be seen from Fig. 3, the subject's inner process can be described as "object information → perceived information → knowledge → intelligent strategy → intelligent action" in which each of the symbol "→" stands for the conversions and creation correspondingly.

As a matter of fact, the subject's inner process of "information conversion and intelligence creation" is the mechanism of intelligence creation and can be pictured like the one in Fig. 4.

Comparing Fig. 4 with Fig. 3, one can find that both the models are in principle the same and the only difference between them is that the subject's functions, which are needed for producing the intelligent action (the embodiment of intelligence), have all been explicitly expressed in Fig. 4 whereas are briefly expressed in Fig. 3.

Briefly speaking, whenever the object information is acting on a subject, the subject will take a series of steps to produce the output (i.e., intelligent action) as its response to the object (see Figs. 3 and 4).

All the inner functions that the subject performs as shown in Fig. 3 can further be explained via a number of points as follows:

Point (1) is for the subject to know the form of the object.

The subject has to know if there is any stimulus (object information) at its input and also has to make a judgment on whether the object information has a relation to the subject's goal implementation. If it has, the subject will take the object information as its form (syntactic) information (knowing the form of the object). If it has not, the subject will then ignore the object information with no action initiated.

Point (2) is for the subject to know the specific value that the object may provide to the subject for implementing his goal.

If the object has a relation to the subject's goal implementation, the subject will then have to take measures to check what kind of utility and how much of the utility the object could provide to the subject's goal implementation. The result of the measuring will become the utility (pragmatic) information that the object provides to the subject.

Point (3) is for the subject to abstract the meaning of the object.

The subject will have to set up an abstract concept — the meaning (semantic) information, via which the subject will be able to understand the object. More specifically, the meaning (semantic) information (denoted by Y) is defined as the name for the joint of the just achieved form (syntactic) information (X) and the utility (pragmatic) information (Z) : $Y = \lambda(X, Z)$, where the symbol "λ" is the operator of mapping and naming.

The trinity of the form (syntactic) information (X), the utility (pragmatic) information (Z), and the meaning (semantic) information (Y) so produced by the subject is termed as the perceived information that the subject has had about the object. The perceived information is just the epistemological information (*refer to* Definition 6). Once the subject has acquired the perceived information, it has also had comprehensive information about the object. That means it has a basic understanding of the object.

All the processes of point (1), point (2), and point (3) constitute the function of "perception", which includes two sub-functions: (a) perceiving the object information and understanding the meaning of the object and (b) making decisions on whether the object information should be selected to be further processed.

Point (4) is for the subject to understand the deep essence of the object.

The perceived information is fed to the unit "cognition" where the perceived information will be converted (abstracted) to the knowledge via which the subject could understand the object much deeper than it has had in point (3). The knowledge will then be stored in comprehensive bases.

Point (5) is for the subject to create the intelligent strategy needed for properly dealing with the object.

The perceived information is also fed to the unit "intelligent strategy creation" where it will gradually be converted (inferred, or deducted) to the intelligent strategy needed for the subject to properly deal with the object, based on the support from the knowledge and under the guidance from the subject's goal.

Point (6) is for the subject to take action for dealing with the object.

The intelligent strategy just created in point (5) will finally be converted to intelligent action for practically dealing with the object via the function of strategy execution.

It can clearly be seen that the points from (1) to (6) are not any others, but the inner processes expressed in Fig. 3. This is exactly the implementation of the methodology of information ecology, specifically, the object information is first converted to the perceived information, and the latter is then converted to the knowledge and further converted to the intelligent strategy and finally to the intelligent action.

Summing up all the steps of the "implementation of the methodology of information ecology", the approach to advanced AI study described above can well be called the **"Principle of Information Conversion and Intelligence Creation"** — the initial stages are the information conversions and the final product is the intelligence (both intelligent strategy and intelligent action) creation.

Considering the fact that all functions performed from points (1) to (6) are by no means case-dependent, the approach is universal. Therefore, the "Principle of Information Conversion and Intelligence Creation" can well legibly be renamed as the "Law of Information Conversion and Intelligence Creation" (Zhong, 2015, 2017c).

This is the panorama of the unified approach to the study of advanced AI theory.

After all, it is the methodology of information ecology that brings out the unified approach to the advanced AI study, which is the "universal mechanism for Intelligence Creation" featured with the "Law of Information Conversion and Intelligence Creation".

The significance of the "universal mechanism for intelligence creation" is to have completely changed the situation in AI: from the three mutually expelled approaches in primary AI studies to the unified approach to the advanced AI studies. This is really the milestone for the study of advanced AI theory.

It is important to note that the "law of information conversion and intelligence creation" in information science is as significant as the "law of matter conversion and matter in-extinguishable" in material science as well as the "law of energy conversion and energy conservation" in energy science. These three laws constitute the complete system of the fundamental laws in modern science and technology.

4.3 Steps 3.1 and 3.2: Making the framework more elaborate

The third contribution that PID can make to the study of advanced AI is to make the framework of advanced AI more elaborate in accordance with the scientific view and the associated methodology.

Since the framework for advanced AI has been created, which includes the global model for advanced AI studies based on the scientific view of PID and the unified approach to advanced AI studies according to the methodology of PID, the next step is to try to make the framework more elaborate and precise.

The elaboration will include two aspects: (1) what is the scientific structure of the knowledge needed for advanced AI? And (2) what are the special, and also the innovative, requirements for the fundamental theory like the mathematical theory and logic theory needed for advanced AI?

Concerning the scientific structure of knowledge needed for advanced AI, there has been a variety of misunderstandings and confusions widely spread for a long time. One of the most serious misunderstandings is the statement saying "AI is an applied branch of computer science only". This has led many AI researchers to consider that AI is just a set of algorithms, supported by computational power and data.

In accordance with the global model of, and the unified approach to, the advanced AI theory, the scientific structure of knowledge needed for advanced AI study should be somewhat inter-disciplinary in nature (Zhong, 2015).

Figure 5 shows some, but not all, of the representatives for the scientific knowledge needed for advanced AI study (Zhong, 1988).

As is seen from Fig. 5, anthropology, social science, humanities, etc. are the reference sciences for advanced AI development; biological neural science, cognitive science, information science, etc.

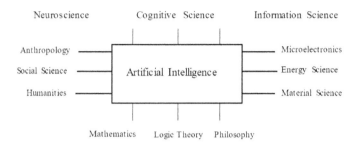

Figure 5. The Scientific Structure of AI.

are the kernel theories for advanced AI study; mathematics, logic theory, philosophy, etc. are the foundational theories for the advanced AI study; microelectronics, energy science, material science, etc. are the technical knowledge for AI implementation.

In accordance with the scientific worldview of PID, as the information process produced by the subject via subject–object interaction, advanced AI is a typical example of complexity sciences — the advanced AI study consists not only of a complex subject and object in environment but also of a very complex interaction between the subject and object. It would be unable to successfully deal with the advanced AI study if only a single branch of sciences, either computer science, automation science, or any other single branch of sciences, is employed. This is an extremely important lesson one must learn and remember in an era in which interdisciplinary studies and complexity science are the major characteristics.

As for point 3.2 in Table 4, the second aspect of making the AI framework more elaborate is to make clear the innovative requirements for foundational theories for the advanced AI study, particularly the innovative requirements for mathematical theory and logic theory for advanced AI.

It is well known that the mathematical methods employed in primary AI research so far, like probability theory, fuzzy set theory, rough set theory, etc., and the logic theory applied in primary AI research, such as the mathematical logic and various non-standard logic, have evident weaknesses due to their features of mutual separation and pure formalism.

In accordance with the scientific methodology of PID, the advanced AI research should observe the rules of information ecology that must include the rule of holistic evolution (not the rule of "divide and conquer") and the rule of form-value-meaning trinity analysis (rather than the rule of pure formalism).

Fortunately, the "Theory of factor space" (Wang, 2021) by Peizhuang Wang, and the "Principles of universal logic" (He, 2001) by Huacan He are respectively the well-satisfied mathematical theory and logic theory for advanced AI studies.

4.4 Steps 4.1 and 4.2: Formulating the advanced AI theory

The fourth contribution that PID can make to the study of advanced AI is to establish the advanced AI theory, including the sets of key concepts and the sets of principles that relate the concepts to one another.

The employment of PID in AI studies has successfully created the global model for advanced AI study (see Fig. 2) and also the universal mechanism of intelligence creation (see Fig. 3).

Figure 4 shows that the universal mechanism of intelligence creation is the real backbone of an advanced AI model. Hence, the fundamental concepts and principles needed for the advanced AI theory would be established through the implementation of the universal mechanism of intelligence creation — the law of information conversion and intelligence creation.

It is also clear from Fig. 4 that the implementation of the pervasive mechanism of intelligence creation should comprise the following five issues:

(1) **Perception:** The first principle and the concepts.
(2) **Cognition:** The second principle and the concepts.
(3) **Knowledge Representation:** The third principle and the concepts.
(4) **Strategy Creation:** The fourth principle and the concepts.
(5) **Strategy Execution:** The fifth principle and the concepts.

These issues are discussed in detail in Section 5.

5 Final Step for Paradigm Change in AI: Formulating the Advanced AI Theory

How to define and establish the fundamental concepts and principles for advanced AI? These are points 4.1 and 4.2 shown in Table 4 for paradigm change in information discipline.

The implementation of a universal mechanism of intelligence creation is just the implementation of the backbone of advanced AI. Therefore, the fundamental concepts and major principles needed for advanced AI theory can be established through the discussion of the implementation of the universal mechanism of intelligence creation in the following five sub-sections.

5.1 *Perception: The first principle and the related concepts*

Referring back to Fig. 4, the first group of key concepts and principles is related to the process of perception defined by definitions 5 and 6. But how are the key concepts and principles harmoniously related to each other? The model of perception is depicted in Fig. 6 (Zhong, 2021).

According to the methodology of information ecology, the function that perception performs is to receive the object information S from the outside world and then convert it to the perceived information as its output. The perceived information has three components

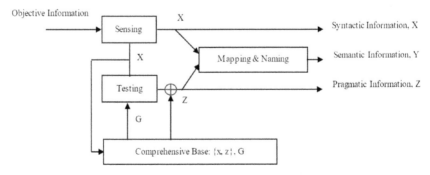

Figure 6. Model for the principle of perception and the equation $Y = \lambda(X, Z)$.

including the form (syntactic) information X, the meaning (semantic) information Y, and the utility (pragmatic) information Z, as is shown in Fig. 6.

Evidently, the form (syntactic) information X can straightly be acquired via the function of the sensing system.

The utility (pragmatic) information may have two different ways to produce.

First, if the mutual relationship between the form (syntactic) information and the utility (pragmatic) information (X, Z) has been stored in the comprehensive knowledge base beforehand and is named as pair set $\{(XZ)\}$, then the utility (pragmatic) information Z can be obtained via retrieval, by using the form (syntactic) information X as the keyword, from the pair set (X, Z) in the base. Whenever the keyword X finds the matching element from the set $\{(X, Z)\}$, then the element Z of $\{X, Z\}$ is just the utility (pragmatic) information related to the form (syntactic) information X.

Second, if the relationship between X and Z, (X, Z), did not exist in $\{(X, Z)\}$ from the comprehensive knowledge base yet, meaning that this is a new object for the perception, the utility (pragmatic) information Z cannot be retrieved from $\{(X, Z)\}$. In this case, the utility (pragmatic) information can be acquired via the measurement of the correlation between the form (syntactic) information X and the subject's goal G. In other words, the correlation between X and G can serve as the utility (pragmatic) information (Zhong, 2021):

$$Z = \mathrm{Cor.}(X, G). \tag{2}$$

As is mentioned before, the three concepts of syntactic information, semantic information, and pragmatic information are not new. In fact, they are the foundational concepts in semiotics (Morris, 1938). They have very frequently been applied in academic literature for decades. But how to correctly define them in theory? How to generate them in practice? And what is the precise relationship among them? All these questions have not been made clear in semiotics study. As a matter of fact, for a very long period of time in history, the relations among the syntactic, semantic, and pragmatic information have been misunderstood as mutually parallel to each other.

But, according to the model in Fig. 6 and the formula in Equation (1), they are not in parallel relations. Both syntactic (form) information X and pragmatic (utility) information Z can be obtained via sensing and testing, respectively, whereas the semantic (meaning) information Y can only be defined based on (X, Z), the pair set of syntactic (form) information and the related pragmatic (utility) information. This is to say that semantic (meaning) information Y cannot be obtained directly via sensing and testing but can only be obtained via the process of abstraction from the pair set (X, Z).

The great significance that the model of the principle of perception expressed in Fig. 6 and the related formula expressed in Equation (1) provide to the academic study, advanced AI in particular, lies in the following two aspects.

First, by utilizing the principle of perception expressed in the model as is shown in Fig. 6, the three components of perceived information not only have been conceptualized precisely but also can practically be realized, both in theory and in technology.

Second, by utilizing the formula described in Equation (1), the big, and old, problem of "what is the ability of understanding" in AI can then be successfully solved as follows (Zhong, 2017d).

For the given three factors: (1) a subject, (2) the goal that the subject wants to reach, and (3) an object, then in accordance with Equation (1) by "the subject has understood the object" is just meant that whenever the subject has acquired the form (syntactic) information X of the object and the utility (pragmatic) information Z that the object could provide to the subject for reaching its goal, the subject is able, based on X and Z, to make decision on how he should handle the object: support or opposite against or just ignored?

Thus, to understand something is just meant to have the meaning (semantic) information Y about that thing because the meaning (semantic) information Y is the appropriate representative of the related form (syntactic) information X and utility (pragmatic) information Z. Or, equivalently, to understand something is just to have the perceived information about the thing concerned. This is a really meaningful principle, making the advanced AI theory able to have the ability of understanding, and the intelligence based on

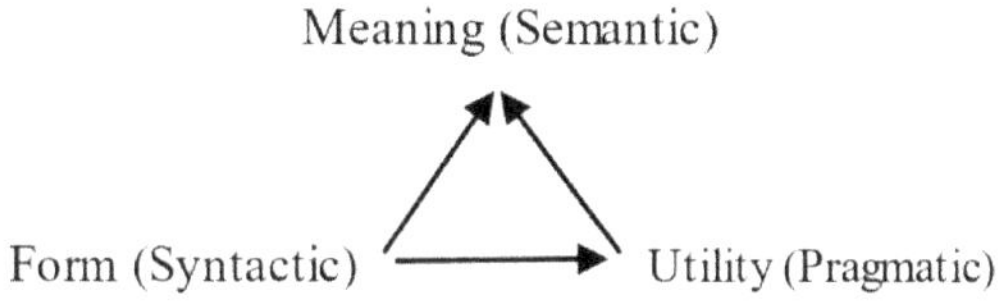

```
                  Meaning (Semantic)
                        ▲▲
                       /  \
                      /    \
         Form (Syntactic) ──────▶ Utility (Pragmatic)
```

Figure 7. Triad for the concept of understanding.

understanding (Zhong, 2017d). So, from the point of view of understanding, we have a new kind of triad as is shown in Fig. 7.

Everybody will accept that the most important performance for any AI system is the high level of intelligence it can have, whereas the well-known fact is that for any AI system to have a high level of intelligence is to have a high level of the ability of understanding.

In primary AI theory, there is only the principle of sensing and the concept of data (Shannon information), but it has no such principle of perception and such concept of perceived information and thus has no concept of understanding, and also has no ability of understanding. This is why the primary AI has no intelligence based on the ability of understanding.

5.2 Cognition: The second principle and the related concepts

The second group of key concepts and principles is related to the cognition process, in which the perceived information is fed to the input of cognition and knowledge is produced at the output of cognition. The question to be asked here is as follows: what is the specific relationship between the perceived information at input and the knowledge at output?

In accordance with the methodology of information ecology (the holistic approach), since the perceived information is a kind of trinity of form (syntactic), utility (pragmatic), and meaning (semantic) information, the knowledge thus generated in advanced AI theory should also be a kind of trinity: (1) the formal knowledge corresponding to the form (syntactic) component of the perceived information, (2) the value knowledge corresponding to the utility (pragmatic) component of the perceived information, and (3) the content knowledge corresponding to the meaning (semantic) component of the perceived information. As a result, the trinity here is named comprehensive knowledge.

As mentioned above, there is a big difference concerning the knowledge in the advanced AI theory and that in the primary AI theory. More specifically, the knowledge in advanced AI theory results from perceived information, while the knowledge in primary AI theory results only from form (syntactic) information (data) (Zhong, 2000). In other words, the knowledge in primary AI theory is merely a kind of formal knowledge.

The model for the principle of cognition in advanced AI (Zhong, 2000) is shown in Fig. 8.

Once again as is seen in Fig. 8, the input of the model is the perceived information, which is just the output of the perception, while the output of the model is the comprehensive knowledge. The function of the comprehensive knowledge base in Fig. 8 is to provide the *a priori* comprehensive knowledge needed for supporting the cognition process. Both the pre-processing unit and the post-processing unit are necessarily needed for the interfaces at the input and the output of the model.

Normally, there should be three levels of the cognition process, that is, levels 1, 2, and 3, for the specific implementation of the principle of cognition in advanced AI theory, as is seen in Fig. 8.

The cognition of level 1 is the primitive, or the lowest, level of cognition, which is frequently termed the level of brute force learning, that simply connects and remembers the corresponding relationships between the form (syntactic) information at the input and the formal

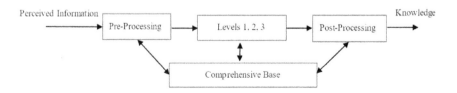

Figure 8. Model for principle of cognition.

knowledge (the name of that object) at the output, without knowing the utility and the meaning, without knowing why.

The primitive level of cognition is exactly the level that young human babies have to adopt. They learn some things from the adults within the family just to remember the form of the things and the related names without understanding the utility and meanings about the things — typical examples of level 1 of cognition. Obviously, level 1 of cognition is very simple and also very basic and really necessary.

In the field of machine learning, brute force learning is the necessary step to provide the seeds of formal knowledge at the very beginning of setting up the formal knowledge base. The reliability of the formal knowledge learned in this way depends basically on the quality of the base builder.

The cognition of level 2 is the intermediate level of the cognition process, which is often named inductive cognition and/or statistical learning in machine learning, that is, to compress the formal knowledge via refining the formal factor in common from the large scale of the samples of form information into the same class. In normal cases, the form factor in common is regarded as the essence of the samples of form information in the class.

It must be pointed out that all of the raw materials provided for cognition, that is, the large set of form (syntactic) information, and all of the results achieved from the cognition of level 2 are purely formal without utility and meaning factors participating in it. Formal knowledge plays a key role in the primary AI theory. People at that stage did not know how the understanding ability could be fostered in machines.

In contrast with the cognition of levels 1 and 2, both of which are purely formal cognition on the whole, however, the cognition of level 3 is no longer purely formal cognition but is comprehensive cognition. This is because of the fact that, by virtue of the model of perception and Equation (1), comprehensive knowledge can be produced. More specifically, within the process of conversion from the perceived information to the comprehensive knowledge, the form (syntactic) information would be converted (be inducted) to formal knowledge, the utility (pragmatic) information would be converted (be inducted) to value knowledge, and the meaning (semantic) information would be converted (be inducted) to content knowledge, and finally the result is that the perceived information as a whole would be converted (be inducted) to the comprehensive knowledge.

Essentially speaking, the cognition of level 1 and level 2 is regarded as the primary level because the products of knowledge produced by both of the two levels are merely formal knowledge. The cognition of level 1 and level 2 can roughly correspond to the learning ability of young babies (brute force learning) and youngsters (statistical learning). The set of knowledge that young babies learned is small in quantity and shallow in quality, while the set of knowledge that youngsters learned is large in quantity, wide in classes, and yet still quite shallow in quality.

The cognition of level 3 is rather different. It is regarded as the advanced level of cognition because the products of knowledge produced by level 3 are no longer formal knowledge but comprehensive knowledge. So, the cognition of level 3 corresponds to the learning ability of adults. They can autonomously build up the ability to understand based on the comprehensive knowledge they learned.

Note that the formal knowledge accumulated via the cognition of levels 1 and 2 can make contributions to the comprehensive cognition of level 3. This is because of the fact that formal knowledge is one of the components of comprehensive knowledge. Put the formal knowledge together with the value knowledge related, then the content knowledge and the related comprehensive knowledge can be formulated.

In fact, the cognition of level 3 can be implemented in two different ways. The first way is to directly produce comprehensive knowledge. The second way is based on levels 1 and 2, that is, to add the related value knowledge to the formal knowledge, making it to be the comprehensive one.

So, all the levels 1, 2, and 3 of the cognition process are useful. The three levels of cognition must mutually support each other in a harmonious manner so as to make all knowledge stored in the knowledge bases become comprehensive knowledge: bottom-up from levels 1 and 2 to level 3 is for producing comprehensive knowledge, while top-down from level 3 to levels 1 and 2 is for adding the value components to the formal knowledge, turning it into the comprehensive knowledge.

The high importance of comprehensive knowledge lies in the fact that formal knowledge reflects only the formal properties of the object, which is only "shallow knowledge", whereas comprehensive knowledge is "deep and complete knowledge" needed to support the ability of understanding in advanced AI.

5.3 Knowledge representation: The third principle and the related concepts

The topic of comprehensive knowledge representation in bases is important because, on the one hand, comprehensive knowledge produced via cognition must be stored in the comprehensive knowledge bases; on the other hand, comprehensive knowledge stored in comprehensive bases must also be easily retrieved for use.

Storage and retrieval are basic issues in computer science and technology. But the principle of storage and retrieval in computer science and technology cannot meet the needs for advanced AI because all the items stored in, and retrieved from, a computer are in the form of data, that is, the form (syntactic) information and formal knowledge.

However, the items stored in, and retrieved from, the bases for advanced AI must be in the format of comprehensive information and comprehensive knowledge. This is because of the fact that it is only the comprehensive information and comprehensive knowledge that can reflect not only the formal properties of the object but also the deep properties of the object. This is why the issues of comprehensive knowledge bases play a fundamental role in advanced AI theory.

The results produced from cognition are the sets of comprehensive knowledge. They should be stored in the comprehensive knowledge base so as to be retrieved from the base effectively and usefully whenever needed. How to realize the requirements concerning the efficient storage and retrieval of comprehensive knowledge in advanced AI theory?

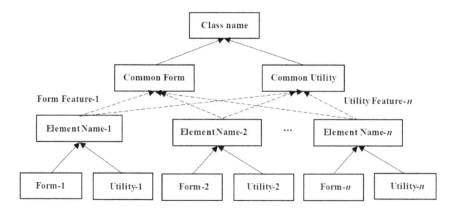

Figure 9. The simplified principle of comprehensive knowledge expression.

This leads to the issues of representation of comprehensive knowledge.

The simplified principle of comprehensive knowledge representation, the organizational architecture, and the working mechanism in the comprehensive knowledge bases are briefly described in Fig. 9 (Zhong, 2000).

As is seen in Fig. 9, at the bottom of the expression system, there is a set of the elementary comprehensive knowledge denoted by name-i $(i = 1, \ldots, n)$ each of which has the formal component and value component of the comprehensive knowledge name-i, where the name-i is just the content knowledge, which is the representative of comprehensive knowledge, and n is an integer, representing the capacity of the elementary comprehensive knowledge concerned in the base.

At one level higher than the elementary comprehensive knowledge of the expression system, there is a name of new comprehensive knowledge that is abstracted from the set of elementary comprehensive knowledge. This new comprehensive knowledge also has two components: one is the new formal component abstracted from those form components, which are mutually in common at the bottom level, and another is the new value component abstracted from the value components, which are mutually in common at the bottom level.

A similar process will be continued from the lower level to the higher level till the highest level reaches the top of the class of comprehensive knowledge. The entire expression of comprehensive knowledge is thus formed.

In a similar way, the expressions of comprehensive knowledge for all of the other classes can also be formed. Finally, the entire architecture of the comprehensive knowledge in comprehensive knowledge bases in advanced AI theory can thus be built up.

This is the biggest difference concerning the principle of knowledge expression in knowledge bases between primary AI theory and advanced AI theory. Without comprehensive knowledge, the subject will almost be unable to satisfactorily understand the problem faced and will thus be unable to make intelligent decisions.

At this point, we would also like to mention that the expression of knowledge in advanced AI theory is also radically different from that in the so-called "knowledge graph" where only the "name" is used for the expressions of the nodes and the links within the graph without formal and value components associated with the name. So, they are "empty names". Surely, the knowledge graph also cannot fully support the needs of advanced AI.

As for the "semantic web", it still cannot meet the needs for advanced AI. This is because of the fact that the ascription "semantic" to the semantic web is not really the true semantic. The "semantic" is not defined by the equation $Y = \lambda(X, Z)$ but by a kind of "pseudo semantic".

It would be a very good idea to reform the "knowledge graph" into the "comprehensive knowledge graph" through the way of changing the expressions of all the nodes from the "empty name" to the comprehensive knowledge. In this case, the comprehensive knowledge graph will not only be able to express the interrelationships among the nodes but also be able to express the comprehensive knowledge for each of the nodes of the graph.

In other words, the comprehensive knowledge graph should have a kind of two-dimensional structure: the interrelationship of the related names of the comprehensive knowledge in the horizontal dimension and the abstraction process (from the most specific one to the most abstract one) for each of the comprehensive knowledge in the vertical dimension.

5.4 Intelligent strategy creation: The fourth principle and the related concepts

As a matter of fact, the discussions about the principle of perception, the principle of cognition, and the principle of comprehensive knowledge representation carried out above are necessary and sufficient preparations. Yet, the real central issue in advanced AI theory is the principle of intelligent strategy creation. That is the issue of how to create an intelligent strategy able to successfully handle and solve the problem.

It is worth pointing out that, in the literature for a very long time, the term "intelligent strategy creation" has been named "decision-making". The problem associated with the term "decision-making" is that one may just consider how to make a choice among a number of alternatives without considering how to produce such a number of alternatives. In fact, the production of the alternatives is the basis of choosing alternatives. In order to emphasize the importance of the alternatives' production, the term "intelligent strategy creation" is adopted here.

Let's put forward a standard statement of the issue in the following way.

Given a set of conditions as follows: (1) The problem for the advanced AI system to solve which is expressed with the perceived information, which in turn is converted from the object information describing the form of the original problem; (2) the comprehensive knowledge for the advanced AI system to be used for problem-solving; (3) the goal the advanced AI system wants to reach as the result of problem-solving.

Then the question is as follows: How does the advanced AI system create the strategy for intelligently solving the given problem with sufficient satisfaction?

By "strategy" here is meant the series of actions that the advanced AI system generates for dealing with, and for solving, the problem

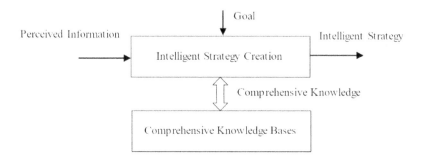

Figure 10. Model of principle for intelligent strategy creation.

given and for achieving the goal that the advanced AI system has to reach, based on the support from the comprehensive knowledge that the advanced AI system could obtain from the comprehensive knowledge bases.

Based on the conditions and question stated above, it is reasonable to build up the model, shown in Fig. 10, for the advanced AI system to solve the given problem and to reach the goal, based on the support from the comprehensive knowledge in the bases.

As can be clearly seen from the model, the perceived information represents the starting point of the problem-solving process on the one hand and the advanced AI system's goal represents the ending point of the process of problem-solving on the other hand. Therefore, the key technique for successfully solving the problem, that is, to go from the starting point to the ending point, lies in how to appropriately (efficient in speed and satisfactory in quality) utilize the comprehensive knowledge provided from the comprehensive knowledge bases.

Note that such a kind of reaction created by an advanced AI system will be named the "intelligent strategy" created and is regarded as the most important specification for advanced AI theory. The reaction is intelligent if it is able to reach the given goal with sufficiently high efficiency in speed and satisfaction in quality.

As is stated above, no matter what kinds of inputs the advanced AI system faces and no matter what kinds of goals that advanced AI system would have to reach, the advanced AI system's reaction will basically be determined by the quality of the comprehensive knowledge that the advanced AI system could use.

In practice, the comprehensive knowledge could roughly be categorized into the following categories:

Category 1 — innate knowledge, the heritage received from genetic source.

Category 2 — common sense knowledge, the one every normal person knows.

Category 3 — experienced knowledge, the one accumulated via practice.

Category 4 — regular knowledge, the knowledge learned via education.

If the subject (stored in the comprehensive knowledge bases) possesses only the knowledge of category 1, the reaction so generated would be named as the kind of subconsciousness (e.g., unconditioned reflections).

If the subject (stored in the comprehensive knowledge bases) possesses the knowledge of category 1 together with category 2, the reaction generated would be the kind of primary consciousness (e.g., conditioned reflections).

If the subject (stored in the comprehensive knowledge bases) possesses the knowledge of category 1, together with categories 2 and 3, the reaction so generated would be named the kind of experienced/emotional intelligence.

If the subject (stored in the comprehensive knowledge bases) possesses the knowledge of category 1, together with categories 2, 3, and 4, the reaction so generated would be regarded as intellectual intelligence.

Note that concerning the comprehensive knowledge bases, it must store not only the very large set of comprehensive knowledge but also the very large set of the rules of inference that infers/deduces one piece of the comprehensive knowledge to another. These kinds of knowledge are called inference rules.

And then the process of intelligent strategy creation can be implemented as follows (Zhong, 2023):

Step 1: Due to the fact that the perceived information contains the form (syntactic) component and the utility (pragmatic) component, it is easy to make a decision that the specific comprehensive knowledge in the base, whose formal knowledge and value knowledge are just respectively corresponding to the form (syntactic) and utility (pragmatic) component of the perceived information, must be taken as the real starting point of the inference (left hand of the inference rule).

Step 2: Among all pieces of comprehensive knowledge related to the one representing the starting point (the left hand of the first inference rule), choosing the one, which has the best value knowledge toward the subject's goal, as the subgoal of the inference (right hand of the first inference rule).

Step 3: Taking the subgoal just reached in step 2 as the new starting point (left hand of the new inference rule to be executed), follow the same criterion stated in step 2 to make the choice of the new right hand of the new inference rule.

Step 4: Continuously using the technique employed in steps 2 and 3, the comprehensive knowledge with the best possible value component toward the subject's goal, among all pieces of the comprehensive knowledge, can finally be reached.

Step 5: The orderly chain, or the orderly sequence, of all the successful rule applications above constitutes the entirety of intelligent strategy.

We can see that the mechanism of intelligent strategy creation very clearly indicates that the comprehensive information, rather than merely the form (syntactic) information, and the comprehensive knowledge, rather than merely the formal knowledge, play the crucial role in the efficient and satisfactory operations in formulating the intelligent strategy for the subject.

Contrarily, in the primary AI theory, the "intelligent strategy", respectively produced via either artificial neural network or expert system or sensor-motor system, is based on the form (syntactic) information as well as formal knowledge and therefore it is not easy to find the best knowledge for the right hand of the rule inference. The creation of the "intelligent strategy" in the primary AI theory would have to be carried on through the statistical approach with lower efficiency and satisfaction.

This clearly indicates the high importance of the comprehensive information. Without comprehensive information, there would be no comprehensive knowledge and there would be no comprehensive intelligence too.

5.5 Strategy execution: The fifth principle and the related concepts

The final issue of the advanced AI theory is about the principle of strategy execution whose function is to convert the intelligent strategy just created into the intelligent action and then to (re-)act on the object so as to reach the goal and to solve the problem.

In the ideal case particularly, this is relatively simple. As can be seen from the discussion in the last sub-section, the execution of the intelligent strategy is to perform orderly rule applications. What the subject should do is to adopt the formal component of the intelligent strategy, which can of course be regarded as a kind of comprehensive knowledge, in accordance with the order by which the successful rules are selected.

In practice, however, more often than seldom, there exist some uncertain and unpredictable factors in the entire process of "information conversion and intelligence creation" such as noise, disturbance, and other non-ideal properties and performances of the systems, which will certainly cause some error between the result of reacting and the goal to be reached. In these cases, the error should be regarded as a kind of new "object information"; this is because of the fact that the error indicates that the "intelligent strategy/action" just applied to the object is insufficiently intelligent and hence should be improved and optimized.

The basic process needed for the strategy/action improvement and optimization can be thought of as follows.

Considering the error as a new kind of "object information", which indicates that the strategy/action just created is not so good due to the lack of some knowledge, let the error feedback to the input of the perception in the advanced AI system. Perform the same process as what the subject did before according to the "mechanism for intelligence creation characterized by information conversion and intelligence creation", as shown in Fig. 4. After the process is performed, the additional knowledge will be learned and added, and the strategy and the action will be improved leading the error to be decreased. This process will be performed at certain times until the error decreased sufficiently small and the goal is satisfactorily reached.

In the case that the above processes fail to reach the goal, this should mean that the goal had not been set reasonably and thus should be reset by the human subject (designers) and then perform the same process of "information conversion and intelligence creation".

Finally, if the strategy/action is successful, it should be stored in the comprehensive knowledge/strategy bases for later use. In this way, the capacity of the comprehensive knowledge base and the comprehensive strategy base as well as the comprehensive information base is continuously expanded and upgraded, meaning that the advanced AI system becomes more and more intelligent. This is a process of learning and evolution for advanced AI.

The entire process of the advanced AI theory establishment is just the process of the implementation of a pervasive mechanism of intelligence creation and is also the realization of the law of "information conversion and intelligence creation". All the processes can be expressed as the model shown in Fig. 4.

With those above, the discussion on "paradigm change" is ended at least for the time being. Reviewing back all the discussions carried out here, a conclusion can be drawn that the advanced AI, thanks to the paradigm change, is the only kind of AI that possesses much superior specifications in the performance of intelligence, understanding ability in particular, compared to all kinds of primary AI existed.

After all, this is clear evidence, indicating that the "paradigm change" in advanced AI study is really necessary and greatly successful. AI research would ever be wandering about within the low stage of its development and there would no advanced AI theory emerge at all if there were no paradigm change in AI.

6 Summary and Remarks

A number of new observations, new findings, new viewpoints, new approaches, a new law, and new perspective results have been presented here, which are of high significance for the further development of AI.

As is seen from the discussions above, almost all new understandings and achievements result from the "paradigm change" in AI. The essence of this change is to give up the borrowed PPD and return it back to the physical discipline on the one hand and, on the other hand, to set up and execute PID in the entire process of its study.

The new and significant understandings and achievements can be briefly and systematically summarized as follows:

(1) There have been two categories of scientific disciplines in natural sciences till the present time. One is the physical discipline, which emerged in the 17th century, or even earlier, and another is the information discipline which appeared in the Second World War in the 1940s.
(2) The information discipline has a number of different branches. But it is only the advanced AI that is able to perform all kinds of information functions. This is due to its taking into account not only the objective factors but also the subjective factors and, even more importantly, the interaction between subject and object in its study. Therefore, advanced AI can serve as the best representative of the entire information discipline. In fact, the full consideration of the subjective, objective, and subject–object interactions is a historical need and is a significant trend for the complexity of science in the complex society of the 21st century.
(3) The paradigm for a scientific discipline refers to the scientific worldview and the associated methodology for the discipline

and is hence the unique supreme guiding force for leading and regulating the research activities for the discipline.
(4) Yet, the formation of the paradigm for a scientific discipline has to be delayed for a very long time after the emergence of the scientific discipline because of the fact that this is a very difficult process with numerous "try and error" steps and puzzles described by the parable "the elephant and seven blind men".
(5) Therefore, the paradigm for any big and newly born scientific discipline will have to be unavailable for the world community of the discipline due to the facts stated above. Information discipline is the first such discipline in recent history. As a result, the studies of information discipline have borrowed the paradigm from the physical discipline for its own use since the very beginning of the discipline till the present day. People have to recognize this reality no matter whether they like it or not and no matter whether they feel happy or not.
(6) The borrowed paradigm has thus requested the researchers in information discipline to follow the paradigm (scientific worldview and the associated methodology) of the physical discipline in their studies. As a result, the study of information discipline has been confined to the lower stage of its development, featuring no general theory for AI (due to the methodology of "divide and conquer") and the absence of high quality intelligence for artificial intelligence (due to the methodology of "pure formalism").
(7) The most reasonable solution to make the studies of AI walk out from the lower stage and enter the higher stage of its development is to summarize, refine, and set up the paradigm for information discipline itself and then to give up the borrowed paradigm.
(8) In this chapter, the paradigm for information discipline has been refined and set up, resulting from our team's long-term investigation for over half a century. Its scientific view says that "the essence of AI is the entire ecological chain of information process generated by the subject (or the agent of the subject)

through the subject-object interaction" and its methodology is characterized by the one of information ecology with neither "divide and conquer" nor "pure formalism". The latter can also be named the holistic approach.

(9) According to the scientific worldview of the paradigm for information discipline, the global model for advanced AI is created as the entire process of the ecological chain of information generated by subject–object interaction activities for problem-solving.

(10) According to the methodology of the paradigm for information discipline, the research approach to advanced AI is the methodology of information ecology featured with holistic, evolution, and global optimization methods, without "divide and conquer" and "pure formalism".

(11) Following the scientific worldview and methodology of the paradigm for information discipline, the universal mechanism of intelligence creation has been explored and discovered, which is characterized by the great "Law of Information Conversion and Intelligence Creation".

(12) Following the paradigm for information discipline, a number of key concepts in the primary AI have been deeply reformed and recreated including comprehensive information, comprehensive knowledge, comprehensive bases for storage and retrieval, primary consciousness, artificial emotion, comprehensive strategy, etc.

(13) Following the paradigm for information discipline, a series of the principles of information conversion have been systematically summarized, such as the principle of perception, the principle of cognition, and the principle of strategy-making, above all the principle and the great law of "information conversion and intelligence creation".

(14) Following the paradigm for information discipline and based on all the results achieved above, the unified theory for advanced AI has eventually been established, namely the "General Theory of AI based on the universal Mechanism of Intelligence Creation" or m-GTAI for simplicity and convenience.

(15) The "m-GTAI" presents highly superior performance compared with all kinds of primary AI theories existed so far. It is the advanced AI theory (m-GTAI) that can have a sufficiently high level of intelligence based on understanding ability.
(16) In summary, the "m-GTAI" is in fact the AI theory in the advanced stage of its development, achieved through the paradigm change in AI.

Finally, the author would also like to point out that the studies of AI are typical examples of complexity science. This should be the nucleus of the contemporary sciences in the 21st century and needs knowledge not only from natural sciences but also from philosophical sciences. Clearly, the studies on the issues of a paradigm do not belong to natural sciences but belong to philosophy. It would not be possible to realize the important issues of "paradigm change in AI" without the support from the studies of information philosophy. So, it is necessary to have a closer collaboration between the studies of information science and that of information philosophy. The author himself has gained many benefits from the philosophical publications (Zhong, 2023; Wu, 2005, 2016).

Acknowledgment

The author would like to express his deep thankfulness to China National Science foundation for a series of research project support (No. 60873001, etc.), to many friends like Professors Claude Shannon, Lotfi Zadeh, Mark Burgin, Wolfgang Hofkirchner, Pedro Marijuan, Gordana Dodig-Crnkovic, Terry Deacon, Ruqian Lu, Yanda Li, Muming Pu, Jianhua Lu, Xuyan Tu, Huacan He, Peizhuag Wang, Zhongzhi Shi, Fuji Ren, Yong Shi, Jiali Feng, Liqun Han, Xiaojie Wang, Yanquan Zhou, Lei Li, Zhicheng Chen, Ruifan Li, Jianxin Lu, Ting Xu, Shiguang Zhang, Dingtao Wang, and many others for their friendly assistance and helpful academic exchanges. My special gratitude goes to Professor Mark Burgin for the very valuable comments he gave to this chapter and the enthusiastic assistance in improving the quality of the English expressions of this chapter.

References

Bengio, Y. and Delalleau, O. (2000). Justifying and generalizing contrastive divergence. *Neural Computation* 21(6), 1601–1621.

Bengio, Y. and LeCun, Y. (2007). Scaling learning algorithms towards AI. *Large Scale Kernel Machines*. MIT Press.

Brooks, R. A. (1989). Engineering approach to building complete, intelligent beings. In *Proceedings of SPIE 1002, Intelligent Robots and Computer Vision VII*, Boston, pp. 618–625.

Brooks, R. A. (1990). Elephant cannot play chess. *Autonomous Robot* 6, 3015.

Brooks, R. A. (1991). Intelligence without representation. *Artificial Intelligence*. 47(1/3), 139–159.

Burgin, M. and Zhong, Y. (2019). Methodology of information ecology in the context of modern academic research (in Chinese). *Philosophy Analysis* 10(1), 119–136.

Delalleau, O. and Bengio, Y. (2011). Shallow vs. deep sum-product networks. In *NIPS*.

Feigenbaum, E. A. (1977). The art of artificial intelligence: Themes and case studies of engineering. In Feigenbaum, E. A. Feldman (eds.), *Proceedings of the 5th International Joint Conference on Artificial Intelligence*, Cambridge: McGraw-Hill.

He, H. (2001). *Principles of Universal Logic* (in Chinese). Science Press.

Hinton, G. E., Osindero, S., and The, Y. (2006). A fast learning algorithm for deep belief nets. *Neural Computation*, 18, 1527–1554.

Hinton, G. E., Deng, L., Yu, D., Dahl, G. E. *et al.* (2012). Deep neural networks for acoustic modeling in speech recognition: The shared views of four research groups. *IEEE Processing Magazine* 20(6), 82–97.

Hopfield, J. J. (1982). Neural networks and physical systems with emergent collective computational abilities. *Proceedings of the National Academy of Sciences of the United States of America* 79(8), 2554–2558.

Kohonen, T. T. (1990). The self-organizing map. *Proceedings of the IEEE*. 78(9), 1464–1480.

Kuhn, T. S. (1962). *The Structure of Scientific Revolution*. University of Chicago Press.

Marcus, G. and Davis, E. (2019). *Rebooting AI: Building Artificial Intelligence We Can Trust*. Pantheon Random House.

McCarthy, J. Minsky, M, Rochester, N. *et al.* (1955). Proposal for the Dartmouth Summer Research Project on Artificial Intelligence. Dartmouth College.

McCulloch, W. and Pitts, W. (1943). A logic calculus of the ideas immanent in nervous activity. *Bulletin of Mathematical Biophysics* 5, 115–133.

Minsky, M. (1986). *The Society of Mind*. Simon and Schuster.

Montufar, G. F., Pascann, R., Cho, K., and Bengio, Y. (2014). On the number of linear regions of deep neural networks. In *NIPS'2014*. https://papers.nips.cc/paper_files/paper/2014.

Morris, W. C. (1938). *Foundation of the Theory of Sign*. The University of Chicago Press.
Newell, A. (1980). Physical symbol systems. *Cognitive Science* 4(2), 152–183.
Newell, A. and Simon, H. A. (1963). GPS, a program that simulate human thoughts. In Feigenbaum, E. A. Feldman, J. (eds.), *Computers and Thought*. McGraw-Hill.
Nilsson, N. J. (1982). *Principles of Artificial Intelligence*. Springer.
Pascann, R., Guleehre, C., Cho, K., and Bengio, Y. (2014). How to construct deep recurrent neural networks. In *ICLR*.
Ranzato, M., Poultney, C., Chopra, S., and LeCun, Y. (2007). Efficient learning of sparse representations with an energy-based model. In B. Scholkopf, J. Platt, and T. Hoffman (Eds.), *Advancs in Neural Information Processing Systems 19* (NIPS'06), MIT Press, pp. 1137–1144.
Rosenblatt, F. (1958). The perceptron: A probabilistic model for information storage and organization in the brain. *Psychological Review* 6(56), 386–408. https://doi.org/10.1037/h0042519.
Rumelhart, D. E. and McClelland, J. L. (1986). *Parallel Distributed Processing*. MIT Press.
Turing, A. M. (1963). Can machine think. In Feigenbaum, E. A. and Feldman, J. (eds.), *Computers and Thought*. McGraw-Hill.
Wang, P. (2021). *Factor-space Theory* (in Chinese). BUPT Press.
Wu, K. (2005). *Information Philosophy: Theory, System, Methods* (in Chinese). Commercial Book Store.
Wu, K. (2016). The interaction and convergence of philosophy and science of information. *Philosophies* 1(3), 228–244.
Zhong, Y. (1988). *Principles of Information Science* (in Chinese). Beijing: BUPT Press.
Zhong, Y. (1992). *Intelligence Theory and Technology: AI and Neural Network* (in Chinese). P&T Press.
Zhong, Y. (2000). A framework of knowledge theory: Information, knowledge, intelligence and the conversions among them (in Chinese) *China Engineering Science* 2(9), 50–64.
Zhong, Y. (2007). *Principles of Cognitive-Cybernetics in Machine* (in Chinese). Beijing: Science Press.
Zhong, Y. (2008). Structuralism? Functionalism? Behaviorism? Or mechanism? Looking for better approach to AI research. *International Journal of Intelligent Computing and Cybernetics* 1(3), 325–336.
Zhong, Y. (2012). The breakthrough in AI due to the innovation in scientific methodology (in Chinese). *Pattern Recognition and Artificial Intelligence* 25(3), 456–461.
Zhong, Y. (2014). *Principles of Advanced AI: New view, New Approach, New Model, and New Theory* (in Chinese). Beijing: Science Press.
Zhong, Y. (2015). Information conversion and intelligence creation: the law that governs the information discipline. *International Journal on Cognitive Informatics and Natural Intelligence* 7(3), 25–41.

Zhong, Y. (2015). *Introduction to Information Science and Technology* (in Chinese). BUPT Press.

Zhong, Y. (2017a). From the methodology of mechanical reductionism to the one of information ecology: A successful approach to AI studies (in Chinese). *Philosophy Analysis* 8(5), 133–144.

Zhong, Y. (2017b). AI: Concept, approaches, and opportunities (in Chinese). *Chinese Science Bulletin* 62(22), 2473–2479.

Zhong, Y. (2017c). The law of information conversion with intelligence creation. In M. Burgin, and W. Hofkirchner (eds.), *Information Studies and the Quest for Transdisciplinarity*, Vol. 9. World Scientific Series in Information Studies, pp. 165–190.

Zhong, Y. (2017d). A theory of semantic information. *China Telecommunications* 14(1), 1–17.

Zhong, Y. (2021). *Mechanism-based Universal Theory of AI* (in Chinese). Beijing: BUPT Press.

Zhong, Y. (2021). Paradigm change in AI and the birth of general theory of intelligence. *Journal of Intelligent System* 16(4), 793–800.

Zhong, Y. (2023). *General Theory of Intelligence* (in Chinese). Beijing: Science Press.

Zurada, J. M. (1992). *Introduction to Artificial Neural Systems*. West Publishing Company.

Chapter 10

Relational Structure of Conceptual Spaces

Mark Burgin[*] and José María Díaz-Nafría[†,‡]

[*]*Department of Computer Science,
UCLA, Los Angeles, CA 90095, USA*
[†]*Department of Telecommunication Engineering,
Madrid Open University, 28400 Collado Villalba, Madrid, Spain*
[‡]*josemaria.diaz.n@udima.es*

Concepts play an important role in human culture, forming explicit and implicit conceptual spaces. The goal of this chapter is to explore the structural organization of information in conceptual spaces.

1 Introduction

Concepts play an important role in human culture as the basic level of our knowledge systems (Burgin, 2016). In essence, all meaningful words are names of concepts. Consequently, dictionaries are collections of concept definitions, while encyclopedias are collections of extended concept descriptions. Thus, it is important to study concepts and their arrangements in the context of knowledge systems.

Concepts are intrinsically related to signs. According to the general understanding, a sign is an object that points to something different than itself, being transcendent to it. At the same time, a concept, or notion, is usually defined as a general idea derived from specific instances, i.e., a concept is a symbolic (usually, linguistic) representation of these instances, and this representation naturally points to these instances. Consequently, a concept is a specific sign.

In terms of their organization, concepts form conceptual spaces, which include relations between and operations with concepts as elements of these spaces. (Doignon and Falmagne,1999; Gärdenfors, 2004; Burgin and Díaz-Nafría, 2019). The goal of this chapter is to study relations in conceptual spaces.

It is important to understand that the conceptual spaces, introduced and explored in Burgin and Díaz-Nafría (2019), are aimed at the study and further development of the structural organization of encyclopedias and encyclopedic dictionaries, while conceptual spaces introduced and studied by other authors are oriented at modeling mental knowledge structures in the mind. The conceptual spaces of the first type are called structural, while the conceptual spaces of the second type are called attributive. Here, we continue our studies of structural conceptual spaces with emphasis on the relations of these spaces, an endeavor aligned with our venture of building up transdisciplinary knowledge (Díaz-Nafría et al., 2019). In this undertaking, the knowledge possessed by interdisciplinary communities is crystalized in a system of interdisciplinary glossaries, as a kind of encyclopedia whose concept has been presented and discussed in previous works (Díaz-Nafría et al., 2018, 2019a, 2019b).

2 Concepts and Their Models

In conceptual spaces, concepts are represented by their models. In our study, we use the most advanced model, which is called the *representational model of a concept* (Díaz-Nafría et al., 2018; Burgin and Gorsky, 1991). On its second level, it can be treated as a synthesis of Russell's model of concept and Peirce's model of sign. In the representational model of a concept, the name of the concept is connected to the concept representative, which consists of three components: *denotat* as the collection of all particular exemplifications or instantiations of this concept, *meaning or connotation*, and

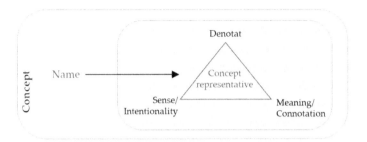

Figure 1. The extended Peircean model of a concept as basic element of conceptual spaces.

sense or intentionality (cf. Fig. 1). We call this special case of the representational model by the name of the extended Peircean model of a concept.

Meaning represents the semantics of the concept. Sense is associated with the pragmatics of the concept. *Denotat* is the generalized syntax of the concept. For a symbolic concept, its denotat consists of syntactic relations of this concept (for more details on the model, cf. Burgin and Semenovich, 2017).

It is possible to treat other known models of concepts as specifications of the representational model of the concept (Burgin and Semenovich 2017; Burgin, 2012; Hampton, 2011).

Thus, the theory of concepts in general and the extended Peircean model of a concept, in particular, bring us to the notion of the *generalized syntax* of the concept, which consists not only of names (words) of objects that constitute the denotat but also include the physical and mental objects from the denotat and relations between all kinds of these objects. While conventional syntax is defined as the arrangement of words (names) in a sentence, generalized syntax is defined as the arrangement of symbolic (e.g., words as names), mental, and physical objects in the denotat of a concept.

Using properties of concepts, which form intermediate relations in conceptual spaces, it is possible to differentiate all concepts in a conceptual space into three groups — general, individual, and impossible concepts — which are defined in the following ways:

- A *general concept* has many instantiations.
- An *individual concept* has only one instantiation.
- An *impossible concept* does not have instantiations.

Note that the membership of a concept in one of these groups is contextual, i.e., it depends on the context. For instance, the name of a person can be an individual concept in one group where there are no other people with this name and a general concept in another group where there are several people with this name.

The concept cat is general. The concept of the Earth is individual. The concept "a ball larger than the Sun and smaller than the Earth" is impossible.

According to the type of the denotat, general concepts can be as follows:

- *set concepts*, in which the denotat is a set,
- *class concepts*, in which the denotat is a class,
- *ensemble concepts*, in which the denotat is an ensemble.

Note that each classification or typology of a concept determines intermediate relations in the conceptual space.

According to the exactness of the denotat, general concepts can be as follows:

- *strict concepts*, in which the denotat is a set,
- *fuzzy concepts*, in which the denotat is a fuzzy set,
- *blurry concepts*, in which the membership in the denotat is not clearly defined.

3 Categorization of Conceptual Spaces

Existential stratification induces three basic types of conceptual spaces:

- *Mental conceptual* spaces consist of mental concepts and relations between them.

- *Symbolic conceptual* spaces consist of symbolic concepts and relations between them being physical representations of ideal conceptual structures.
- *Substantial conceptual* spaces include physical components of concepts in addition to symbolic conceptual representations.

Mental conceptual spaces are in the heads of people and are studied by psychologists and educators.

Encyclopedia is an example of a symbolic conceptual space.

If the concept *tree* belongs to a substantial conceptual space, then this space also includes some number of trees.

Here, we study symbolic conceptual spaces with the goal of explicating the structural organization of information in conceptual spaces. We speak of information in conceptual spaces because it is the interaction among concepts that create meaning in the knowledge system based upon the concepts of such spaces, and the interaction can be expressed in terms of information.

4 Relations in Conceptual Spaces

Studying conceptual spaces, we consider relations of two types: inner and inter-mediate relations (Burgin, 2012). On the first level of the space structure, inner relations of a conceptual space are relations between elements of this space (cf. Fig. 2), while intermediate relations of a conceptual space are relations between elements of this space and some other objects (see Fig. 3). Examples of intermediate

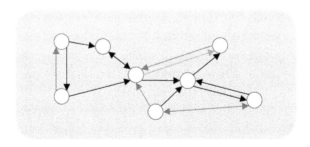

Figure 2. Conceptual space with inner relations between concepts.

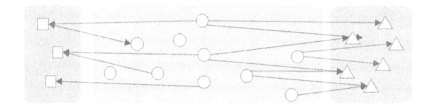

Figure 3. Conceptual space with intermediate relations of concepts.

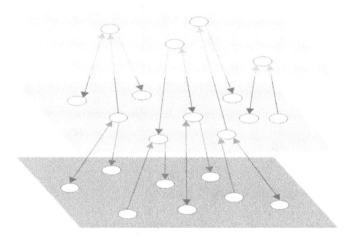

Figure 4. Conceptual spaces of different abstraction levels with intermediate relations among their respective concepts.

relations are abstract properties of concepts. Examples of inner relations are relations "to be more (less) general" or "to be more (less) abstract".

There is a particular type of intermediate relation of special interest for our endeavor to integrate knowledge, as discussed in the following section. This is the case in which the relations are established between concepts of different conceptual spaces each one corresponding to a different abstraction level (cf. Fig. 4).

Usually, a knowledge system incorporates concepts of different abstraction levels that enable us to provide a more detailed description or a more abstract one depending on the issues we are dealing with, but in some cases, we develop conceptual spaces in which the intention is to be detached from the level below. This is, for instance, the case of network theory in which the universe of discourse is simplified significantly. The intermediate relations among conceptual systems of different abstractions (Fig. 4) and between concepts and objects (Fig. 3) ultimately enable us to reattach conceptual spaces — no matter at what abstraction level they are — and other objects. By these means, we can solve problems at the highest abstraction level that can be generally applied to a large number of situations. Therefore, the conceptual spaces represented in Fig. 4 at different levels should rather be understood as subspaces.

Abstraction levelism is particularly relevant to the purpose of knowledge integration the authors have referred to in a previous work through a conceptual integration diagram (Díaz-Nafría, 2019b; diagram 4) that is supposed to happen in a single step, i.e., between only two levels. Therefore, the current representation corresponds to a more general case in which knowledge integration happens throughout multiple levels.

In addition to the previous typology of conceptual relations, it is also important to consider another fundamental distinction that divides conceptual relations into three groups:

- *Semantic* relations explicate the meaning of the concept.
- *Pragmatic* relations display the intentionality of the concept.
- *Syntactic* relations form the (symbolic) denotat of the concept.

Let us consider some basic inner relations in conceptual spaces of utmost importance for our purposes.

A concept C is more abstract than a concept D if the properties that define the concept C form a part of the properties that define the concept D.

Proposition 1. *The relation "to be more abstract" is transitive.*

A concept C is more general than a concept D if the denotat of the concept D is a subset of the denotat of the concept C.

For instance, the concept real number is more general than the concept whole number. The concept love is more general than friendship.

Proposition 2. *The relation "to be more general" is transitive.*

Often, but not always, more general concepts are also more abstract. For instance, the concept *tree* is more abstract and more general than the concept *pine*. At the same time, the concept *real number* is more general than the concept *whole number*, but it is not more abstract than the concept *whole number*. Both concepts are situated on the same level of abstraction. In a similar way, the concept *red* is more abstract than *red hat* but is not more general.

5 Creating Knowledge Structures for Building Up Transdisciplinary Knowledge Through GlossaLAB

Using the previous characterization of conceptual spaces, we are in a better situation to describe our venture of creating transdisciplinary knowledge through the articulation of Interdisciplinary Glossaries (ID-G), as open platforms for the clarification of concepts by interdisciplinary communities, with the scope of progressively maximizing its generality. As discussed in Díaz-Nafría *et al.* (2018, 2019a, 2019b), ID-G offers a fundamental tool in the process of knowledge integration, "bringing together the different understanding of terms from the summoning of various disciplinary perspectives" (Díaz-Nafríam, 2019b) and at the same time, they are used as proxies to assess the performance achieved in the process the knowledge integration in terms of the conceptual relations discussed above, and particularly the definitions concerning abstraction and generality, which are complementary to the definitions of intensional performance as the "capacity of a conceptual system to refer a knowledge field" given in Díaz-Nafría *et al.* (2019a).

System of federated ID-G: GlossaLAB (Díaz-Nafría, 2022) is conceived as a system of federated ID-G, represented in Fig. 5, each one oriented to different object fields. These ID-G evolve

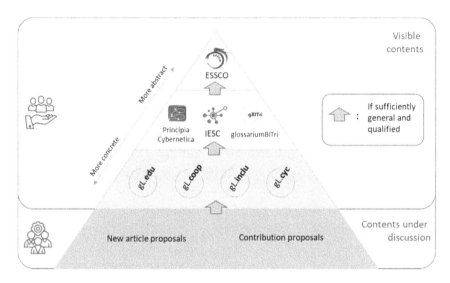

Figure 5. GlossaLAB as a federated system of ID-G, which can be treated as conceptual spaces at different abstraction levels. The more abstract and qualified contents are located at the higher level, ESSCO being at the top, while the lowest level is composed of problem-oriented elucidations with stronger bonds to real-world objects and problems.

autonomously but cooperate in the development of a network of transdisciplinary concepts whose intensional performance (defined in Díaz-Nafría et al., 2019a) is intended to grow over time (Díaz-Nafría et al., 2021). Since the scope, abstraction level, and quality criteria of each ID-G are different, we distinguish different levels of abstraction. The more abstract (and usually more qualified) the contents are, the higher they are located. The lowest level is composed of problem-oriented ID-G and therefore more directly linked to objects, i.e., with stronger intermediate relations. Figure 5 shows some project-specific ID-G named with the prefix gL- and a suffix characterizing the target project, while ESSCO (standing for "Encyclopedia of Systems Science and Cybernetics Online" which inherits and elaborates upon a large corpus of several encyclopedic projects) occupies the highest level of abstraction, responding also to a high-quality level. Below the lowest visible level (i.e., accessible to the public), there is a fundamental ground of proposals where knowledge co-creation is carried out through content proposals and interdisciplinary discussion. The vertical dynamics corresponds to the building up of the transdisciplinary conceptual framework composed of a network of concepts sufficiently abstract. The vertical arrows represent the escalation between levels, provided that the contribution involved satisfies the quality and abstraction criteria.

The vertical dynamics, i.e., the process in which knowledge integration is achieved and abstraction relations — as the ones represented in Fig. 4 — are developed, further explained, and discussed in Díaz-Nafría et al. (2021). Therein, the authors present a category tree elaborated to qualify in an integrative manner, on the one hand, the reliability of intermediate conceptual relations — as the ones represented in Fig. 3 — and on the other, the participation of knowledge domains (using the classification first presented in Díaz-Nafría et al. (2019b). This categorization offers a fundamental asset to evaluate the integration achieved in terms of the diversity and generality assessments using the methodologies defined and discussed in Díaz-Nafría et al. (2018, 2019a, 2019b).

References

Burgin, M. (2012). *Structural Reality.* New York, NY, USA: Nova Science Publishers.
Burgin, M. (2016). *Theory of Knowledge: Structures and Processes.* Singapore: World Scientific.
Burgin, M. and Díaz-Nafría, J. M. (2019). Introduction to the mathematical theory of knowledge conceptualization: conceptual systems and structures. *Commun. Comput. Inform. Sci.* 1051, 469–482. https://doi.org/10.1007/978-3-030-32475-9_34.
Burgin, M. S. and Gorsky, D. P. (1991). Towards the construction of a general theory of concept. In K. Lehrer and E. Sosa (eds.) *The Opened Curtain. A US-Soviet Philosophy Summit.* Boulder, CO, USA: Westview Press, pp. 167–195.
Burgin, M. S. and Semenovich, M. (2017). *Theory of Knowledge: Structures and Process.* Singapore: World Scientific Publishing.
Díaz-Nafría, J. M., Guarda, T., and Coronel, I. (2018). A network theoretical approach to assess knowledge integration in information studies. *Smart Innov. Syst. Technol.* 94, 360–371. https://doi.org/10.1007/978-3-319-78605-6_31.

Díaz-Nafría, J. M., et al. (2019a). GlossaLAB: co-creating interdisciplinary knowledge. *Commun. Comput. Inform. Sci.* 1051, 423–437. https://doi.org/10.1007/978-3-030-32475-9_31.

Díaz-Nafría, J. M., Burgin, M., and Rodríguez-Bravo, B. (2019b). Evaluation of knowledge integration through knowledge structures and conceptual networks. In G. Dodig-Crnkovic and M. Burgin (eds.), *Philos. Methodol. Inform.* Singapore: Word Scientific Publishing, pp. 457–489.

Díaz-Nafría, J. M. et al. (2021). GlossaLAB: enabling the co-creation of interdisciplinary knowledge through the reviving of long-term conceptual elucidation. *Commun. Comput. Inform. Sci.* 1485, 18–33. https://doi.org/10.1007/978-3-030-90241-4_2.

Díaz-Nafría, J. M. (2022). *GlossaLAB: Co-creating Interdisciplinary Knowledge* retrieved March 11, from: http://glossalab.org.

Doignon, J.-P. and Falmagne, J.-C. L. (1999). *Knowledge Spaces.* Germany: Springer, Berlin/Heidelberg.

Gärdenfors, P. (2004). Conceptual spaces as a framework for knowledge representation. *Mind Matt.* 2, 9–27.

Hampton, J. A. (2011). Concepts and natural language. In *Concepts and Fuzzy Logic.* Cambridge, MA, USA: MIT Press, pp. 233–258.

Chapter 11

Extended Probabilities and their Application to Statistical Inference[*]

Michele Caprio[†] and Sayan Mukherjee[‡]

*University of Pennsylvania and
Max Planck Institute for Mathematics in the Sciences,*
[†]*caprio@seas.upenn.edu*
[‡]*sayan.mukherjee@mis.mpg.de*

We propose a new, more general definition of extended probability measures. We study their properties and provide a behavioral interpretation. We put them to use in an inference procedure, whose environment is canonically represented by the probability space $(\Omega, \mathcal{F}, P)$, when both P and the composition of Ω are unknown. We develop an *ex ante* analysis — taking place before the statistical analysis requiring knowledge of Ω — in which the true composition of Ω is progressively learned. We describe how to update extended probabilities in this setting and introduce the concept of lower extended probabilities. We apply our findings to a species sampling problem and to the study of the boomerang effect (the empirical observation that sometimes persuasion yields the opposite effect: the persuaded agent moves their opinion away from the opinion of the persuading agent).

---

[*]This work is dedicated to the memory of Professor Mark Burgin. Without his profound contributions to the field of extended probabilities, this chapter would not have existed.

1 Introduction

Researchers use the terminology "extended probabilities" to refer to set functions whose codomain is either a superset of $[0,1]$ or defined using entirely different number types, such as p-adic numbers (Khrennikov, 2009). They first came up in physics (noted by Burgin (2013) and Heisenberg (1931)), where they are still studied today (Ferrie, 2011; Hartle, 2008; Kronz, 2007). They then became popular in other fields too, including economics and finance (Burgin and Meissner, 2011; Jarrow and Turnbull, 1995), machine learning (Lowe, 2004/2007), stochastic processes (Wiewiora, 2008), and queuing theory (Tijms and Staats, 2007). Of course, they have been extensively studied in mathematics (see, e.g., Allen 1976; Bartlett 1945 for early works, and; Khrennikov, 2009 for more recent ones). A complete account is given in Burgin (2012).

"Extended probabilities" have been given differing definitions and interpretations across — and even within — fields of study. In quantum theory, for example, Benavoli *et al.* (2019) point out that "negative probabilities" do not have intrinsic meaning beyond the fact that they constitute a probabilistic model compatible with quantum events. In Gell-Mann and Hartle (2012) and Hartle (2004, 2008), instead, the authors interpret them as being associated with unsettleable bets (see Section 2.3).

In this chapter, we aim to give a foundational definition of extended probabilities. We give two properties that a set function must have in order to be called an extended probability, irrespective of the field the scholar works in. This represents a great improvement with respect to previous works on the matter, in which definitions depend on the area of study.

We explain how extended probabilities are different from regular Kolmogorovian probabilities, and we characterize some of their more interesting properties. We also give a behavioral interpretation of extended probabilities taking on negative values that complements the frequentist one given in Burgin (2012) to "negative probabilities". Finally, we relate our definition and interpretation to the ones existing in the literature, and we illustrate how these latter can be reconciled with the ones we provide.

After that, we present what is to the best of our knowledge the first application of extended probabilities to an inference procedure. Consider a generic statistical experiment; let Ω be a finite or countable space, and endow it with the sigma-algebra $\mathcal{F} = 2^\Omega$. The space usually adopted to express uncertainty around the elements of $\mathcal{F}$ is the probability space $(\Omega, \mathcal{F}, P)$, for some probability measure $P : \mathcal{F} \to [0,1]$. Now, suppose we want to express further uncertainty regarding either the composition of Ω, or which P to consider on $(\Omega, \mathcal{F})$; we can do so by using lower probabilities. In particular, in the first case, we consider the probability space $(\Omega, \mathcal{F}, P)$ and we follow the example in Gong (2018). There, as a consequence of survey nonresponse (that is, subjects not answering to the questions in a survey), the author is forced to consider $\check{\Omega} = 2^\Omega$. In this case, probabilities cannot be computed exactly: only lower and upper bounds to precise probabilities — lower and upper probabilities, respectively — are available. In the second case, we consider the triple $(\Omega, \mathcal{F}, \mathcal{P})$, where $\mathcal{P}$ is a set of probability measures, and we proceed, e.g., as in Caprio and Gong (2022) and Caprio and Mukherjee (2022).

To the best of our knowledge, there is no cogent way of expressing uncertainty on both the composition of Ω and which P to consider on $(\Omega, \mathcal{F})$. In this chapter, we aim to fill this gap by using extended probabilities. We describe an *ex ante* analysis, meaning one that takes place before the actual statistical analysis that requires the knowledge of the state space. In the most general case, we start from the number type we believe we are working with (naturals, wholes, integers, or rationals) and call it Ω. Then, at time 0 we divide it into an actual space Ω_0^+ that we deem a plausible state space for our experiment, and a latent one Ω_0^-, which we do not know about; notationally, $\Omega = \Omega_0^- \sqcup \Omega_0^+$, where $\sqcup$ denotes a disjoint union of sets. We assign negative extended probabilities to the subsets of Ω_0^-. To capture the uncertainty around which P to consider, we specify a set $\mathcal{P}^{ex}$ of extended probabilities supported on the whole Ω, instead of a single one. We then describe how we progressively discover the true composition of the state space associated with our experiment (which may be the whole Ω we initially specified, or a proper subset $\Omega' \subsetneq \Omega$) and how to update extended probabilities accordingly. We conclude our analysis by discovering the state space associated with our experiment.

In addition, we show that the limiting set of the sequence $(\mathcal{P}_t^{ex})$ of updates of the initially specified set of extended probability measures must be one of the following. It is either a set of regular probability measures, if the state space associated with our experiment is the whole Ω, or a set of extended probability measures, if the state space associated with our experiment is a proper subset Ω' of Ω. In this latter case, the limiting set induces a set of regular probability measures on Ω'.

We also develop the concept of lower and upper extended probabilities. They represent the "boundaries" of a generic set $\mathcal{P}^{ex}$ of extended probabilities and more in general allow for an imprecise elicitation of extended probabilities. They are extremely important because under a mild assumption, knowing lower probability $\underline{P}^{ex}$ is enough to be able to retrieve the whole set $\mathcal{P}^{ex}$. We provide bounds for the lower extended probability $\underline{P}_t^{ex}(A)$ of any element $A \in \mathcal{F}$, at any time t in our *ex ante* analysis. For the sake of completeness, we give the definition of an extended Choquet capacity, and we show how lower and upper extended probabilities are extended Choquet capacities.

In Example 1, we provide a simple application to the field of ecology. We illustrate how the analysis we describe in this chapter can be put to use in a species sampling problem, specifically to retrieve the number of birds that inhabit a certain region throughout the year.

We also provide an example — adapted from the one in Allahverdyan and Galstyan (2014) — in the field of opinion dynamics; in particular, we describe the boomerang effect. We have a persuading agent acting on a persuaded agent, but the latter perceives the former as having low credibility. This can be modeled so that the persuaded agent does not know the composition of the entire state space, while she suspects the persuading agent does: she thinks the persuading agent may be hiding something form her. As she discovers the true composition of the state space, the credibility of the persuading agent is restored.

This chapter is organized as follows: in Section 2, we give the foundational definition of an extended probability, its properties, and its behavioral interpretation. In Section 3, we use extended probabilities in an inference procedure. Section 4 deals with lower extended probabilities, and Section 5 concludes our work. In Appendix A.1, we give the proofs of our results. In Appendix A.2 we give the opinion dynamics example, and in Appendix A.3, we report two interesting quotes from de Finetti (1974).

Remark 1. To deal with uncertainty in the composition of Ω, one could proceed as in Walley (1991, Section 4.3.3). The agent could begin the elicitation by specifying the state space Ω_0 and then, as they analyze the problem in greater detail, could realize that a refinement to a finer-grained Ω_1 is needed. This corresponds to specifying what Dempster (1967) calls a multivalued mapping from Ω_0 to Ω_1; that is,

$$A : \Omega_0 \rightrightarrows \Omega_1, \quad \omega_0 \mapsto A(\omega_0) \subset \Omega_1.$$

So Ω_0 corresponds to a partition of Ω_1, and each state $\omega_0 \in \Omega_0$ can be identified with the set $A(\omega_0)$ of "refined possibilities" in Ω_1. To illustrate the complication deriving from this approach, let Ω_0 and Ω_1 be finite or countable, and call $\mathcal{F}_0 = 2^{\Omega_0}$ and $\mathcal{F}_1 = 2^{\Omega_1}$. If the agent specifies a probability measure P_0 on $(\Omega_0, \mathcal{F}_0)$, they then need to come up with a probability measure P_1 on $(\Omega_1, \mathcal{F}_1)$ such that

$$P_0(\{\omega_0\}) = \sum_{\omega_1 \in A(\omega_0)} P_1(\{\omega_1\})$$

holds for all $\omega_0 \in \Omega_0$. This means that a new (subjective) probability elicitation must take place once the agent refines the state space to Ω_1. This can be avoided using extended probabilities, as we shall argue in Section 3.

It is worth noting that while this multivalued mapping approach is pointed out in Dempster (1967) and Walley (1991), the authors sidestep its associated complications by means of probability bounding. Furthermore, in Walley (1996) too the author tries to solve the incompletely specified possibility space problem by probability bounding; this enables progressive learning of the sample space in a way that representation invariance as well as symmetry are satisfied. We verify whether the model that we introduce in Section 3 meets these conditions in future work.

2 Extended Probability Measures

In this section, we first illustrate the philosophical reason to introduce extended probabilities, and then we dive into more technical details. We conclude with a thorough analysis on how our interpretation of extended probabilities relates to the existing literature.

2.1 *Philosophical motivation for extended probabilities*

We give a behavioral interpretation of extended probabilities. Consider a generic event A, and suppose we want to express our belief about the likelihood of it taking place. Suppose we can enter a bet about A that gives us \$1 if event A happens and \$0 if it does not happen. Then, we say that the probability we attach to A is given by $p \geq 0$, the supremum buying price as well as the infimum selling price for the bet about A (we call it *fair price*). That is, for every $\varepsilon > 0$, we are willing to enter a bet which gives $1 - p + \varepsilon$ if A happens, and $-p + \varepsilon$ if A does not happen, as well as the bet which gives $p - 1 + \varepsilon$ if A happens and $p + \varepsilon$ if A does not happen. This interpretation is inspired by the classical subjective probability interpretation given by de Finetti (1974, 1975). As pointed out in Nau (2001), the other two fathers of subjective probability theory, Ramsey (1964) and Savage Savage (1954), simultaneously introduced measurement schemes for utility. They tied their definitions of probability to bets in which the payoffs were effectively measured in utiles rather than dollars. In this way, they obtained probabilities that were interpretable as measures of pure belief, uncontaminated by marginal utilities for money. De Finetti later admitted that it might have been better to adopt the seemingly more general approach of Ramsey and Savage since it leads to a theory of decision-making that does not rely on monetary values. Nevertheless, he found other reasons for preferring the money bet approach. In particular, he maintained that it would be extremely difficult to settle bets based on utiles because the monetary sums needed to settle them would need to be adjusted to the complex variations in a unit of measure (utiles) that is unobservable.[1]

---

[1] The complete quotes from de Finetti (1974) can be found in Appendix A.3.

This is why we retain the de Finettian interpretation of (subjective) probability, and more in general why we adopt a betting scheme approach.

Suppose now that we are not given the possibility to enter a bet like the aforementioned one, so we cannot assess a probability for A as before. Instead, such a possibility is given to our doppelgänger, who tells us that their subjective assessment for the probability of A is some $q \in [0,1]$. Here, it is assumed that the doppelgänger assigns the same probabilities as ourselves to all the events we both can enter a bet about. This procedure of asking the doppelgänger is equivalent to setting the probability of A ourselves; we introduce the doppelgänger — a logical artifice — because, given our interpretation of probability, it is impossible to elicit the probability of event A if we cannot enter a bet similar to the one described before. We conclude that if we were given the opportunity to enter the bet about A, the amount of money we would deem fair to pay would be q dollars. Therefore we are prepared to lose $\$q$ in the case A^c happens.[2] We express this by setting $p = -q$.

If after a while we are given the opportunity to enter a bet about an event A whose extended probability we deemed to be $-q < 0$, the price we consider fair to pay need not be $p = |-q|$. If $p \neq |-q|$, it means that we changed idea about how likely event A is once given the possibility of entering the bet. If instead $p = |-q|$, we are obeying to Allen's principle of conservation of knowledge Allen (1976). It states that, like the law of conservation of mass, knowledge is neither created nor destroyed, but only transformed; its total possible amount is constant. By having us choose the probability equal in absolute value, we comply with conservation of knowledge. As we can see, this principle allows us to mechanically retrieve positive probabilities starting from negative ones. It also allows us to work backwards to negative probabilities starting from positive ones. Suppose we have an event $A \in \mathcal{F}$ to which we attach probability $p \geq 0$. Then, we know that if we were not given the opportunity to enter the bet that allowed us to indicate p as the probability that event A

---

[2]Provided that we gain $\$(1-q)$ if A happens.

happens, then we would have expressed our uncertainty by assigning A probability $-p \leq 0$.

A Dutch book is, informally, the possibility of constructing a bet such that the bookmaker always profits, while the punter always loses money (see Remark 2 for a formal definition). In de Finetti (1937), a coherent subjective probability is defined as one that does not allow for a Dutch book to be made against the punter, however a bet is made. Necessary and sufficient conditions for coherence require that subjective probabilities satisfy the Kolmogorovian axioms of probability (with only finite additivity). As we shall see in Section 2.2, this does not hold for extended probabilities. This is not a problem though, in light of the interpretation we give to negative probabilities. The events whose attached probabilities are negative are events for which the punter cannot enter a bet; hence, they cannot be used to build a Dutch book. We give the definition of coherence in the context of extended probabilities and the formal statement that extended probabilities are always coherent in Remark 2. In addition, as we show in Section 3, the inferential procedure we consider is such that, starting with extended probabilities, we recover — at the end of our analysis — regular probabilities. This also ensures that no Dutch books can be created.

2.2 *Technical definition and properties*

Consider a measurable space $(\Omega, \mathcal{F})$, where $\mathcal{F} \subset 2^\Omega$ is a sigma-algebra. An extended probability $P^{ex} : \mathcal{F} \to \mathbb{R}$ is a set function on $\mathcal{F}$ such that we have the following:

(i*) $P^{ex}(A) \in [-1, 1]$, for all $A \in \mathcal{F}$.

(ii*) If $\{A_j\}_{j \in I}$ is a countable collection of disjoint events such that $\cup_{j \in I} A_j \in \mathcal{F}$, then

$$P^{ex}\left(\bigcup_{j \in I} A_j\right) = \sum_{j \in I} P^{ex}(A_j).$$

(iii*) Given any partition $\mathcal{E} = \{E\}$ of Ω such that $E \in \mathcal{F}$, for all $E \in \mathcal{E}$, $\sum_{E \in \mathcal{E}} |P^{ex}(E)| = 1$.

Condition (iii*) is needed in light of the interpretation we gave in Section 2.1 to extended probabilities: if (iii*) were not to hold, the doppelgänger's opinion would be represented by a signed measure, not a probability measure.

Let $\mathcal{F} \subset 2^\Omega$ be the same sigma-algebra as before. The Kolmogorovian axioms for any regular probability measure $P : \mathcal{F} \to \mathbb{R}$ are the following:

(K1) $P(A) \in [0,1]$, for all $A \in \mathcal{F}$.
(K2) If $\{A_j\}_{j \in I}$ is a countable collection of disjoint events such that $\cup_{j \in I} A_j \in \mathcal{F}$, then $P(\cup_{j \in I} A_j) = \sum_{j \in I} P(A_j)$.[3]
(K3) $P(\Omega) = 1$.

It is easy to see, then, how conditions (i*) and (iii*) are more general than (K1) and (K3), respectively. To this extent, extended probabilities are a generalization of the concept of regular probabilities. From its definition, we see that an extended probability is a finite signed measure. We call the triple $(\Omega, \mathcal{F}, P^{ex})$ an *extended probability space*.

Since we do not require $P^{ex}(\Omega) = 1$, the extended probability of the complement of an event A is given by

$$P^{ex}(A^c) = P^{ex}(\Omega \setminus A) = P^{ex}(\Omega) - P^{ex}(A). \tag{1}$$

Equation (1) comes from (ii*); indeed, consider the disjoint sets $\Omega \setminus A$ and $\Omega \cap A = A$. Then,

$$P^{ex}(\Omega) = P^{ex}\left([\Omega \setminus A] \sqcup A\right) = P^{ex}(\Omega \setminus A) + P^{ex}(A)$$
$$\iff P^{ex}(\Omega \setminus A) = P^{ex}(\Omega) - P^{ex}(A).$$

Equation (1) ensures us that $P^{ex}(\emptyset) = P^{ex}(\Omega^c) = P^{ex}(\Omega) - P^{ex}(\Omega) = 0$. We also have that $P^{ex}(\Omega) = \sum_{E_j \in \mathcal{E}} P^{ex}(E_j)$, where $\mathcal{E} = \{E_j\}$ is any finite or countable partition of Ω.

Proposition 1. *Extended probabilities have the adequacy property: for all $A, B \in \mathcal{F}$ such that $A = B$, then $P^{ex}(A) = P^{ex}(B)$.*

We also have that extended probabilities retain a version of the monotonic property of regular probabilities.

---

[3]We consider the countably additive version of Kolmogorov axioms.

Proposition 2. *For all $A, B \in \mathcal{F}$ such that $A \subset B$, if $P^{ex}(A) \geq 0$, $P^{ex}(B) \geq 0$, and $P^{ex}(B \cap A^c) \geq 0$, then $P^{ex}(A) \leq P^{ex}(B)$. If instead $P^{ex}(A) \leq 0$, $P^{ex}(B) \leq 0$, and $P^{ex}(B \cap A^c) \leq 0$, then $P^{ex}(A) \geq P^{ex}(B)$.*

This version of the monotonic property implies a version of the continuity property enjoyed by regular probabilities.

Corollary 1. *If we have a collection $\{A_j\}$ of elements of $\mathcal{F}$ such that*

- *it is nested (i.e., $A_1 \supset A_2 \supset \cdots \supset A_n \supset \cdots$),*
- *$\cap_j A_j = \emptyset$,*
- *$P^{ex}(A_j) \geq 0$, for all A_j,*

then $\lim_{j \to \infty} P^{ex}(A_j) = P^{ex}(\cap_j A_j) = 0$.

Extended probabilities satisfy the inclusion–exclusion principle.

Proposition 3. *The following is true:*

$$P^{ex}(A \cup B) = P^{ex}(A) + P^{ex}(B) - P^{ex}(A \cap B),$$

for all $A, B \in \mathcal{F}$.

This implies immediately that, for any $A, B, C \in \mathcal{F}$,

$$P^{ex}((A \cup B) \cap C) = P^{ex}((A \cap C) \cup (B \cap C))$$
$$= P^{ex}(A \cap C) + P^{ex}(B \cap C) - P^{ex}(A \cap B \cap C),$$

by Proposition 3 and the De Morgan laws.

We define the extended conditional probability of A given B to be the extended counterpart of regular conditional probabilities:

$$P^{ex}(A \mid B) := \frac{P^{ex}(A \cap B)}{P^{ex}(B)} \in [-1, 1], \tag{2}$$

for all $A, B \in \mathcal{F}$ such that $P^{ex}(B) \neq 0$.[4]

---

[4]Note that in de Finetti's approach de Finetti (1974, 1975), a similar formula follows from the "called off bet" interpretation of (regular) conditional probability. In the future, we plan to provide a direct behavioral motivation for (2), in the spirit of de Finetti's work.

To avoid confusion arising from the sign, we say that $A, B \in \mathcal{F}$ are independent if and only if

$$|P^{ex}(A \cap B)| = |P^{ex}(A) \times P^{ex}(B)|.$$

This way of describing independence corresponds to the one given in Allen (1976, Section 2).

Remark 2. Consider any extended probability space $(\Omega, \mathcal{F}, P^{ex})$. Call $\mathscr{B}$ the sigma-algebra generated by the events B in Ω that we can enter a bet about, that is, the sigma-algebra generated by the collection $\{B \subset \Omega : P^{ex}(B) \geq 0 \text{ and } B \cap C = \emptyset, \forall C \in \mathcal{F}, P^{ex}(C) < 0\}$. Then, a *Dutch book* is a finite collection $\{B_j\}_{j=1}^n \subset \mathscr{B}$ along with numbers $\{s_j\}_{j=1}^n \subset \mathbb{R}$ such that

$$\sup_{\omega \in \Omega} \mathfrak{f}(\omega) := \sup_{\omega \in \Omega} \left\{ \sum_{j=1}^n s_j \left[\mathbb{1}_{B_j}(\omega) - P^{ex}(B_j) \right] \right\} < 0. \quad (3)$$

Note that the s_j's are the payout for a winning bet on B_j, and $s_j P^{ex}(B_j)$ is the bet's fair buy-in. The inequality in (3) suggests that a bet can be built such that the bookmaker always gets a profit, while the punter always loses money.

Definition 1. Extended probability P^{ex} is coherent if no Dutch books can be made against the punter.

Then, we have the following important result.

Theorem 1. P^{ex} is always coherent.

In our definition of Dutch book, we do not take into account the events $C \in \mathcal{F}$ for which $P^{ex}(C) < 0$ since those are events for which the punter cannot enter a bet. Hence, they cannot be used to build a Dutch book.

2.3 Related literature

Let us now inspect how the definition and the interpretation of extended probabilities we have given so far relates to the existing literature.

Our framework is related to the one studied in Burgin (2012). Burgin studies a static environment (as opposed to the dynamic one we inspect in this work in the next section) where the state space Ω can be divided in two irreducible parts Ω^+ and Ω^- such that $\#\Omega^+ = \#\Omega^-$, where $\#$ denotes the cardinality operator. The elements of Ω^- are called anti-events, and they are usually connected to negative objects: encountering a negative object is a negative event. The author then states that an example of a negative object is given by anti-particles, the anti-matter counterpart of quantum particles. Anti-events are given negative probabilities. If we require that the following conditions hold, we obtain a framework very similar to Burgin's one:

- Extended probabilities are finitely additive (instead of countably additive).
- Our state space can be divided in two irreducible parts Ω^+ and Ω^-.
- $P^{ex}(A) \geq 0$ if and only if $A \subset \Omega^+$.
- There exists a function $\alpha : \Omega \to \Omega$ such that $\alpha(\omega) = -\omega$ and $\alpha^2(\omega) = \omega$.
- $P^{ex}(\Omega^+) = 1$.
- $\{v_i, \omega, -\omega : v_i, \omega \in \Omega, i \in I\} = \{v_i : v_i \in \Omega, i \in I\}$, for all $\omega \in \Omega$ and all set of indices I.

The main difference is that in Burgin (2012) $P^{ex}(\Omega^+) + |P^{ex}(\Omega^-)| = 2$. Our interpretation of negative probabilities reconciles with Burgin's one if we consider anti-events as events for which we cannot enter a bet, which seems a reasonable assumption. He also gives a frequentist interpretation of negative probabilities that complements our interpretation of negative extended probabilities.

In a subsequent paper (Burgin, 2013), Burgin generalizes his own setup, mainly by not requiring $\#\Omega^+ = \#\Omega^-$ and by allowing any event to have either positive or negative probability, depending on external conditions. If we require that the following conditions hold, we obtain Burgin's generalized framework:

- Extended probabilities are finitely additive.
- $A \in \mathcal{F}$ implies $-A := \{-\omega : \omega \in A\} \in \mathcal{F}$.
- For all $A \in \mathcal{F}$, $P^{ex}(-A) = -P^{ex}(A)$.

Note that this last condition is very similar to Allen's principle of conservation of knowledge. The interpretation he gives for negative probabilities is similar to the one given in his previous work, with the peculiarity that now negative probabilities are not assigned exclusively to anti-events. Our interpretation can be seen as being a step behind Burgin's one: we give a specific reason for why an event $\tilde{A}$ is assigned a negative probability, namely that we cannot enter a bet on it. Then, Burgin states that the anti-event of $\tilde{A}$, $-\tilde{A}$, will have a positive (regular) probability. Of course, the vice versa holds: if an event is assigned a positive probability (because we can enter the bet), then according to Burgin, its anti-event will have a negative probability.

Another interesting interpretation is given by Székely in Székely (2005). The author proves that we can encounter a negative probability if we work with a random variable having a signed distribution. In addition, if X has a signed distribution, then there exist two random variables Y, Z having an ordinary (not signed) distribution such that $X + Y = Z$ in distribution. Therefore, X can be seen as a "difference" of two ordinary random variables Z and Y. We reconcile our interpretation and Székely's one as follows. A signed distribution P^s is simply an extended probability on the space of outcomes of a random variable. A pullback argument can then be used to define an extended probability P^{ex} on Ω: $P^{ex}(X^{-1}(I)) = P^s(I)$, for all subsets I of the outcome space of random variable X. So there are going to be events in Ω having negative probabilities: those are events we cannot enter a bet about.

The interpretation Kronz (2007) gives of negative probabilities is the following: he calls negative probabilities inferred probabilities that can only be obtained indirectly by inference from operational (regular) probabilities. The associated events (that is, events having negative probabilities) are called virtual events in that they are non-operational and so do not give rise to directly accessible relative frequencies. Actual events have non-negligible effect on them, and the reverse is also true. As it appears clear, virtual events are equivalent to latent events, e.g., in the psychology (Bollen, 2002), economics (Hu, 2017) and medicine (Rabe-Hesketh and Skrondal, 2008) literatures: Kronz assigns negative probabilities to latent events. These latter are events we do not observe, so it is fair to think we cannot enter a bet involving them.

In Wigner (1932), Wigner allows probabilities to go negative to sidestep the uncertainty principle of quantum mechanics. This latter states that given a particle that moves in one dimension, there is a limit to the precision with which position x and momentum p can be determined simultaneously in a given state. The observer can know the distribution of x and that of p, but there is no joint probability distribution of (x,p) with the correct marginals of x and p. To overcome this shortcoming, Wigner came up with a quasiprobability distribution $W_\psi(x,p)$, that is, a countably additive function on the measurable subsets of state space $\Omega = \mathbb{R}^2$ such that $W_\psi(\Omega) = 1$. Wigner's function $W_\psi(x,p)$ is a signed probability distribution; all values of $W_\psi(x,p)$ are real, but some values may be negative. In Blass and Gurevich (2022), the authors show that Wigner's quasiprobability distribution is the unique signed probability distribution yielding the correct marginal distributions for position and momentum and all their linear combinations, and in Gurevich and Vovk (2021), the authors show that Wigner's function can be tested using the canonical approach of game-theoretic probability. We can give Wigner's function two interpretations. The first one is purely "utilitarian": because it is the unique quasiprobability distribution that gives the correct marginal distributions for x and p, scientists should use it even if it allows for probabilities being negative-valued. The second one can be reconciled with the interpretation we give in the present work to negative extended probabilities. Because a scientist cannot observe x and p simultaneously, it is fair to think that they cannot enter a bet about events involving both position and momentum of a particle; hence, they resort to negative probabilities.

In Benavoli et al. (2021), the authors study bounds on the algorithmic capabilities of a mathematical theory and analyze their implications. In particular, they require that the theory is logically consistent — that is, it has to be based on a few axioms and rules from which mathematical truths can be unambiguously derived — and that inferences in the theory should be computable in polynomial time — that is, there should be an efficient way to execute the theory. The postulates of consistency and computation are apparently in conflict with each other: intuitively, if only polynomial time computations are allowed, the theory is consistent only up to what polynomial calculus allows, an instance that the authors call external–internal clash. They formalize such a clash via a "weirdness theorem", which shows that any theory obeying their two postulates necessarily departs in a very peculiar way from the probabilistic point of view. In particular, the theorem proves that all models compatible with the theory will present some negative probabilities (quasiprobabilities à la Wigner). They also show that quantum paradoxes are a consequence of the weirdness theorem. To develop their results, the authors rely on a characterization of probability in terms of lotteries (or gambles, a particular type of bets). They then provide a subjective foundation, à la de Finetti, of so-called generalized probability theories. Negative probabilities arise when an agent would like to enter a given bet involving an event, but computational limitations related to the event prevent them from doing so. This interpretation coincides with the one we give in the present chapter: the authors give a reason for which an agent is denied the possibility of entering a bet.

Finally, the interpretation of negative probabilities in Gell-Mann and Hartle (2012) and Hartle (2004, 2008) is very similar — although not identical — to the one in this chapter. We give the summary of their way of interpreting negative probabilities as reported in Feintzeig and Fletcher (2017). First, they point out that the probabilities dictated by a physical theory instruct (rational) agents on how to bet on the outcomes of phenomena. Standard arguments from subjective Bayesian probability theory — the Dutch book arguments — demand that the Kolmogorovian axioms for classical probability theory must hold for the degrees of belief of any rational agent, which determine which bets the agent regard as fair. However, they point out that these arguments only apply to bets which are settleable, that is, bets about events the agent will certainly know at some point. They then argue that only bets on sufficiently coarse-grained alternative histories will be settleable, where this coarse-graining guarantees that the alternatives' probabilities will lie in the unit interval. The main difference between their interpretation and the one presented in this chapter is that for Gell-Mann and Hartle the bets on events that have negative probabilities cannot be settled, whereas we deem those bets to be settleable. The agent simply cannot enter them at the moment, but they may be given the opportunity in the future.

3 Extended Probabilities in Statistical Inference

In this section, we are going to consider an *ex ante* analysis in which we progressively learn the composition of the state space. It is *ex ante* in that it takes place before the actual statistical analysis. As time goes by, we collect new observations via a learning procedure specified in advance, e.g., an urn with or without replacement.

Consider a measurable space $(\Omega, \mathcal{F})$ with Ω at most countable and $\mathcal{F} = 2^\Omega$. At any time t, we have $\Omega = \Omega_t^- \sqcup \Omega_t^+$. For all $t \in \mathbb{N}_0 := \mathbb{N} \cup \{0\}$, Ω_t^- represents the "latent" part of Ω at time t, that is, the part that we have not yet observed. Ω_t^+ represents the "actual" part of Ω, that is, the portion of Ω that we have observed at time t (at time $t = 0$, Ω_0^+ is the portion of Ω that we know *ex ante*, which is assumed non-empty). This approach is similar to the "*ex ante* humility" one introduced in Allen (1976, Section 1), where latent and actual portions of the state space are first introduced, and a dynamic is described. A graphical representation of $\Omega = \Omega_t^- \sqcup \Omega_t^+$ is given in Fig. 1. At time $t = 0$, we specify the finest possible partition of Ω, $\mathcal{E} = \{\{\omega\}\}_{\omega \in \Omega}$. It is such that $\mathcal{E} = \mathcal{E}_0^- \sqcup \mathcal{E}_0^+$; in this notation,

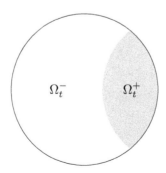

Figure 1. Graphical representation of $\Omega = \Omega_t^- \sqcup \Omega_t^+$.

$\mathcal{E}_0^-$ partitions Ω_0^-, and $\mathcal{E}_0^+$ partitions Ω_0^+. The most general way of proceeding is by specifying the number type (natural numbers $\mathbb{N}$, whole numbers $\mathbb{N}_0$, integers $\mathbb{Z}$, or rationals $\mathbb{Q}$); we subjectively believe we are working with, and setting it to be Ω. To this extent, we give to Ω the apparently possible interpretation of Walley (1991, Section 2.1.2): Ω is the space of apparently possible states if it contains all the states ω that we believe are logically consistent with our available information. To give an example, a space of apparently possible states associated with a coin toss is

$\Omega = \{$heads, tails, coin landing on its edge, coin breaking into pieces on landing, coin disappearing down a crack in the floor$\}$.

Cosidering the space of apparently possible states, then, amounts to considering the broadest state space associated with the statistical experiment of interest.

After that, we subjectively specify what we initially think the state space is, and we set it to be Ω_0^+; for example, we may have previous information coming from similar (but not equal) experiments.

Given the sequences (Ω_t^+) and (Ω_t^-), we require that, for all t,

$$\Omega_t^+ \subset \Omega_{t+1}^+ \quad \text{and} \quad \Omega_t^- \supset \Omega_{t+1}^-.$$

This means that (Ω_t^+) is monotone nondecreasing, and (Ω_t^-) is monotone nonincreasing, which implies that the limits of both exist and are well defined. In particular, we have that

$$\lim_{t \to \infty} \Omega_t^+ = \bigcup_{t \in \mathbb{N}_0} \Omega_t^+ \quad \text{and} \quad \lim_{t \to \infty} \Omega_t^- = \bigcap_{t \in \mathbb{N}_0} \Omega_t^-.$$

We then have two possible scenarios. In the first one, $\bigcup_{t \in \mathbb{N}_0} \Omega_t^+ = \Omega$, which implies that $\bigcap_{t \in \mathbb{N}_0} \Omega_t^- = \emptyset$. In the second one, $\bigcup_{t \in \mathbb{N}_0} \Omega_t^+ \equiv \Omega' \subsetneq \Omega$, which implies that $\bigcap_{t \in \mathbb{N}_0} \Omega_t^- \equiv \Omega'' := \Omega \setminus \Omega' \neq \emptyset$.

As we collect more and more observations, we progressively discover the composition of our sample space. In the first scenario, in the limit we discover that the sample space associated with our experiment corresponds to the whole set Ω we initially specified. In this scenario, at time $t+1$ when an element $\tilde{\omega}$ is observed that belongs to Ω_t^-, the element $E_{t,\tilde{\omega}}^- = \{\tilde{\omega}\}$ of the partition $\mathcal{E}_t^-$ consistent with $\tilde{\omega}$ becomes an element of $\mathcal{E}_{t+1}^+$. We use the following notation: $E_{t+1,\tilde{\omega}}^+ = E_{t,\tilde{\omega}}^-$. So the partition of the latent space loses an element, while the partition of the actual space gains one. This means that, for all t, $\mathcal{E}_t^+ \subset \mathcal{E}_{t+1}^+$ and $\mathcal{E}_t^- \supset \mathcal{E}_{t+1}^-$. We abuse the notation: we write $\mathcal{E}_t^+ \subset \mathcal{E}_{t+1}^+$ to indicate that $\#\mathcal{E}_t^+ \leq \#\mathcal{E}_{t+1}^+$, where $\#\mathcal{E}_t^+$ denotes the number of elements of the partition $\mathcal{E}_t^+$, for all t, and we write $\mathcal{E}_t^- \supset \mathcal{E}_{t+1}^-$ to indicate that $\#\mathcal{E}_t^- \geq \#\mathcal{E}_{t+1}^-$. Hence, in the limit, the partition $\mathcal{E}_t^+$ of the "actual" space coincides with the partition $\mathcal{E}$ of the whole set Ω. This mirrors the behavior of Ω_t^+, that converges to Ω.

In the second scenario, Ω' can be finite or countable. In the former case, there exists a $T \in \mathbb{N}$ after which the observations we collect at time $T+i$ already belong to Ω_T^+, for all $i \in \mathbb{N}$. If that is the case, we write

$$\Omega' \equiv \Omega_T^+. \tag{4}$$

This corresponds to discovering that the actual sample space is finite and smaller than the whole set Ω that we specified at the beginning of our analysis. If Ω' is countable, we can only say that it is a proper subset of Ω, and that we discover its composition in the limit. This may happen, for example, if we begin our analysis by setting $\Omega = \mathbb{N}_0$ but then realize that the state space associated with our experiment is actually $\Omega' = \mathbb{N}$.

In general, when the second scenario takes place, we have $\Omega_t^+ \uparrow \Omega'$, and $\Omega_t^- \downarrow \Omega'' := \Omega \setminus \Omega'$. We also have that, in the limit, the partition $\mathcal{E}_t^+$ of the "actual" space coincides with the (finest possible) partition $\mathcal{E}'$ of set Ω'. We call $\mathcal{E}''$ the partition whose elements belong to $\mathcal{E}$ but not to $\mathcal{E}'$; again abusing notation, we write $\mathcal{E}'' := \mathcal{E} \setminus \mathcal{E}'$. We call $\mathcal{F}' = 2^{\Omega'}$ and $\mathcal{F}'' = 2^{\Omega''}$.

Remark 3. In this chapter, our *ex ante* analysis is concluded by finding the true composition of the state space. We can interpret this using the concept of a benevolent bookmaker. While our doppelgänger is always able to enter a bet on all the events in the space Ω of apparently possible states, we end up only able to bet on the events of Ω that are crucial to the statistical analysis taking place after our *ex ante* analysis. This is akin to a benevolent bookmaker preventing us from entering bets on irrelevant events, that, if we were to bet on, would certainly make us lose money.[5]

However, there may be cases in which the state space does not get fully discovered. For example, this may happen if we have an urn with a timer attached; once the time runs out, the urn is sealed so that we do not discover its entire composition. The following statistical analysis, which would require the use of extended probabilities, is explored in a future work. This case corresponds to the existence of a malevolent bookmaker that does not allow us to enter bets that are crucial to the statistical analysis.[6] This setting is especially important when studying events about which we will never be able to enter a bet, for example latent events.

Remark 4. If at any time t we collect an observation $\breve{\omega}$ that does not belong to Ω, this means that the space Ω we initially specified is not rich enough. We have then to specify a richer, larger set $\breve{\Omega} \supset \Omega$, and start our analysis over. We can either consider $\breve{\Omega} = \Omega \cup \{\breve{\omega}\}$, or define $\breve{\Omega}$ as a larger number type. Let us give an example. Suppose we begin our analysis by setting $\Omega = \mathbb{N}$, and after a while we observe $\breve{\omega} = 1/2$. Then, we have to restart our analysis and we can either let $\breve{\Omega}$ be $\mathbb{N} \cup \{1/2\}$ or $\breve{\Omega} = \mathbb{Q}$.

It may also happen that our true sample space is $\breve{\Omega} \subsetneq \Omega_0^+$. In this case, Equation (4) holds with $T = 0$, and our analysis would still be valid. The drawback is that, if that happens, we are not respecting one of the conditions listed in Tsitsiklis (2018) for a sample space to be valid. In particular, that the sample space Ω must have the right granularity depending on what we are interested in. This means that we must remove irrelevant information from the sample space. In other words, we must choose the right abstraction and forget irrelevant information. This issue can be avoided by initially specifying Ω_0^+ so that it has the fewest possible elements.

---

[5]A more correct expression would be "almost benevolent" since we are considering only one-way bets.

[6]Here, "malevolence" has to be understood in terms of restricting the menu of available options to the agent.

On top of dealing with uncertainty on the composition of the sample space, we also address the problem of not being able to specify a unique (extended) probability measure on Ω. That is why we are going to work with sets of extended probability measures. We call this approach *extended sensitivity analysis*, since it corresponds to the extended probabilities counterpart of Bayesian sensitivity analysis (Berger, 1984). We begin by considering a set $\mathcal{P}^{ex} \equiv \mathcal{P}^{ex}_0$ of extended probabilities that represent the agent's initial beliefs, and we update it as described in Section 3.2. We denote the sequence of successive updates of $\mathcal{P}^{ex}_0$ as $(\mathcal{P}^{ex}_t)_{t \in \mathbb{N}}$. By working with sets of extended probability measures, we represent the condition of a researcher facing a decision under ambiguity (Ellsberg, 1961). As we see in Section 3.2, sequence $(\mathcal{P}^{ex}_t)_{t \in \mathbb{N}_0}$ converges in the Hausdorff metric.

Now, fix any $t \in \mathbb{N}_0$, and consider any $P^{ex}_t \in \mathcal{P}^{ex}_t$. We require the following:

(i) $P^{ex}_t(A) \in [0,1]$ if $A \subset \Omega^+_t$;
(ii) $P^{ex}_t(A) \in [-1,0]$ if $A \subset \Omega^-_t$;
(iii) $P^{ex}_t(A) \in [-1,1]$ if $A \cap \Omega^+_t \neq \emptyset \neq A \cap \Omega^-_t$.

In particular, we compute the latter as follows.

Proposition 4. *The following is true:*

$$P^{ex}_t(A) = \sum_{E^+_{t,j} \in \mathcal{E}^+_t : P^{ex}_t(E^+_{t,j}) \neq 0} P^{ex}_t(A \mid E^+_{t,j}) P^{ex}_t(E^+_{t,j}) \\ + \sum_{E^-_{t,j} \in \mathcal{E}^-_t : P^{ex}_t(E^-_{t,j}) \neq 0} P^{ex}_t(A \mid E^-_{t,j}) P^{ex}_t(E^-_{t,j}). \quad (5)$$

Note that we need to consider the elements of $\mathcal{E}^+_t$ and $\mathcal{E}^-_t$ whose extended probabilities are not 0 otherwise the conditional extended probabilities $P^{ex}_t(A \mid E^+_{t,j})$ and $P^{ex}_t(A \mid E^-_{t,j})$ may result in an indeterminate form of the $\frac{0}{0}$ kind. This requirement yields no loss of generality since if an element of a partition is assigned extended probability 0, then it does not convey any information around event A. Note also that, for the elements of $\mathcal{E}^+_t$ and $\mathcal{E}^-_t$ whose extended probabilities are not 0, $P^{ex}_t(A \mid E^+_{t,j}) = \frac{P^{ex}_t(A \cap E^+_{t,j})}{P^{ex}_t(E^+_{t,j})} \geq 0$ by (i), and

$P_t^{ex}(A \mid E_{t,j}^-) = \frac{P_t^{ex}(A \cap E_{t,j}^-)}{P_t^{ex}(E_{t,j}^-)} \geq 0$ because it is the ratio of two negative quantities; once multiplied by $P_t^{ex}(E_{t,j}^-)$, which is negative by (ii), it gives us a negative value. So the sign of $P_t^{ex}(A)$ is not predetermined when $A \cap \Omega_t^+ \neq \emptyset \neq A \cap \Omega_t^-$. The interpretation of these conditions is straightforward: we assign negative extended probabilities to events that belong to the latent space at time t (meaning that at time t we cannot enter a bet about them), while we assign positive extended probabilities to events that are in the actual, observed space (meaning that at time t we can enter a bet about them). If a given event is only partially known, its extended probability has not a predetermined sign: it will depend on whether we know enough about it (then the probability will be positive), or not (vice versa). This means that we can enter a bet about "sub-event" $A \cap \Omega_t^+$, but not about $A \cap \Omega_t^-$. For example, let $A = \{$tomorrow there will be a thunderstorm$\}$. Then, for some t, suppose that

$A \cap \Omega_t^+ = \{$tomorrow will rain$\}$,

$A \cap \Omega_t^- = \{$tomorrow there will be a dry thunderstorm$\}$.

Then we can place a bet on $A \cap \Omega_t^+$ but not on $A \cap \Omega_t^-$, so the extended probability we assign to A does not have a predetermined sign.

Note also that we can define a set of events $\mathscr{C}_{P_t^{ex}} := \{A \in \mathcal{F} : P_t^{ex}(A \cap \Omega_t^-) = -P_t^{ex}(A \cap \Omega_t^+)\}$, which we call critical events according to P_t^{ex}, with the property that $P_t^{ex}(A) = 0$, for all $A \in \mathscr{C}_{P_t^{ex}}$ (immediate from the definition). We assign the sub-event we can enter a bet about the same probability that our doppelgänger assigns to the sub-event that we are not allowed to bet on. This means that we deem their "actual" portion (the one we know/we have observed so far) to be just as likely than their "latent" portion (the one we do not know/we have not yet observed). The set of critical events according to the whole set of extended probability measures $\mathcal{P}_t^{ex}$ is given by

$$\mathscr{C}_t := \bigcap_{P_t^{ex} \in \mathcal{P}_t^{ex}} \mathscr{C}_{P_t^{ex}}.$$

3.1 Properties of this environment

We now give some results concerning the environment we depicted so far.

Proposition 5. Let $A_t^+ := A \cap \Omega_t^+$, $A_t^- := A \cap \Omega_t^-$, $B_t^+ := B \cap \Omega_t^+$, and $B_t^- := B \cap \Omega_t^-$. Then, $A \cup B = (A_t^+ \cup B_t^+) \cup (A_t^- \cup B_t^-)$ and $A \cap B = (A_t^+ \cap B_t^+) \cup (A_t^- \cap B_t^-)$, for all t. Also, $A \setminus B = (A_t^+ \setminus B_t^+) \cup (A_t^- \setminus B_t^-)$, for all t.

Recall now that a set ring is a system of sets $\mathbb{B}$ such that $A, B \in \mathbb{B}$ implies $A \cap B \in \mathbb{B}$ and $(A \setminus B) \cup (B \setminus A) =: A \triangle B \in \mathbb{B}$. A set ring $\mathbb{B}$ with a unit element, i.e., $E \in \mathbb{B}$ such that for all $A \in \mathbb{B}$, $A \cap E = A$, is called a set algebra. Let $\mathcal{F}_t^+ = 2^{\Omega_t^+}$ and $\mathcal{F}_t^- = 2^{\Omega_t^-}$.

Proposition 6. $\mathcal{F}_t^+$ and $\mathcal{F}_t^-$ are set algebras, for all t.

We also point out that if $A = \{\omega_1, \ldots, \omega_k\}$ and $\omega_1, \ldots, \omega_k \in \Omega$, then

$$P^{ex}(A) = \sum_{j=1}^{k} P^{ex}(\{\omega_j\}).$$

This is immediate from the countable additivity of extended probabilities.

Another property is the following. Fix any t and let $A, B \in \mathcal{F}_t^+$ such that $A \subset B$. Then, $P_t^{ex}(A) \leq P_t^{ex}(B)$. Let then $C, D \in \mathcal{F}_t^-$ such that $C \subset D$. Then, $P_t^{ex}(C) \geq P_t^{ex}(D)$. Both these results come from Proposition 2.

The following is also interesting.

Proposition 7. Consider $A \subset \cup_{j \in \mathbb{N}_0} A_j$, where $A_j \in \mathcal{F}$ for all j, and also $A \in \mathcal{F}$. Then $P_t^{ex}(A) \leq \sum_{j \in \mathbb{N}_0} P_t^{ex}(A_j)$ if $\cup_{j \in \mathbb{N}_0} A_j \in \mathcal{F}_t^+$, and $P_t^{ex}(A) \geq \sum_{j \in \mathbb{N}_0} P_t^{ex}(A_j)$ if $\cup_{j \in \mathbb{N}_0} A_j \in \mathcal{F}_t^-$. If instead some of the A_j's are in $\mathcal{F}_t^+$ and some are in $\mathcal{F}_t^-$, then $P_t^{ex}(A) \gtreqless \sum_{j \in \mathbb{N}_0} P^{ex}(A_j)$, that is, $P_t^{ex}(A)$ can be larger, smaller, or equal to $\sum_{j \in \mathbb{N}_0} P^{ex}(A_j)$.

In the setting we have outlined so far, there is a way of operationalizing Equation (1). Pick any $A \in \mathcal{F}$; we have

$$P_t^{ex}(A^c) = P_t^{ex}\left(\Omega_t^+ \setminus [\Omega_t^+ \cap A]\right) + P_t^{ex}\left(\Omega_t^- \setminus [\Omega_t^- \cap A]\right) \quad (6)$$

We retain the fact that $P_t^{ex}(\emptyset) = 0$; indeed,

$$P_t^{ex}(\Omega^c) = P_t^{ex}(\emptyset) = P_t^{ex}\left(\Omega_t^+ \setminus [\Omega_t^+ \cap \Omega]\right) + P_t^{ex}\left(\Omega_t^- \setminus [\Omega_t^- \cap \Omega]\right)$$
$$= P_t^{ex}(\emptyset) + P_t^{ex}(\emptyset)$$
$$\iff P_t^{ex}(\emptyset) = 0.$$

We can also write $P_t^{ex}(A^c) = P_t^{ex}(A^c \cap \Omega_t^+) + P_t^{ex}(A^c \cap \Omega_t^-)$ because, as we know, $A^c = (A^c \cap \Omega_t^+) \sqcup (A^c \cap \Omega_t^-)$.

3.2 Interpretation and updating procedure

Let us now discuss the probability assigned to the whole sample space Ω. From (ii*), we know that since, for all t, $\Omega_t^+ \sqcup \Omega_t^- = \Omega$, then $P_t^{ex}(\Omega) = P_t^{ex}(\Omega_t^+) + P_t^{ex}(\Omega_t^-)$. Also, from (i) and (ii), we know that $P_t^{ex}(\Omega_t^+) \geq 0$ and $P_t^{ex}(\Omega_t^-) \leq 0$. So,

$$P_t^{ex}(\Omega) = \begin{cases} p > 0 & \text{if } P_t^{ex}(\Omega_t^+) > |P_t^{ex}(\Omega_t^-)| \\ p = 0 & \text{if } P_t^{ex}(\Omega_t^+) = |P_t^{ex}(\Omega_t^-)| \\ p < 0 & \text{if } P_t^{ex}(\Omega_t^+) < |P_t^{ex}(\Omega_t^-)| \end{cases}.$$

What does it mean, then, for $P_t^{ex}(\Omega)$ to be equal to 0? And to be negative? And to be positive, but not 1? Given the benevolent bookmaker interpretation, we have the following. If $P_t^{ex}(\Omega) = 0$, it means that we have no sufficient information to say whether the sample space associated with our experiment is in fact Ω. If $P_t^{ex}(\Omega) < 0$, it means that, for the time being, the sample space associated with our experiment appears to be some $\check{\Omega} \subsetneq \Omega$. If $P_t^{ex}(\Omega) \in (0,1)$, it means that there is evidence that the sample space associated with our experiment could be in fact Ω, but we cannot state it with certainty.

The natural question that one might ask now is, for some $t \in \mathbb{N}_0$, how we come up with negative numbers to assign to events that belong to the latent space Ω_t^-. Or, for that matter, how do we come up with positive numbers to assign to events that belong to the actual space Ω_t^+. Call $\Delta(\Omega, \mathcal{F})$ the set of probability measures on $(\Omega, \mathcal{F})$, and $\Delta^{ex}(\Omega, \mathcal{F})$ the set of extended probability measures on $(\Omega, \mathcal{F})$. The latter is a linear space, as shown in Rao and Rao (1983). In a future work, we will argue that it is a Dedekind complete Banach lattice with respect to the norm induced by the total variation of an element P^{ex} of $\Delta^{ex}(\Omega, \mathcal{F})$. We also show that if Ω is a compact separable space, then the subset $\Delta^{ex}_{Baire}(\Omega, \mathcal{F}) \subset \Delta^{ex}(\Omega, \mathcal{F})$ of Baire extended probability measures is the dual of the real Banach space of all continuous real-valued functions on Ω.

Consider any $P \in \Delta(\Omega, \mathcal{F})$ such that, for all $\tilde{A} \in \mathcal{F}_0^+$, $P(\tilde{A}) = p \in [0, 1]$ is the amount we deem fair to pay to enter a bet about $\tilde{A}$; we call it the *oracle probability measure*. Then, for a generic $A \in \mathcal{F}$, we have that

$$P_0^{ex}(A \cap \Omega_0^+) = P(A \cap \Omega_0^+) \tag{7}$$

and

$$P_0^{ex}(A \cap \Omega_0^-) = -P(A \cap \Omega_0^-). \tag{8}$$

It is easy to see that P_0^{ex} satisfies (i*) and (ii*). Note also that

$$\sum_{E \in \mathcal{E}} |P_0^{ex}(E)| = \sum_{E_0^+ \in \mathcal{E}_0^+} P_0^{ex}(E_0^+) + \sum_{E_0^- \in \mathcal{E}_0^-} |P_0^{ex}(E_0^-)| = 1, \tag{9}$$

so P_0^{ex} satisfies (iii*) as well. It is hence a properly defined extended probability measure. Clearly, (9) holds for all $t \in \mathbb{N}_0$, not just for $t = 0$.

This way of assessing initial extended probabilities well reconciles with the interpretation we gave in general for extended probabilities: we cannot enter bets about events $A \notin \mathcal{F}_0^+$ (in this case, because we cannot observe them for the time being). So we assess the probabilities of the events $A \in \mathcal{F}_0^+$ as specified in Section 2, and then we ask our doppelgänger the probabilities $P_{0,D}(B) \geq 0$ they assign to the elements $B \in \mathcal{F}_0^-$.[7] Then, we flip the sign to those, that is,

---

[7] Subscript D stands for "doppelgänger".

$P_0^{ex}(B) = -P_{0,D}(B)$, for all $B \in \mathcal{F}_0^-$. In this notation, $P_{0,D}(B)$ is the probability the doppelgänger assigns to event B at time $t = 0$.

Since the agent faces ambiguity, they need to specify a set $\mathcal{P} \subset \Delta(\Omega, \mathcal{F})$ of probability measures. Every element $P \in \mathcal{P}$ induces an extended probability P_0^{ex} as we just described. In this way, we build the set $\mathcal{P}_0^{ex}$.

Let us now discuss how to update extended probabilities, that is, how to update $P_t^{ex}(A)$ to $P_{t+1}^{ex}(A)$, for all $A \in \mathcal{F}$, for all $P_t^{ex} \in \mathcal{P}_t^{ex}$, for all $t \in \mathbb{N}_0$.[8] We first consider a procedure to discover the components of the sample space that is equivalent to an urn without replacement. That is, after specifying Ω and Ω_0^+, we start our analysis with an urn whose content is unknown and possibly countable; it represents the true sample space associated with our experiment. At any time point t, we extract a ball (an element ω of the sample space). Once we learn about that element, our knowledge about the composition of the urn increases. We do not put the ball back into the urn; we discuss the case in which the discover procedure is equivalent to an urn with replacement later in this chapter.

Let us begin with the updating procedure from $P_0^{ex} \in \mathcal{P}_0^{ex}$ to $P_1^{ex} \in \mathcal{P}_1^{ex}$.

We collect a new observation $\omega \in \Omega$. If $\omega \in \Omega_0^-$, this means that, at time $t = 1$, we learn a new element of the true sample space. Notice that ω is consistent with an element $E_{0,\omega}^-$ of $\mathcal{E}_0^-$ (that is, $E_{0,\omega}^- = \{\omega\}$), which — given that we observe such ω — at time $t = 1$ becomes an element of $\mathcal{E}_1^+$; in formulas, $E_{0,\omega}^- = E_{1,\omega}^+$. Then, $\mathcal{E}_1^+ \supset \mathcal{E}_0^+$, $\mathcal{E}_1^- \subset \mathcal{E}_0^-$, and we update the extended probability assigned to $E_{0,\omega}^- = E_{1,\omega}^+$ as follows:

$$P_1^{ex}(E_{1,\omega}^+) = \left| P_0^{ex}(E_{0,\omega}^-) \right|. \tag{10}$$

The extended probabilities assigned to the other elements of $\mathcal{E}$ (the finest possible partition of the whole Ω) are held constant. From (5),

---

[8]It is worth mentioning that in Walley (1996) too the author discusses a process of learning the state space combined with learning about the probability of the elements of $\mathcal{F}$.

the updated extended probability associated with event A is given by

$$P_1^{ex}(A) = \sum_{E_{1,j}^+ \in \mathcal{E}_1^+ : P_1^{ex}(E_{1,j}^+) \neq 0} P_1^{ex}(A \mid E_{1,j}^+) P_1^{ex}(E_{1,j}^+)$$

$$+ \sum_{E_{1,j}^- \in \mathcal{E}_1^- : P_1^{ex}(E_{1,j}^-) \neq 0} P_1^{ex}(A \mid E_{1,j}^-) P_1^{ex}(E_{1,j}^-), \quad (11)$$

where

$$P_1^{ex}(A \mid E_{1,j}^+) = \frac{P_1^{ex}(A \cap E_{1,j}^+)}{P_1^{ex}(E_{1,j}^+)}$$

and

$$P_1^{ex}(A \mid E_{1,j}^-) = \frac{P_1^{ex}(A \cap E_{1,j}^-)}{P_1^{ex}(E_{1,j}^-)}.$$

Note that we are working with the finest possible partition of Ω, that is, $\mathcal{E} = \{\{\omega\}\}_{\omega \in \Omega}$. Then, for any $A \in \mathcal{F}$ and any $E \in \mathcal{E}$, $A \cap E = A \cap \{\omega\}$, which is equal to the empty set if $\omega \notin A$, and it is equal to $\{\omega\} = E$ if $\omega \in A$. Hence, for all $E \in \mathcal{E}$ such that $P_1^{ex}(E) \neq 0$,

$$P_1^{ex}(A \mid E) = \frac{P_1^{ex}(A \cap E)}{P_1^{ex}(E)} = \begin{cases} \frac{P_1^{ex}(\emptyset)}{P_1^{ex}(E)} = 0 & \text{if } A \cap E = \emptyset \\ \frac{P_1^{ex}(E)}{P_1^{ex}(E)} = 1 & \text{if } A \cap E = \{\omega\} \end{cases},$$

for all $E \in \mathcal{E}$.

So, Equation (11) can be rewritten as

$$P_1^{ex}(A) = \sum_{E_{1,j}^+ \in \mathcal{E}_1^+ : A \cap E_{1,j}^+ \neq \emptyset,\, P_1^{ex}(E_{1,j}^+) \neq 0} P_1^{ex}(E_{1,j}^+)$$

$$+ \sum_{E_{1,j}^- \in \mathcal{E}_1^- : A \cap E_{1,j}^- \neq \emptyset,\, P_1^{ex}(E_{1,j}^-) \neq 0} P_1^{ex}(E_{1,j}^-). \quad (12)$$

Of course, this holds for all $P_1^{ex} \in \mathcal{P}_1^{ex}$. Notice that we are implicitly assuming Allen's principle of conservation of knowledge. Indeed, suppose at time $\mathbf{t}$ the event A is entirely in the latent portion of Ω, that is, $A \cap \Omega_{\mathbf{t}}^- = A$, and at time $\mathbf{t} + k$ it is entirely in the actual portion of Ω, that is, $A \cap \Omega_{t+k}^+ = A$. Then, by (10) and (12), we have that $P_{t+k}^{ex}(A) = |P_t^{ex}(A)|$. This has the practical advantage of making the updating procedure mechanical (we do not need to reassess any subjective extended probability when new observations become available) and of preserving the initial opinion of the researcher. The importance of this last point is discussed in Remark 5.

If instead we observe $\omega \in \Omega_0^+$, this means that we made a good job in specifying Ω_0^+, and so we keep the extended probability constant

$$P_1^{ex}(E_{1,\omega}^+) = P_0^{ex}(E_{0,\omega}^+) \geq 0. \tag{13}$$

Of course, the extended probabilities assigned to the other elements of $\mathcal{E}$ remain constant as well.

We follow the procedures in (10) and (13) to update P_t^{ex} to P_{t+1}^{ex} for all $t \in \mathbb{N}_0$, not just from $t = 0$ to $t = 1$. We call (P_t^{ex}) the sequence of updates of initial extended probability P_0^{ex}. Now consider the extended probability $P_\infty^{ex} \in \Delta^{ex}(\Omega, \mathcal{F})$ such that

$$P_\infty^{ex}\left(A \cap \bigcup_{t \in \mathbb{N}_0} \Omega_t^+\right) = P\left(A \cap \bigcup_{t \in \mathbb{N}_0} \Omega_t^+\right)$$

and

$$P_\infty^{ex}\left(A \cap \bigcap_{t \in \mathbb{N}_0} \Omega_t^-\right) = -P\left(A \cap \bigcap_{t \in \mathbb{N}_0} \Omega_t^-\right),$$

for all $A \in \mathcal{F}$, where P is the oracle probability measure we used in (7) and (8) to specify P_0^{ex}. Let us denote by d_{ETV} the extended total variation distance,

$$d_{ETV}(P^{ex}, Q^{ex}) := \sup_{A \in \mathcal{F}} |P^{ex}(A) - Q^{ex}(A)|.$$

It is routine to check that d_{ETV} is a metric: the proof goes along the lines of showing that the total variation distance is a metric. Then, the following holds.

Proposition 8. $P_t^{ex} \to P_\infty^{ex}$ as $t \to \infty$ in the extended total variation metric.

The following claim is especially important.

Proposition 9. *If $\Omega_t^+ \uparrow \Omega$, then $P_\infty^{ex}(\Omega) = 1$.*

This result implies that if $\Omega_t^+ \uparrow \Omega$, then P_∞^{ex} is a regular probability measure. Indeed, it is easy to see that it satisfies the (countably additive) Kolmogorovian axioms for regular probability measures.

Now call $\mathcal{P}_\infty^{ex} \subset \Delta^{ex}(\Omega, \mathcal{F})$ the following set:

$$\mathcal{P}_\infty^{ex} := \left\{ P_\infty^{ex} \in \Delta^{ex}(\Omega, \mathcal{F}) : d_{ETV}(P_t^{ex}, P_\infty^{ex}) \xrightarrow[t \to \infty]{} 0, \ P_t^{ex} \in \mathcal{P}_t^{ex} \right\}.$$

That is, $\mathcal{P}_\infty^{ex}$ is the set of limits (in the extended total variation metric) of the sequences (P_t^{ex}) whose elements P_t^{ex} belong to $\mathcal{P}_t^{ex}$.

The Hausdorff distance between an element of $(\mathcal{P}_t^{ex})_{t \in \mathbb{N}_0}$ and $\mathcal{P}_\infty^{ex}$ is given by

$$d_H(\mathcal{P}_t^{ex}, \mathcal{P}_\infty^{ex}) = \max \left\{ \sup_{P_t^{ex} \in \mathcal{P}_t^{ex}} \inf_{P_\infty^{ex} \in \mathcal{P}_\infty^{ex}} d_{ETV}(P_t^{ex}, P_\infty^{ex}), \right. \tag{14}$$
$$\left. \sup_{P_\infty^{ex} \in \mathcal{P}_\infty^{ex}} \inf_{P_t^{ex} \in \mathcal{P}_t^{ex}} d_{ETV}(P_t^{ex}, P_\infty^{ex}) \right\}.$$

Then, the sequence $(\mathcal{P}_t^{ex})_{t \in \mathbb{N}_0}$ of successive updates of set $\mathcal{P}_0^{ex}$ representing the initial beliefs of the agent facing ambiguity converges in the Hausdorff distance to $\mathcal{P}_\infty^{ex}$. This result is an immediate consequence of Proposition 8: every element of $\mathcal{P}_t^{ex}$ converges to an element of $\mathcal{P}_\infty^{ex}$, so the distance between the "borders" of these two sets — measured by the Hausdorff metric — converges to 0.

Now, there are two possible scenarios: one in which we continue discovering the elements of the state space until we retrieve the full Ω we specified *ex ante* (this corresponds to $\Omega_t^+ \uparrow \Omega$), and another one in which we discover that the actual sample space associated with our experiment is $\Omega' \subsetneq \Omega$. In the first scenario, any $P_\infty^{ex} \in \mathcal{P}_\infty^{ex}$ is a regular probability measure, a consequence of Proposition 9. So, after discovering the composition of the state space, we have that $(\Omega, \mathcal{F}, P_\infty^{ex})$ is a regular probability space, for all $P_\infty^{ex} \in \mathcal{P}_\infty^{ex}$.

In the second scenario, Ω' can be finite or countable. In the former case, we have that, for some $T \in \mathbb{N}$, $\Omega_T^+ \equiv \Omega' \subsetneq \Omega$, so

$\sum_{E_T^+ \in \mathcal{E}_T^+} P_T^{ex}(E_T^+) = q < 1$ because, by (9) and (10),

$$\sum_{E_T^+ \in \mathcal{E}_T^+} P_T^{ex}(E_T^+) + \sum_{E_T^- \in \mathcal{E}_T^-} |P_T^{ex}(E_T^-)| = 1.$$

This may seem problematic: $\mathcal{P}_t^{ex}$ converges to $\mathcal{P}_T^{ex}$ (that is, $\mathcal{P}_\infty^{ex}$ coincides with $\mathcal{P}_T^{ex}$), which is not a set of regular probability measures. To solve this issue, we need to describe the regular probability measure induced by every $P_T^{ex} \in \mathcal{P}_T^{ex}$. It would be desirable to find $\tilde{P}$ such that $\tilde{P}(A) = cP_T^{ex}(A)$, for all $A \in \mathcal{F}_T^+$. This because such a regular probability measure preserves the ratios between extended probabilities of the elements of $\mathcal{F}_T^+$, that is,

$$\frac{\tilde{P}(A)}{\tilde{P}(B)} = \frac{cP_T^{ex}(A)}{cP_T^{ex}(B)} = \frac{P_T^{ex}(A)}{P_T^{ex}(B)},$$

for all $A, B \in \mathcal{F}_T^+$ such that $P_T^{ex}(B) \neq 0$.[9] To find such a c, the following needs to hold:

$$\sum_{E_T^+ \in \mathcal{E}_T^+} cP_T^{ex}(E_T^+) = 1,$$

which happens if and only if

$$c = \frac{1}{\sum_{E_T^+ \in \mathcal{E}_T^+} P_T^{ex}(E_T^+)} = \frac{1}{P_T^{ex}(\Omega_T^+)}.$$

Hence, $\tilde{P} = cP_T^{ex}$ is the regular probability measure induced by P_T^{ex} that preserves the ratios between extended probabilities of elements of $\mathcal{F}_T^+$. Clearly, this holds for all $P_T^{ex} \in \mathcal{P}_T^{ex}$, so $(\Omega_T^+, \mathcal{F}_T^+, \tilde{P})$ is a regular probability space, for all $\tilde{P} \in \tilde{\mathcal{P}}$, where $\tilde{\mathcal{P}}$ is the set of regular probabilities induced by the elements of $\mathcal{P}_T^{ex}$.

If Ω' is countable, we proceed in a similar way. We still want to preserve the ratios between extended probabilities of the elements of $\mathcal{F}'$,

---

[9]Note that if the ratio is greater than 1, we deem A more likely than B, and vice versa if the ratio is smaller than 1. We deem A and B equally likely if the ratio is exactly 1. Since we want to preserve this behavioral interpretation when we move to the regular probabilities induced by the elements of $\mathcal{P}_T^{ex}$, we want the ratio to be maintained.

so we have to find c such that $\sum_{E \in \mathcal{E}'} cP^{ex}_\infty(E) = 1$. This happens when

$$c = \frac{1}{\sum_{E \in \mathcal{E}'} P^{ex}_\infty(E)} = \frac{1}{P^{ex}_\infty(\Omega')},$$

so $\tilde{P} = cP^{ex}_\infty$ is the regular probability measure we were looking for. This holds for all $P^{ex}_\infty \in \mathcal{P}^{ex}_\infty$, so $(\Omega', \mathcal{F}', \tilde{P})$ is a regular probability space, for all $\tilde{P} \in \tilde{\mathcal{P}}$.

Remark 5. Note that, in the first scenario $(\Omega^+_t \uparrow \Omega)$, every element $P^{ex}_\infty \in \mathcal{P}^{ex}_\infty$ coincides with its corresponding oracle probability measure $P \in \mathcal{P}$ we expressed *ex ante* on the whole Ω. Indeed, we have that $P^{ex}_\infty(E) = |P^{ex}_\infty(E)|$, for all $E \in \mathcal{E}$, which implies $P^{ex}_\infty(A) = |P^{ex}_\infty(A)|$, for all $A \in \mathcal{F}$. But then, by (9), we know that $\sum_{E \in \mathcal{E}} P^{ex}_\infty(E) = 1$, and that $P^{ex}_\infty(A) = P(A)$, for all $A \in \mathcal{F}$. This should not surprise: the updating procedure we described earlier is not based on collecting new data like the Bayesian one, nor on repeating the experiment many times like the frequentist one. Rather, it is based on discovering the true composition of the state space associated with our inference procedure. Then, it is natural to retrieve the opinion we expressed *ex ante* on the state space Ω once we get the confirmation that the true state space is indeed Ω. This also reconciles well with the interpretation we gave to negative extended probabilities: once we are given the possibility to enter the bets we were denied before, we tend to agree with our doppelgänger who had the opportunity to bet on those events in the first place (Allen's principle of conservation of knowledge holds).

Note also that $\tilde{P}$, albeit not equal, is proportional to P^{ex}_∞, $\tilde{P} = cP^{ex}_\infty$, for every $\tilde{P} \in \tilde{\mathcal{P}}$ and its corresponding $P^{ex}_\infty \in \mathcal{P}^{ex}_\infty$. The interpretation is immediate: we maintain our opinion on the elements of the sample space $\Omega' \subsetneq \Omega$ that pertains to our experiment.

If the procedure for discovering the composition of the state space is equivalent to an urn with replacement, everything we discussed so far still holds with just two differences:

- If we extract twice or more times the same element, its extended probability flips sign the first time, and then stays constant.

- The convergence may be slower because we may need more extractions from the urn to learn the true composition (because we can extract twice or more times the same element).

Let us give a simple example that illustrates one of the possible situations in which the analysis depicted so far can be put to use.

Example 1. We consider a species sampling problem in the field of ecology. Suppose we want to know the number of bird species that inhabit a certain region throughout the year. What we can do is to start our analysis by letting $\Omega = \mathbb{N}$ and $\Omega_0^+ = \{1, \ldots, n\}$, where n is the number of species that inhabit a region similar to the one of interest throughout the year. Then, we specify a set $\mathcal{P}$ of probability measures on $\mathbb{N}$, e.g., a collection $\{\text{Geom}(p)\}_{p \in [0,1]}$ of geometric distributions having parameter $p \in [0, 1]$. We specify a set of probabilities because we are not able to express our initial opinion via a unique probability measure (we face ambiguity). We specify the probability measures on the whole number field $\mathbb{N}$ because we do not know exactly the composition of our state space. Then, after eliciting the set $\mathcal{P}_0^{ex}$ of extended probability measures induced by (the elements of) $\mathcal{P}$, we begin the *ex ante* analysis described in this chapter. After collecting observations for an entire year, we end up discovering that the state space associated with our experiment is $\Omega' = \{1, \ldots, m\} \subsetneq \mathbb{N}$, where $m \geq n$ is the number of species we counted during the year. We recover also the set $\tilde{\mathcal{P}}$ of regular probability measures induced by $\mathcal{P}_\infty^{ex}$. Now, the "real" statistical analysis can take place. Indeed, we know that m is the maximum number of species that live in the region during the year, but it does not take into account migrations to or from the region itself. Hence, the number of bird species that inhabit the region throughout the year may well be smaller than m. So, every $\tilde{P} \in \tilde{\mathcal{P}}$ will be such that $\tilde{P}(\{\omega_k\}) \in [0, 1]$, and $\sum_{k=1}^m \tilde{P}(\{\omega_k\}) = 1$, where $\omega_k = k$, for all $k \in \{1, \ldots, m\}$. Gathering data $y_1, \ldots, y_\ell$ and updating these $\tilde{P}$'s via Bayesian conditioning, we obtain the set $\{\tilde{P}(\cdot \mid y_1, \ldots, y_\ell)\}_{\tilde{P} \in \tilde{\mathcal{P}}}$ of posterior (regular) probability measures. This gives a robust analysis: for all $\omega_k \in \Omega'$, the posterior probability of ω_k being the correct number of species belongs to the interval

$$\left[\underline{\tilde{P}}(\{\omega_k\} \mid y_1, \ldots, y_\ell), \overline{\tilde{P}}(\{\omega_k\} \mid y_1, \ldots, y_\ell)\right],$$

where the lower bound is a lower (regular) probability, and the upper bound is an upper (regular) probability. The narrower the interval, the less imprecise our beliefs resulting from the analysis. △

A criticism that can be made of our *ex ante* analysis is the following: why should the scholar be concerned with the exact composition of the state space? It should be enough to specify it as richly as possible, and then proceed to a regular statistical analysis. There are two responses to such a critique. First, as pointed out in Remark 4, in doing so the scholar would not be respecting the condition listed in Tsitsiklis (2018) that the state space must have the right granularity depending on the statistical experiment they are interested in. Second, in the case that the state space associated with our statistical experiment is $\Omega' \subsetneq \Omega$, working with probability measures supported on the whole space Ω of apparently possible states may be computationally costly. Our *ex ante* analysis allows the scholar to focus on the "minimal" state space — the one containing only the necessary states — that is more meaningful to the analysis.

4 Upper and Lower Extended Probabilities

Fix any $t \in \mathbb{N}_0$, and considering the "boundary elements" of the set $\mathcal{P}_t^{ex}$,

$$\underline{P}_t^{ex}(A) = \inf_{P^{ex} \in \mathcal{P}_t^{ex}} P^{ex}(A) \tag{15}$$

and

$$\overline{P}_t^{ex}(A) = \sup_{P^{ex} \in \mathcal{P}_t^{ex}} P^{ex}(\Omega) - \underline{P}_t^{ex}(A^c) = \sup_{P^{ex} \in \mathcal{P}_t^{ex}} P^{ex}(A), \tag{16}$$

for all $A \in \mathcal{F}$. They are not extended probabilities; we call $\underline{P}_t^{ex}(A)$ a lower extended probability measure, and $\overline{P}_t^{ex}(A)$ an upper extended probability measure. Notice that upper extended probability measures differ from upper regular probability measures. These latter are defined as 1 minus the lower regular probability of the complement of the event we are interested in. In (16), we give a similar conjugate type of definition, but we cannot write 1, because we do not require that the lower extended probability of Ω is 1.

Given a generic lower extended probability $\underline{P}^{ex}$ and a generic event $A \in \mathcal{F}$, we interpret $\underline{P}^{ex}(A)$ as follows. If $\underline{P}^{ex}(A) = p > 0$, then p represents the supremum price that we are willing to pay to enter a bet on A that gives us \$1 if A takes place and \$0 otherwise. If $\underline{P}^{ex}(A) = 0$, then in the most conservative of our mental states we deem A impossible to take place. Finally, if $\underline{P}^{ex}(A) = q < 0$, then $|q|$ represents the infimum selling price at which our doppelgänger takes bets on A that pay \$1 if A takes place and \$0 otherwise. As we can see, this interpretation captures the ideas of worst case scenario and of prudent behavior. It can be seen as a betting scheme analogous to the one in Walley (1991, Section 2.3.1), but with monetary instead of utiles outcomes, and extended to probabilities that can take on negative values as well.

Both lower and upper extended probabilities are extended Choquet capacities. A generic extended Choquet capacity is defined as a set function $\nu^{ex} : \mathcal{F} \to \mathbb{R}$ such that

(EC1) $\nu^{ex}(\emptyset) = 0$,
(EC2) $\nu^{ex}(A) \in [-1, 1]$, for all $A \in \mathcal{F}$,
(EC3) for any $A, B \in \mathcal{F}$ such that $A \subset B$, if $\nu^{ex}(A) \geq 0$, $\nu^{ex}(B) \geq 0$, and $\nu^{ex}(B \cap A^c) \geq 0$, then $\nu^{ex}(A) \leq \nu^{ex}(B)$. If instead $\nu^{ex}(A) \leq 0$, $\nu^{ex}(B) \leq 0$, and $\nu^{ex}(B \cap A^c) \leq 0$, then $\nu^{ex}(A) \geq \nu^{ex}(B)$.

As we can see, we do not require countable additivity (ii*) to hold for extended capacities. An extended probability measure is an additive extended capacity.

It is easy to see that upper and lower extended probability measures satisfy (EC1)–(EC3); in addition, lower extended probabilities are superadditive, while upper extended probabilities are subadditive. That is, for all $A, B \in \mathcal{F}$,

$$\underline{P}_t^{ex}(A \sqcup B) \geq \underline{P}_t^{ex}(A) + \underline{P}_t^{ex}(B) \qquad (17)$$

and

$$\overline{P}_t^{ex}(A \sqcup B) \leq \overline{P}_t^{ex}(A) + \overline{P}_t^{ex}(B). \qquad (18)$$

These inequalities come immediately from the properties of the infimum and supremum operators.

Remark 6. Note that the behavioral interpretation that we gave to lower extended probabilities entails that they can be specified even without eliciting a set of extended probability measures first. To this extent, our behavioral interpretation can be called minimal, similarly to Walley (1991, Section 2.3.1). An immediate question the reader may ask is as follows: "If we were to specify a lower extended probability without resorting to a set of extended probabilities, are we sure it is subadditive?" The answer is yes, under a mild assumption: Example 2 — based on the example in Walley (1991, Section 1.6.4) — shows that if lower extended probabilities avoid sure loss, then they are superadditive, and Theorem 2 shows that if $\underline{P}^{ex}$ can be obtained as the infimum of a set of extended probabilities, then it avoids sure loss.

We first define *sure loss* for lower extended probabilities. It is the immediate lower counterpart of Definition 1.

Definition 2. Lower extended probability $\underline{P}^{ex}$ avoids sure loss if no Dutch books can be made against the punter, that is, if we cannot find a finite collection $\{B_j\}_{j=1}^n \subset \mathscr{B}'$ along with numbers $\{s_j\}_{j=1}^n \subset \mathbb{R}_+$ such that

$$\sup_{\omega \in \Omega} \left\{ \sum_{j=1}^n s_j \left[\mathbb{1}_{B_j}(\omega) - \underline{P}^{ex}(B_j) \right] \right\} < 0, \qquad (19)$$

where $\mathscr{B}'$ is the sigma-algebra generated by the collection $\{B \subset \Omega : \underline{P}^{ex}(B) \geq 0 \text{ and } B \cap C = \emptyset, \forall C \in \mathcal{F}, \underline{P}^{ex}(C) < 0\}$.

We have the following interesting result, that is a version of the lower envelope theorem in Walley (1991, Corollary 2.8.6).[10]

---

[10]Note that the lower envelope theorem in Walley (1991, Corollary 2.8.6) comprises an envelope of probabilities that are merely finitely additive, and it states a necessary and sufficient condition. We plan to prove the opposite direction of Theorem 2 in future work.

Theorem 2. *Consider a generic lower extended probability $\underline{P}^{ex}$ defined on a measurable space $(\Omega, \mathcal{F})$. If there exists a nonempty set $\mathcal{P}^{ex}$ of extended probabilities on $(\Omega, \mathcal{F})$ such that $\underline{P}^{ex}(A) = \inf_{P^{ex} \in \mathcal{P}^{ex}} P^{ex}(A)$, for all $A \in \mathcal{F}$, then $\underline{P}^{ex}$ avoids sure loss.*

The following example shows that if $\underline{P}^{ex}$ avoids sure loss, then it is superadditive.

Example 2. Pick two mutually exclusive events A and B. By our behavioral interpretation, the highest amount we are willing to pay to get \$1 if A occurs (or the highest amount we would deem reasonable to lose if we were given the possibility to enter the bet) is $\underline{P}^{ex}(A)$. The same holds for $\underline{P}^{ex}(B)$. Since we want to avoid sure loss, the net outcome is equivalent to paying $\underline{P}^{ex}(A) + \underline{P}^{ex}(B)$ to get \$1 if $A \sqcup B$ occurs. Now, $\underline{P}^{ex}(A \sqcup B)$ is the highest price we are willing to pay to obtain \$1 if $A \sqcup B$ occurs. So we recover the superadditivity constraint

$$\underline{P}^{ex}(A \sqcup B) \geq \underline{P}^{ex}(A) + \underline{P}^{ex}(B).$$

Similarly, an upper lower probability $\overline{P}^{ex}$ should satisfy the subadditivity constraint $\overline{P}^{ex}(A \sqcup B) \leq \overline{P}^{ex}(A) + \overline{P}^{ex}(B)$. △

An important concept worth introducing is the core of a lower extended probability measure.

Definition 3. *Given a generic lower extended probability $\underline{P}^{ex}$, we call core of $\underline{P}^{ex}$ the set*

$$\operatorname{core}(\underline{P}^{ex}) := \big\{ P^{ex} \in \Delta^{ex}(\Omega, \mathcal{F}) : \ P^{ex}(A) \geq \underline{P}^{ex}(A), \forall A \in \mathcal{F} \\ \text{and } P^{ex}(\Omega) = \underline{P}^{ex}(\Omega) \big\}.$$

This definition tells us that the core of $\underline{P}^{ex}$ is the set of all suitably normalized extended probabilities that setwise dominate $\underline{P}^{ex}$. Note that in general it may be empty and that

$$\operatorname{core}(\underline{P}^{ex}) = \big\{ P^{ex} \in \Delta^{ex}(\Omega, \mathcal{F}) : \ \underline{P}^{ex} \leq P^{ex} \leq \overline{P}^{ex} \big\} \\ = \big\{ P^{ex} \in \Delta^{ex}(\Omega, \mathcal{F}) : \ \underline{P}^{ex}(A) \leq \overline{P}^{ex}(A), \forall A \in \mathcal{F} \\ \text{and } P^{ex}(\Omega) = \underline{P}^{ex}(\Omega) \big\},$$

so the core can be seen as the set of extended probabilities "sandwiched" between lower extended probability $\underline{P}^{ex}$ and upper extended probability $\overline{P}^{ex}$, as well as the set of extended probabilities setwise dominated by $\overline{P}^{ex}$. The following is a corollary to Theorem 2.

Corollary 2. *If* $core(\underline{P}^{ex}) \neq \emptyset$, *then* $\underline{P}^{ex}$ *avoids sure loss.*

A crucial property of the core is the following.

Proposition 10. *Given a generic lower extended probability* $\underline{P}^{ex}$, *its core is convex and weak*-compact.*

Being compact and convex, the core of $\underline{P}^{ex}$ is completely characterized by $\underline{P}^{ex}$. This means that it is enough to know $\underline{P}^{ex}$ to be able to retrieve every element in its core. So in our analysis we can focus on updating $\underline{P}^{ex}_t$ to $\underline{P}^{ex}_{t+1}$, and then require that $\mathcal{P}^{ex}_{t+1} = core(\underline{P}^{ex}_{t+1})$, instead of updating $\mathcal{P}^{ex}_t$ to $\mathcal{P}^{ex}_{t+1}$ elementwise. This justifies our focus in the remainder of this section on studying how to update lower extended probabilities. The procedure to update upper extended probability measures is going to be similar (their relation is described by Equation (16)).

Recall that the conditions to perform the update in the additive case are given by (10) and (13).

Proposition 11. *The sublinear counterpart of* (10) *is the following*:

$$\underline{P}^{ex}_{t+1}(E^+_{t+1,\omega}) = \left|\overline{P}^{ex}_t(E^-_{t,\omega})\right| \quad \text{and} \quad \overline{P}^{ex}_{t+1}(E^+_{t+1,\omega}) = \left|\underline{P}^{ex}_t(E^-_{t,\omega})\right|. \tag{20}$$

It holds when we learn a new element ω of our sample space, that is, when the new observation ω belongs to the latent space at time t, but "moves" to the actual space at time $t+1$. In formulas, $\omega \in \Omega^-_t$, but $\omega \in \Omega^+_{t+1}$. The lower extended probabilities assigned to the other elements of $\mathcal{E}$ are held constant.

Note that we are working with the finest possible partition of Ω, so, as before, the intersection $A \cap E$ between any $A \in \mathcal{F}$ and any

$E = \{\omega\} \in \mathcal{E}$ is either the empty set, $A \cap E = \emptyset$, if $\omega \notin A$, or the element ω itself, $A \cap E = \{\omega\} = E$, if $\omega \in A$. Hence, for any $t \in \mathbb{N}_0$, for any $E = \{\omega\} \in \mathcal{E}$, and for any $A \in \mathcal{F}$, we have that

$$\underline{P}_t^{ex}(A \cap E) = \begin{cases} \underline{P}_t^{ex}(E) & \text{if } \omega \in A \\ \underline{P}_t^{ex}(\emptyset) = 0 & \text{if } \omega \notin A \end{cases}. \tag{21}$$

Given any $A, B \in \mathcal{F}$, if $\underline{P}_t^{ex}(B) \neq 0$, we define

$$\underline{P}_t^{ex}(A \mid B) := \frac{\inf_{P^{ex} \in \mathcal{P}_t^{ex}} P^{ex}(A \cap B)}{\inf_{P^{ex} \in \mathcal{P}_t^{ex}} P^{ex}(B)} = \frac{\underline{P}_t^{ex}(A \cap B)}{\underline{P}_t^{ex}(B)}, \tag{22}$$

for all $t \in \mathbb{N}_0$. We call it the *extended Geometric rule*. We note immediately that, combining (21) and (22), we get

$$\underline{P}_t^{ex}(A \mid E) = \frac{\underline{P}_t^{ex}(A \cap E)}{\underline{P}_t^{ex}(E)} = \begin{cases} \frac{\underline{P}_t^{ex}(E)}{\underline{P}_t^{ex}(E)} = 1 & \text{if } \omega \in A \\ \frac{\underline{P}_t^{ex}(\emptyset)}{\underline{P}_t^{ex}(E)} = 0 & \text{if } \omega \notin A \end{cases}, \tag{23}$$

for all $E = \{\omega\} \in \mathcal{E}$ such that $\underline{P}_t^{ex}(E) \neq 0$ and all $A \in \mathcal{F}$. So, we have that $\underline{P}_t^{ex}(A \mid E) = P_t^{ex}(A \mid E)$, for all A, all E, and all $P_t^{ex} \in \mathcal{P}_t^{ex}$. This is true only because we are working with the finest possible partition of Ω, and because we endorse the extended Geometric rule.

The sublinear counterpart of (13) is the following:

$$\underline{P}_{t+1}^{ex}(E_{t+1,\omega}^+) = \underline{P}_t^{ex}(E_{t,\omega}^+) \geq 0, \tag{24}$$

for all $t \in \mathbb{N}_0$. This comes from the fact that we draw an element already belonging to the actual space, so we do not need to update its lower extended probability (similarly to what is described in Equation (13)). The lower extended probabilities assigned to the other elements of $\mathcal{E}$ are held constant.

At this point, a natural question one may ask is how to compute $\underline{P}_t^{ex}(A)$, for any t, for any $A \in \mathcal{F}$. It would be tempting to write that

$$\underline{P}_t^{ex}(A) = \sum_{E_{t,j}^+ \in \mathcal{E}_t^+ : \underline{P}_t^{ex}(E_{t,j}^+) \neq 0} \underline{P}_t^{ex}(A \mid E_{t,j}^+) \underline{P}_t^{ex}(E_{t,j}^+)$$

$$+ \sum_{E_{t,j}^- \in \mathcal{E}_t^- : \underline{P}_t^{ex}(E_{t,j}^-) \neq 0} \underline{P}_t^{ex}(A \mid E_{t,j}^-) \underline{P}_t^{ex}(E_{t,j}^-).$$

This would mimic exactly (11), with lower extended probabilities in place of additive extended probabilities. Alas, that would not be true, since lower extended probabilities are not additive. Instead, we have the following.

Fix any $t \in \mathbb{N}_0$. Call $\mathfrak{E}_t \equiv \{E_{t,A}\}$ the collection of elements of $\mathcal{E}$ such that $A \cap E_{t,A} \neq \emptyset$. We index $\mathfrak{E}_t$ to time t to highlight the fact that although its elements stay the same, some of them may "move" from the latent to the actual space as time goes by and we collect more observations. Since we are working with the finest possible partition of Ω, we can write

$$A = \bigsqcup_{E_{t,A} \in \mathfrak{E}_t} E_{t,A}. \tag{25}$$

In the most general case, some of these $E_{t,A}$'s belong to the actual space and some to the latent space. Let us denote the former by $E_{t,A}^+$'s and the latter by $E_{t,A}^-$'s. Formally, we have that $E_{t,A}^+ \cap \Omega_t^+ \neq \emptyset$ and $E_{t,A}^+ \cap \Omega_t^- = \emptyset$, and vice versa for the $E_{t,A}^-$'s. Hence, (25) can be rewritten as

$$A = \bigsqcup_{E_{t,A}^+ \in \mathfrak{E}_t} E_{t,A}^+ \sqcup \bigsqcup_{E_{t,A}^- \in \mathfrak{E}_t} E_{t,A}^-. \tag{26}$$

Now, from (20) and (24), we know how to update the lower extended probabilities of all the elements of the partition $\mathcal{E}$ of Ω, which implies that we know the value of $\underline{P}_{t+1}^{ex}(E_{t+1,A})$, for all $E_{t+1,A} \in \mathfrak{E}_{t+1}$. Then, consider now the summation

$$\sum_{E_{t+1,A} \in \mathfrak{E}_{t+1}} \underline{P}_{t+1}^{ex}(E_{t+1,A}).$$

From Equation (17), we have that

$$\underline{P}_{t+1}^{ex}(A) \geq \sum_{E_{t+1,A} \in \mathfrak{E}_{t+1}} \underline{P}_{t+1}^{ex}(E_{t+1,A})$$

$$= \sum_{E_{t+1,A}^+ \in \mathfrak{E}_{t+1}} \underline{P}_{t+1}^{ex}(E_{t+1,A}^+) + \sum_{E_{t+1,A}^- \in \mathfrak{E}_{t+1}} \underline{P}_{t+1}^{ex}(E_{t+1,A}^-).$$

From Equation (18), we can give an upper bound for the upper extended probability of A:

$$\overline{P}_{t+1}^{ex}(A) \leq \sum_{E_{t+1,A} \in \mathfrak{E}_{t+1}} \overline{P}_{t+1}^{ex}(E_{t+1,A})$$

$$= \sum_{E_{t+1,A}^+ \in \mathfrak{E}_{t+1}} \overline{P}_{t+1}^{ex}(E_{t+1,A}^+) + \sum_{E_{t+1,A}^- \in \mathfrak{E}_{t+1}} \overline{P}_{t+1}^{ex}(E_{t+1,A}^-).$$

5 Conclusion

In this chapter, we give a definition of extended probability measures that does not depend on the field a scholar works in. We give some of their more interesting properties, and a behavioral interpretation to positive and negative values of extended probabilities. We then apply extended probabilities to statistical inference. Given the probability space $(\Omega, \mathcal{F}, P)$ associated with the experiment we want to conduct, we use extended probabilities to express uncertainty about both the composition of the state space Ω, and which probability measure P to select. We develop an *ex ante* analysis; our method describes how the researcher progressively discovers the true composition of the state space so that, at the end of the process, a regular statistical analysis (that requires the knowledge of Ω) can take place. We introduce the concept of extended Choquet capacities and, in particular, of upper and lower extended probabilities that represent the "borders" of sets of extended probabilities, and we give bounds for the lower extended probability of any element $A \in \mathcal{F}$. We also apply our model to the fields of opinion dynamics and species sampling models. This chapter is important because it gives a foundational definition of extended probability measures and makes these latter relevant to the field of statistics.

In the future, we will deal with the possibility that in our *ex ante* analysis we are not able to discover the whole composition of the state space associated with our experiment. In that case, the statistical analysis itself will have to be carried out using extended probabilities. We also plan to relax the assumption we made about working with a finite or countable state space. We will also provide direct behavioral motivation for the properties of extended probabilities so as to be able to relate our approach to those of de Finetti (1974, 1975) and Walley (1991). Furthermore, since it is one of the main benefits of a betting approach, we will study partially specified extended probabilities (that is, extended probabilities defined not on a sigma-algebra, but rather on a generic collection of events) and their coherent extension. Finally, we would like to deepen the study of lower extended probabilities.

Acknowledgments

This chapter is supported in part by the following grants: NSF CCF-193496, NSF DEB-1840223, NIH R01 DK116187-01, HFSP RGP0051/2017, NIH R21 AG055777-01A, NSF DMS 17-13012, NSF ABI 16-61386, and ARO MURI W911NF2010080. The authors would like to thank Aaditya Ramdas, Mark Burgin, Shounak Chattopadhyay, Roberto Corrao, Ruobin Gong, Vittorio Orlandi, Glenn Shafer, Peter Walley, Marco Zaffalon, and Alessandro Zito for helpful comments.

Appendix

A.1 *Proofs*

Proof of Proposition 1. Suppose for the sake of contradiction that for some $A = B$,

$$P^{ex}(A) \neq P^{ex}(B). \tag{A.1}$$

Consider then $A \setminus B = B \setminus B = \emptyset$; we have that

$$P^{ex}(A \setminus B) \neq P^{ex}(B \setminus B) = P^{ex}(\emptyset) = 0. \tag{A.2}$$

This is true because $A = \{A \setminus B\} \sqcup \{A \cap B\}$ and $B = \{B \setminus B\} \sqcup \{B \cap B\}$; by (ii*), this implies that $P^{ex}(A \setminus B) = P^{ex}(A) - P^{ex}(A \cap B)$ and $P^{ex}(B \setminus B) = P^{ex}(B) - P^{ex}(B \cap B)$. Then, we have that $P^{ex}(A \setminus B) =$

$P^{ex}(B \setminus B)$ if and only if

$$P^{ex}(A) = P^{ex}(B) - P^{ex}(B \cap B) + P^{ex}(A \cap B),$$

but $B \cap B = B$, and $A \cap B = B$ because we assumed $A = B$, so $P^{ex}(A \setminus B) = P^{ex}(B \setminus B)$ if and only if $P^{ex}(A) = P^{ex}(B)$. But by (A.1), $P^{ex}(A) \neq P^{ex}(B)$, so we have that the inequality in (A.2) holds.

By assumption, we know that $A = B$, so $A \setminus B = \emptyset$, and so $P^{ex}(A \setminus B) = P^{ex}(\emptyset)$. Then, by (A.2), we have that $P^{ex}(\emptyset) = P^{ex}(A \setminus B) \neq P^{ex}(B \setminus B) = P^{ex}(\emptyset)$, a contradiction. $\square$

Proof of Proposition 2. Let $A \subset B$; then, we can write B as $B = A \sqcup (B \cap A^c)$. By (ii*), this implies that $P^{ex}(B) = P^{ex}(A) + P^{ex}(B \cap A^c)$. Then, if $P^{ex}(B)$, $P^{ex}(A)$, and $P^{ex}(B \cap A^c)$ are all non-negative, then $P^{ex}(B) = P^{ex}(A) + P^{ex}(B \cap A^c) \geq P^{ex}(A)$. If instead they are all nonpositive, then $P^{ex}(B) = P^{ex}(A) + P^{ex}(B \cap A^c) \leq P^{ex}(A)$. $\square$

Proof of Proposition 3. Pick $A, B \in \mathcal{F}$; if they are disjoint, the equation follows immediately from (ii*). If they are not, consider $A \cup B = \{A \setminus B\} \sqcup \{A \cap B\} \sqcup \{B \setminus A\}$. Then, again by (ii*), we have that

$$P^{ex}(A \cup B) = P^{ex}(A \setminus B) + P^{ex}(A \cap B) + P^{ex}(B \setminus A). \quad (A.3)$$

Note then that $P^{ex}(A \setminus B) = P^{ex}(A) - P^{ex}(A \cap B)$; indeed,

$$P^{ex}(A) = P^{ex}(\{A \cap B\} \sqcup \{A \setminus B\}) = P^{ex}(A \cap B) + P^{ex}(A \setminus B).$$

Similarly, $P^{ex}(B \setminus A) = P^{ex}(B) - P^{ex}(B \cap A)$. So, (A.3) becomes

$$P^{ex}(A \cup B) = P^{ex}(A) - P^{ex}(A \cap B) + P^{ex}(A \cap B)$$
$$+ P^{ex}(B) - P^{ex}(B \cap A)$$
$$= P^{ex}(A) + P^{ex}(B) - P^{ex}(A \cap B). \quad \square$$

Proof of Theorem 1. Note that P^{ex} restricted to $\mathscr{B}$ is a finite signed measure. By the Hahn-Jordan decomposition theorem Fischer (2012), there exists a unique decomposition of P^{ex} into a difference $P^{ex} = P^{ex}_+ - P^{ex}_-$ of two finite positive measures. Pick any $\{B_j\}_{j=1}^n \subset \mathscr{B}$, $\{s_j\}_{j=1}^n \subset \mathbb{R}$. We have that, for any payoff

$$\mathfrak{f}(\omega) = \sum_{j=1}^n s_j \left[\mathbb{1}_{B_j}(\omega) - P^{ex}(B_j)\right]$$

$$= \sum_{j=1}^n s_j \left[\mathbb{1}_{B_j}(\omega) - P^{ex}_+(B_j) + P^{ex}_-(B_j)\right],$$

the following holds:

$$\int_\Omega \mathfrak{f} dP^{ex} = \sum_{j=1}^n s_j \left[\int_\Omega \mathbb{1}_{B_j} dP^{ex} - P^{ex}_+(B_j) + P^{ex}_-(B_j)\right]$$

$$= \sum_{j=1}^n s_j \left[\int_\Omega \mathbb{1}_{B_j} dP^{ex}_+ - \int_\Omega \mathbb{1}_{B_j} dP^{ex}_- - P^{ex}_+(B_j) + P^{ex}_-(B_j)\right]$$

$$= \sum_{j=1}^n s_j \left[\int_\Omega \mathbb{1}_{B_j} dP^{ex}_+ - P^{ex}_+(B_j)\right]$$

$$- \sum_{j=1}^n s_j \left[\int_\Omega \mathbb{1}_{B_j} dP^{ex}_- - P^{ex}_-(B_j)\right]$$

$$= 0.$$

So $\mathfrak{f}$ cannot have a negative supremum. $\square$

Proof of Proposition 4. We have that

$$P^{ex}_t(A) = P^{ex}_t(A \cap \Omega^+_t) + P^{ex}_t(A \cap \Omega^-_t)$$

$$= \sum_{E^+_{t,j} \in \mathcal{E}^+_t} P^{ex}_t(A \cap E^+_{t,j}) + \sum_{E^-_{t,j} \in \mathcal{E}^-_t} P^{ex}_t(A \cap E^-_{t,j})$$

$$= \sum_{E^+_{t,j} \in \mathcal{E}^+_t : P^{ex}_t(E^+_{t,j}) \neq 0} P^{ex}_t(A \mid E^+_{t,j}) P^{ex}_t(E^+_{t,j}) \qquad (A.4)$$

$$+ \sum_{E^-_{t,j} \in \mathcal{E}^-_t : P^{ex}_t(E^-_{t,j}) \neq 0} P^{ex}_t(A \mid E^-_{t,j}) P^{ex}_t(E^-_{t,j}),$$

where the first equality comes from $A = (A \cap \Omega_t^+) \sqcup (A \cap \Omega_t^-)$ and (ii*), the third equality comes from (2), and the second equality comes from $\Omega_t^+ = \sqcup_{E_{t,j}^+ \in \mathcal{E}_t^+} E_{t,j}^+$, so

$$A \cap \Omega_t^+ = A \cap (\sqcup_{E_{t,j}^+ \in \mathcal{E}_t^+} E_{t,j}^+) = \sqcup_{E_{t,j}^+ \in \mathcal{E}_t^+} (A \cap E_{t,j}^+),$$

where the last equality comes from De Morgan laws. □

Proof of Proposition 5. Pick any $A, B \in \mathcal{F}$. Since $\Omega_t^+ \cap \Omega_t^- = \emptyset$, we have that

$$\begin{aligned} A \cup B &= ((A \cup B) \cap \Omega_t^+) \sqcup ((A \cup B) \cap \Omega_t^-) \\ &= ((A \cap \Omega_t^+) \cup (B \cap \Omega_t^+)) \sqcup ((A \cap \Omega_t^-) \cup (B \cap \Omega_t^-)) \\ &= (A_t^+ \cup B_t^+) \sqcup (A_t^- \cup B_t^-). \end{aligned}$$

A similar argument shows that $A \cap B = (A_t^+ \cap B_t^+) \cup (A_t^- \cap B_t^-)$, for all t.

For the second part, note that $A \setminus B = \{\omega \in \Omega : \omega \in A, \omega \notin B\}$. Then, $A \setminus B \cap \Omega_t^+ \equiv A_t^+ \setminus B_t^+ = \{\omega \in \Omega_t^+ : \omega \in A, \omega \notin B\}$; similarly, $A \setminus B \cap \Omega_t^- \equiv A_t^- \setminus B_t^- = \{\omega \in \Omega_t^- : \omega \in A, \omega \notin B\}$. But $\Omega_t^+ \sqcup \Omega_t^- = \Omega$, so the claim follows. □

Proof of Proposition 6. Let us focus on $\mathcal{F}_t^+$; it is closed with respect to countable intersections because it is a sigma-algebra. Then, pick $A, B \in \mathcal{F}_t^+$ such that $A \neq B$. If $A \cap B = \emptyset$, then $A \setminus B = A \in \mathcal{F}_t^+$; if $A \cap B \neq \emptyset$, then $A \setminus B = A \cap B^c$. Then, $B \in \mathcal{F}_t^+$ implies $B^c \in \mathcal{F}_t^+$ because $\mathcal{F}_t^+$ is a sigma-algebra; also, $A \cap B^c \in \mathcal{F}_t^+$ because $\mathcal{F}_t^+$ is closed with respect to countable intersections. So $A \setminus B \in \mathcal{F}_t^+$. Finally, the unit element is Ω_t^+: for any $A \in \mathcal{F}_t^+$, $A \cap \Omega_t^+ = A$ because $A \in \mathcal{F}_t^+$. Hence, $\mathcal{F}_t^+$ is a set algebra. We show in a similar fashion that $\mathcal{F}_t^-$ is a set algebra as well. □

Proof of Proposition 7. To ease notation, let $C \equiv \cup_{j \in \mathbb{N}_0} A_j$. C belongs to $\mathcal{F}$ because $\mathcal{F}$ is a sigma-algebra. Then, let $C \in \mathcal{F}_t^+$.

This implies that $P_t^{ex}(C) \geq 0$ but also that $P^{ex}(A) \geq 0$ and that $P^{ex}(C \cap A^c) \geq 0$, since $C = A \sqcup (C \cap A^c)$. Then,

$$P_t^{ex}(A) \leq P_t^{ex}(C) \leq \sum_{j \in \mathbb{N}_0} P_t^{ex}(A_j),$$

where the first inequality comes from Proposition 2 and the second one from Proposition 3.

If $C \in \mathcal{F}_t^-$, then $P_t^{ex}(C) \leq 0$, and also that $P^{ex}(A) \leq 0$ and that $P^{ex}(C \cap A^c) \leq 0$. Then,

$$P_t^{ex}(A) \geq P_t^{ex}(C) \geq \sum_{j \in \mathbb{N}_0} P_t^{ex}(A_j),$$

where again the first inequality comes from Proposition 2 and the second one from Proposition 3.

If $C \equiv \cup_{j \in \mathbb{N}_0} A_j$ is such that $C \cap \Omega_t^+ \neq \emptyset \neq C \cap \Omega_t^-$, then we cannot say anything general about the relation between $P_t^{ex}(A)$ and $\sum_{j \in \mathbb{N}_0} P_t^{ex}(A_j)$. □

Proof of Proposition 8. Note that by (7) and (8), we have that, for all t,

$$P_t^{ex}(A \cap \Omega_t^+) = P(A \cap \Omega_t^+)$$

and

$$P_t^{ex}(A \cap \Omega_t^-) = -P(A \cap \Omega_t^-),$$

for all $A \in \mathcal{F}$, where $P \in \Delta(\Omega, \mathcal{F})$. Now, to save some notation, let us denote by $\Omega^+ \equiv \cup_{t \in \mathbb{N}_0} \Omega_t^+$ and by $\Omega^- \equiv \cap_{t \in \mathbb{N}_0} \Omega_t^-$. Then, we have the following:

$$d_{ETV}(P_t^{ex}, P_\infty^{ex}) = \sup_{A \in \mathcal{F}} |P_t^{ex}(A) - P_\infty^{ex}(A)| \equiv |P_t^{ex}(\mathbf{A}) - P_\infty^{ex}(\mathbf{A})|.$$

Let us denote by $\mathbf{A}^+ \equiv \mathbf{A} \cap \Omega^+$ and by $\mathbf{A}^- \equiv \mathbf{A} \cap \Omega^-$, so $\mathbf{A} = \mathbf{A}^+ \sqcup \mathbf{A}^-$. Note that even though in the limit we fully discover $\mathbf{A}^+$, there may be some $t \in \mathbb{N}_0$ such that $\mathbf{A}^+ \cap \Omega_t^+ \neq \emptyset \neq \mathbf{A}^+ \cap \Omega_t^-$. So we have

$$|P_t^{ex}(\mathbf{A}) - P_\infty^{ex}(\mathbf{A})| = |P_t^{ex}(\mathbf{A}^+) + P_t^{ex}(\mathbf{A}^-) - P_\infty^{ex}(\mathbf{A}^+) - P_\infty^{ex}(\mathbf{A}^-)| \quad (A.5)$$

$$= |P_t^{ex}(\mathbf{A}^+) - P_\infty^{ex}(\mathbf{A}^+)| \quad (A.6)$$

$$= |P(\mathbf{A}^+ \cap \Omega_t^+) - P(\mathbf{A}^+ \cap \Omega_t^-) - P(\mathbf{A}^+ \cap \Omega_t^+)$$
$$- P(\mathbf{A}^+ \cap \Omega_t^-)| \quad (A.7)$$

$$= 2P(\mathbf{A}^+ \cap \Omega_t^-).$$

Equation (A.5) comes from $\mathbf{A} = \mathbf{A}^+ \sqcup \mathbf{A}^-$ and (ii*). Equation (A.6) comes from the fact that, for all t, $P_t^{ex}(\mathbf{A}^-) = P_\infty^{ex}(\mathbf{A}^-)$ because $\mathbf{A}^-$ is the portion of $\mathbf{A}$ that never leaves the latent space. Equation (A.7) comes from the updating procedure described in Section 3.2, from the countable additivity of P, and from $\mathbf{A}^+ = (\mathbf{A}^+ \cap \Omega_t^+) \sqcup (\mathbf{A}^+ \cap \Omega_t^-)$.

Now, note that the limit as $t \to \infty$ of $\mathbf{A}^+ \cap \Omega_t^-$ is

$$\mathbf{A}^+ \cap \bigcap_{t \in \mathbb{N}_0} \Omega_t^- = \mathbf{A}^+ \cap \Omega^- = \emptyset,$$

where the last equality is by construction. Then, by the continuity of P we have that

$$\lim_{t \to \infty} d_{ETV}(P_t^{ex}, P_\infty^{ex}) = 2 \lim_{t \to \infty} P(\mathbf{A}^+ \cap \Omega_t^-) = 2P(\emptyset) = 0,$$

which concludes the proof. □

Proof of Proposition 9. Suppose that $P_\infty^{ex}(\Omega) \neq 1$. This of course implies that $P_\infty^{ex}(\Omega) < 1$ because, from the definition of extended probabilities, $P_\infty^{ex}(A) \in [-1, 1]$, for all $A \in \mathcal{F}$. Then, this means that there exists a set $A \in \mathcal{F}$ such that $P_\infty^{ex}(A) \leq 0$, which implies $A \subset \Omega_\infty^-$. But we know that, if $\Omega_t^+ \uparrow \Omega$, then $\Omega_t^- \downarrow \Omega_\infty^- = \emptyset$, which contradicts $A \subset \Omega_\infty^-$. □

Proof of Theorem 2. Pick any lower extended probability $\underline{P}^{ex}$ and suppose there exists $\emptyset \neq \mathcal{P}^{ex} \subset \Delta^{ex}(\Omega, \mathcal{F})$ such that $\underline{P}^{ex}(A) = \inf_{P^{ex} \in \mathcal{P}^{ex}} P^{ex}(A)$, for all $A \in \mathcal{F}$. Now, by Theorem 1, we know that P^{ex} is coherent, for all $P^{ex} \in \mathcal{P}^{ex}$. This means that for all $\{B_j\}_{j=1}^n \subset \mathcal{B}'$, all $\{s_j\}_{j=1}^n \subset \mathbb{R}$, and all $P^{ex} \in \mathcal{P}^{ex}$,

$$\sup_{\omega \in \Omega} \left\{ \sum_{j=1}^n s_j \left[\mathbb{1}_{B_j}(\omega) - P^{ex}(B_j) \right] \right\} \geq 0. \tag{A.8}$$

Note that we use $\mathcal{B}'$ because that is the sigma-algebra generated by the events we can enter a bet about. Then, (A.8) entails that

$$\sup_{\omega \in \Omega} \left[\sum_{j=1}^n s_j \mathbb{1}_{B_j}(\omega) \right] \geq \sum_{j=1}^n s_j P^{ex}(B_j). \tag{A.9}$$

The inequality in (A.9) implies that

$$\sup_{\omega \in \Omega} \left[\sum_{j=1}^n s_j \mathbb{1}_{B_j}(\omega) \right] \geq \inf_{P^{ex} \in \mathcal{P}^{ex}} \sum_{j=1}^n s_j P^{ex}(B_j)$$

$$\geq \sum_{j=1}^n s_j \inf_{P^{ex} \in \mathcal{P}^{ex}} P^{ex}(B_j) \tag{A.10}$$

$$= \sum_{j=1}^n s_j \underline{P}^{ex}(B_j).$$

The results in (A.10) hold if and only if $\sup_{\omega \in \Omega} \{ \sum_{j=1}^n s_j [\mathbb{1}_{B_j}(\omega) - \underline{P}^{ex}(B_j)] \} \geq 0$. The claim follows. $\square$

Proof of Corollary 2. Pick a generic lower extended probability $\underline{P}^{ex}$, and suppose $\text{core}(\underline{P}^{ex})$ is non-empty. Then, $\underline{P}^{ex}$ avoids sure loss by Theorem 2. $\square$

Proof of Proposition 10. We first show that $\text{core}(\underline{P}^{ex})$ is convex. Pick any P_1^{ex}, P_2^{ex} in the core of $\underline{P}^{ex}$, any $\alpha \in (0,1)$, and any $A \in \mathcal{F}$. We have

$$\alpha P_1^{ex}(A) + (1-\alpha) P_2^{ex}(A) \geq \alpha \underline{P}^{ex}(A) + (1-\alpha)\underline{P}^{ex}(A) = \underline{P}^{ex}(A),$$

so $\alpha P_1^{ex} + (1-\alpha) P_2^{ex} \in \text{core}(\underline{P}^{ex})$.

We then show that core($\underline{P}^{ex}$) is weak*-compact. Recall that, in the weak* topology, a net $(P^{ex}_\alpha)_{\alpha \in I}$ converges to P^{ex} if and only if $P^{ex}_\alpha(A) \to P^{ex}(A)$, for all $A \in \mathcal{F}$. This proof is similar to the proof of Marinacci and Montrucchio (2004, Proposition 3), where the authors prove the same claim for the core of a bounded game. Pick any $P^{ex} \in \text{core}(\underline{P}^{ex})$, and let $k := 2\sup_{A\in\mathcal{F}}|\underline{P}^{ex}(A)|$. For any $A \in \mathcal{F}$, it holds that $P^{ex}(A) \geq \underline{P}^{ex}(A) \geq -k$. On the other hand, for all $A \in \mathcal{F}$, we have that

$$P^{ex}(A) = P^{ex}(\Omega) - P^{ex}(A^c) \leq \underline{P}^{ex}(\Omega) - \underline{P}^{ex}(A^c) \leq 2\sup_{A\in\mathcal{F}}|\underline{P}^{ex}(A)|.$$

This implies that $|P^{ex}(A)| \leq k$, for all $A \in \mathcal{F}$. By Dunford and Schwartz (1958, page 94), we have that

$$\|P^{ex}\| := \sup \sum_{j=1}^n |P^{ex}(A_j) - P^{ex}(A_{j-1})| \leq 2k, \quad (A.11)$$

where the supremum is taken over all finite chains $\emptyset = A_0 \subset A_1 \subset \cdots \subset A_n = \Omega$. Then, (A.11) implies that

$$\text{core}(\underline{P}^{ex}) \subset \{P^{ex} \in \Delta^{ex}(\Omega, \mathcal{F}) : \|P^{ex}\| \leq 2k\}.$$

By the Alaoglu theorem Dunford and Schwartz (1958, Theorem 2, page 424), we know that $\{P^{ex} \in \Delta^{ex}(\Omega, \mathcal{F}) : \|P^{ex}\| \leq 2k\}$ is weak*-compact. Hence, to complete the proof, we are left to show that core($\underline{P}^{ex}$) is weak*-closed. Let then $(P^{ex}_\alpha)_{\alpha\in I}$ be a net in core($\underline{P}^{ex}$) that weak*-converges to $P^{ex} \in \Delta^{ex}(\Omega, \mathcal{F})$. Using the properties of the weak* topology, it is easy to see that $P^{ex} \in \text{core}(\underline{P}^{ex})$. Hence, core($\underline{P}^{ex}$) is weak*-closed. The claim follows. $\square$

Proof of Proposition 11. Equation (20) comes from the fact that, for all t,

$$P^{ex}_{t+1}(E^+_{t+1,\omega}) = |P^{ex}_t(E^-_{t,\omega})|,$$

that is, $P^{ex}_{t+1}(E^+_{t+1,\omega}) = -P^{ex}_t(E^-_{t,\omega})$, and that

$$-\overline{P}^{ex}_t(E^-_{t,\omega}) \leq -P^{ex}_t(E^-_{t,\omega}) \leq -\underline{P}^{ex}_t(E^-_{t,\omega}). \quad \square$$

A.2 Application to opinion dynamics

This example comes from the model in Allahverdyan and Galstyan (2014). There, opinion dynamics between a persuaded and a persuading agents is studied, in particular when the persuaded agent evaluates new information in a way that is consistent with her own preexisting belief. In this example, we are going to adoperate extended probabilities to model the boomerang effect. This phenomenon corresponds to the empirical observation that sometimes persuasion yields the opposite effect: the persuaded agents move her opinion away from the opinion of the persuading agent. That is, she enforces her old opinion.

In Allahverdyan and Galstyan (2014), the authors assume that the state of the world does not change, that the agents are aware of this fact, and that the persuaded agent changes her opinion only under the influence of the opinion of the persuading agent. They model this dynamic in a linear fashion. The first iteration is the following:

$$\hat{P}_1(\{\omega_k\}) = \epsilon P_0(\{\omega_k\}) + (1-\epsilon)Q(\{\omega_k\}), \quad \epsilon \in [0,1], \qquad (A.12)$$

where $\omega_k \in \Omega$, a finite sample space, and P_0 and Q are regular probability measures that represent the initial opinions of the persuaded and the persuading agents, respectively. Q is not indexed to time because they make the simplifying assumption that the persuading agent does not change his mind: he tries to persuade the other agent of the same thing at every iteration. ϵ is a weight, and several qualitative factors contribute to its subjective assessment: egocentric attitude of the persuaded agent, the fact that the persuaded agent has access to internal reasons for choosing her opinion, while she is not aware of the internal reasons of the persuading agent, and many more. The authors relate ϵ to the credibility of the persuading agent: the higher the ϵ, the less credible he is. The successive iterations are modeled as follows:

$$\hat{P}_{t+1}(\{\omega_k\}) = \epsilon \hat{P}_t(\{\omega_k\}) + (1-\epsilon)Q(\{\omega_k\}), \qquad (A.13)$$

for $t \geq 1$. This means that at every iteration, the persuading agent tries to shift the persuaded agent's opinion closer to his own. This continues until either the persuaded agent is content with her opinion (and hence does not further change her beliefs) or the persuading agent completely convinces the other agent.

Now, one way to model the boomerang effect is to consider $\epsilon > 1$. This conveys the idea that the persuading agent has an extremely low credibility. This results in the updated opinion of the persuaded agent to be an extended probability measure. Indeed, for $\epsilon > 1$, $\hat{P}_{t+1}(\{\omega_k\})$ is negative whenever $\epsilon \hat{P}_t(\{\omega_k\}) < |1-\epsilon| Q(\{\omega_k\})$. In Allahverdyan and Galstyan (2014), the authors consider the induced regular probability measure to avoid working with extended probabilities.

We modify slightly the linear model in Allahverdyan and Galstyan (2014). The main differences are three: we allow the use of extended probabilities, we describe a state space that is divided in latent and known, and whose composition is gradually discovered as the time passes by, and we do not let ϵ be a free parameter. It depends on both the element ω_k of the state space we examine and on the iteration t we are considering.

Mathematically, we can describe the low credibility of the persuading agent through hidden states: the persuaded agent may think that she does not know the state space well enough, that is, that there are some hidden portions of Ω she is not (yet) aware of.

We assume that the state space $\Omega = \{\omega_1, \ldots, \omega_N\}$ is a finite set (a common simplifying assumption in opinion dynamics). Note that the finest possible partition of Ω is $\mathcal{E} = \{\{\omega_j\}\}_{j=1}^N$. At the beginning of the interaction between persuading and persuaded agents, the former is fully aware of the composition of the state space, while the latter is only aware of the composition of $\Omega_0^+ \subsetneq \Omega$, but suspects that the actual state space is larger. This corresponds to having suspects on the persuading agent hiding some pieces of information. In particular, she correctly guesses that the true state space is Ω. This correct guess is without loss of generality for our analysis: we are in the $\Omega_t^+ \uparrow \Omega$ case; we also assume that the discovering procedure is equivalent to an urn without replacement. She then defines an extended probability on Ω the way we explained in Section 3. That is, she specifies the oracle probability distribution P on the whole Ω and then flips the sign to the probabilities of the latent events: $P_0^{ex}(\{\omega_k\}) = P(\{\omega_k\})$ if $\omega_k \in \Omega_0^+$, and $P_0^{ex}(\{\omega_k\}) = -P(\{\omega_k\})$ if $\omega_k \notin \Omega_0^+$.

To obtain the influenced extended probability at any time $t \geq 0$, we modify slightly Equation (A.13) to get

$$\hat{P}_t^{ex}(\{\omega_k\}) = \epsilon_{k,t} P_t^{ex}(\{\omega_k\}) + (1 - \epsilon_{k,t}) Q(\{\omega_k\}). \qquad (A.14)$$

The non-negative $\epsilon_{k,t}$'s have to be chosen such that $\hat{P}_t^{ex}$ is an extended probability measure, that is, $\hat{P}_t^{ex}(\{\omega_k\}) \in [-1, 1]$ for all k, for all t, and

$$\hat{P}_t^{ex}(\Omega) = \sum_{k=1}^{N} \hat{P}_t^{ex}(\{\omega_k\}) \leq 1.$$

Here, $\hat{P}_t^{ex}$ denotes the influenced extended probability at time t. Note that $\hat{P}_t^{ex}$ is analytically similar to an ϵ-contaminated probability measure. There are of course two major differences: for some ω_k, $\epsilon_{k,t}$ is greater than 1, and also one of the elements of the mixture is an extended probability measure (rather than a regular one). Note also that Q is not indicized to time because we too make the simplifying assumption that the persuading agent does not change his mind. Another characteristic worth noting is that in our model, at every iteration t, the persuaded agent combines her updated belief (expressed via the extended probability P_t^{ex}) with the other agent's belief to obtain $\hat{P}_t^{ex}$. This is different from the model in Allahverdyan and Galstyan (2014) where at every iteration t the persuaded agent combines her influenced belief at iteration $t - 1$ (expressed through $\hat{P}_{t-1}$) with the other agent's belief to obtain $\hat{P}_t$. This because in their model the world does not change, so the only way of describing an opinion dynamics is the one the authors illustrate.

The persuaded agent updates P_t^{ex} as specified in Section 3. That is, when at time t she observes ω_k that used to belong to the latent space, $P_t^{ex}(\{\omega_k\}) = |P_{t-1}^{ex}(\{\omega_k\})|$, while the extended probabilities for $\omega_s \neq \omega_k$ are kept constant. Let us be more precise about the differences with Equation (A.12); $\epsilon_{k,t}$ depends on both k and t. The persuaded agent has a different perception of the opinion of the persuading agent depending on whether she is not sure the topic they are debating about belongs to the state space Ω, so for ω_k in the latent space, $\epsilon_{k,t}$ is greater than 1. In addition, as time passes by, the hidden elements of the state space become known, so (part of) the credibility of the persuading agent is restored. The $\epsilon_{k,t}$ associated with ω_k observed at time t becomes smaller than 1, for all ω_k.

Note that, for $t \geq N$, P_t^{ex} is a regular probability measure because the persuaded agent discovers the composition of the whole state space, so the latent space shrinks to the empty set.

Throughout this section we made the tacit assumption that $\mathcal{P}$, the set of probability measures on Ω that induces the set of extended probabilities $\mathcal{P}_0^{ex}$ at time $t = 0$, is the singleton $\{P\}$ so that $\mathcal{P}_0^{ex} = \{P_0^{ex}\}$. This simplifying assumption can be dropped, and the analysis stays the same. The only difference is that we have to repeat it for all the elements of $\mathcal{P}_t^{ex}$, for all $t \in \mathbb{N}_0$. Every set $\mathcal{P}_t^{ex}$ of extended probabilities induces a set $\hat{\mathcal{P}}_t^{ex}$ of influenced extended probabilities. For all $t \geq N$, $\mathcal{P}_t^{ex}$ is a set of regular probability measures, and $\hat{\mathcal{P}}_t^{ex}$ is a set of influenced regular probability measures.

A.3 De Finettian interpretation of subjective probability

De Finetti admits that it might have been better to adopt the seemingly more general approach of Ramsey and Savage of defining bets whose payoffs are in utils (de Finetti, 1974, page 79):

> The formulation [...] could be made watertight [...] by working in terms of the utility instead of with monetary value. This would undoubtedly be the best course from the theoretical point of view, because one could construct, in an integrated fashion, a theory of decision-making [...] whose meaning would be unexceptionable from an economic viewpoint, and which would establish simultaneously and in parallel the properties of probability and utility on which it depends.

Nevertheless, he found "other reasons for preferring" the money bet approach (de Finetti, 1974, page 81):

> The main motivation lies in being able to refer, in a natural way to combinations of bets, or any other economic transactions, understood in terms of monetary value (which is invariant). If we referred ourselves to the scale of utility, a transaction leading to a gain of amount S if the event E occurs would instead appear as a variety of different transactions, depending on the outcome of other random transactions. These, in fact, cause variations in one's fortune, and therefore in the increment

of utility resulting from the possible additional gain S: conversely, suppose that in order to avoid this one tried to consider bets, or economic transactions, expressed, let us say, in "utiles" (units of utility, definable as the increment between two fixed situations). In this case, it would be practically impossible to proceed with the transactions, because the real magnitudes in which they have to be expressed (monetary sums or quantities of goods, etc.) would have to be adjusted to the continuous and complex variations in a unit of measure that nobody would be able to observe.

References

Allahverdyan, A. E. and Galstyan, A. (2014). Opinion dynamics with confirmation bias, *PLoS One* **9**(7).

Allen, E. H. (1976). Negative probabilities and the uses of signed probability theory, *Philosophy of Science* **43**(1), 53–70.

Bartlett, M. S. (1945). Negative probability, *Mathematical Proceedings of the Cambridge Philosophical Society* **41**, 71–73.

Benavoli, A. Facchini, A. and Zaffalon, M. (2019). Computational complexity and the nature of quantum mechanics (extended version), *Available at arXiv:1902.03513*.

Benavoli, A. Facchini, A., and Zaffalon, M. (2021). The weirdness theorem and the origin of quantum paradoxes, *Foundations of Physics* **51**(95).

Berger, J. O. (1984). The robust Bayesian viewpoint. In ed. Kadane, J. B. *Robustness of Bayesian Analyses*. Amsterdam: North-Holland.

Blass, A. and Gurevich, Y. (2022). Negative probabilities: What they are and what they are for, *arXiv:2009.10552*.

Bollen, K. A. (2002). Latent variables in psychology and the social sciences, *Annual Review of Psychology* **53**, 605–634.

Burgin, M. and Meissner, G. (2011). Negative probabilities in modeling random financial processes, *Integration: Mathematical Theory and Applications* **2**(3), 305–322.

Burgin, M. (2012). Integrating random properties and the concept of probability, *Integration: Mathematical Theory and Applications* **3**(2), 137–181.

Burgin, M. (2013). Negative probability in the framework of combined probability, *arXiv:1306.1166*.

Caprio, M. and Gong, R. (2022). Dynamic precise and imprecise probability kinematics, *arXiv:2110.04382*.

Caprio, M. and Mukherjee, S. (2022). Ergodic theorems for dynamic imprecise probability kinematics, *arXiv:2003.06502*.

de Finetti, B. (1937). La prévision: ses lois logiques, ses sources subjectives, *Annales de l'institut Henri Poincaré* **7**(1), 1–68.

de Finetti, B. (1974). *Theory of Probability*. vol. 1, New York: Wiley.

de Finetti, B. (1975). *Theory of Probability*. vol. 2, New York: Wiley.

Dempster, A. P. (1967). Upper and lower probabilities induced by a multivalued mapping, *The Annals of Mathematical Statistics* **38**(2), 325–339.

Dirac, P. A. M. (1930). Note on exchange phenomena in the Thomas atom, *Mathematical Proceedings of the Cambridge Philosophical Society* **26**, 376–395.

Dunford, N. and Schwartz, J. T. (1958). *Linear Operators, Part I: General Theory*. London: Wiley Interscience.

Ellsberg, D. (1961). Risk, ambiguity, and the Savage axioms, *The Quarterly Journal of Economics* **75**(4), 643–669.

Feintzeig, B. H. and Fletcher, S. C. (2017). On noncontextual, non-Kolmogorovian hidden variable theories, *Foundations of Physics* **47**, 294–315.

Ferrie, C. (2011). Quasi-probability representations of quantum theory with applications to quantum information science, *Reports on Progress in Physics* **74**(11), 116001.

Fischer, T. (2012). Existence, uniqueness, and minimality of the Jordan measure decomposition, *arXiv:1206.5449*.

Gell-Mann, M. and Hartle, J. B. (2012). Decoherent histories quantum mechanics with one "real" fine-grained history, *Physical Review A* **85**, 062120.

Gong, R. (2018). Modeling uncertainty with sets of probabilities. In *Foundations of Probability Seminar Series*, Rutgers University.

Gurevich, Y. and Vovk, V. (2021). Betting with negative probabilities, *The Game-Theoretic Probability and Finance Project, Working Paper #58*.

Hartle, J. B. (2004). Linear positivity and virtual probability, *Phisical Review A* **70**, 022104.

Hartle, J. B. (2008). Quantum mechanics with extended probabilities, *Physical Review A* **78**, 012108.

Heisenberg, W. K. (1931). Über die inkohärente Streuung von Röntgenstrahlen, *Physikalische Zeitschrift* **32**, 737–740.

Hu, Y. (2017). The econometrics of unobservables: Applications of measurement error models in empirical industrial organization and labor economics, *Journal of Econometrics* **200**(2), 154–168.

Jarrow, R. A. and Turnbull, S. M. (1995). Pricing derivatives on financial securities subject to credit risk, *Journal of Finance* **L**(1), 53–85.

Khrennikov, A. (2009). *Interpretations of probability*. Berlin/New York: Walter de Gruyter.

Kronz, F. (2007). Non-monotonic probability theory and photon polarization, *Journal of Philosophical Logic* **36**, 449–472.

Lowe, D. (2004/2007). Machine learning, uncertain information, and the inevitability of negative "probabilities", *Available at Machine Learning Workshop 2004*.

Marinacci, M. and Montrucchio, L. (2004). Introduction to the mathematics of ambiguity. In ed. Gilboa, I. *Uncertainty in Economic Theory: A Collection of Essays in Honor of David Schmeidler's 65th Birthday*, London: Routledge.

Nau, R. F. (2001). De Finetti was right: Probability does not exist, *Theory and Decision* **51**, 89–124.

Ramsey, F. P. (1964). Truth and Probability. In eds. H. E. K. Jr. and Smokler, H. E. *Studies in Subjective Probability*, pp. 61–92. New York: Wiley.

Rabe-Hesketh, S. and Skrondal, A. (2008). Classical latent variable models for medical research, *Statistical Methods in Medical Research* **17**(1), 5–32.

Rao, K. P. S. B. and Rao, M. B. (1983). *Theory of Charges, a Study of Finitely Additive Measures*. London: Academic Press.

Savage, L. J. (1954). *The Foundations of Statistics*. New York: John Wiley and Sons.

Székely, G. J. (2005). Half of a coin: Negative probabilities, *Wilmott Magazine*. pp. 66–68.

Tijms, H. C. and Staats, K. (2007). Negative probabilities at work in the M/D/1 queue, *Probability in the Engineering and Informational Sciences* **21**(1), 67–76.

Tsitsiklis, J. (2018). Sample spaces. In *Introduction to Probability (online course)*. Massachusetts Institute of Technology.

Walley, P. (1991). *Statistical Reasoning with Imprecise Probabilities*, vol. 42, *Monographs on Statistics and Applied Probability*. London: Chapman and Hall.

Walley, P. (1996). Inferences from multinomial data: Learning about a bag of marbles, *Journal of the Royal Statistical Society. Series B (Methodological)* **58**(1), 3–57.

Wiewiora, E. W. (2008). *Modeling Probability Distributions with Predictive State Representations*. PhD thesis, University of California, San Diego.

Wigner, E. P. (1932). On the quantum correction for thermodinamic equilibrium, *Physical Review* **40**, 749–759.

Chapter 12

Infoautopoiesis and the Mind–Body Problem

Jaime F. Cárdenas-García

Department of Mechanical Engineering University of Maryland, Baltimore County, 1000 Hilltop Circle Baltimore, MD 21250, USA

jfcg@umbc.edu

Since the beginning of time, the matter/energy in the universe has existed in constant flux. The result after billions of years is the materialization of life, a phenomenon that remains unexplained. Continuous development allowed life to attain its countless forms of expression that we find today. All of these multifarious expressions of life share the ability to interact with their environment to satisfy their physiological and/or relational needs and make effective their unavoidable metabolic connection with the environment. Thus resolving the mind–body problem or how organisms coordinate the world and their images of the world to create an environment in their own self-image, to make it amenable for their development. This is accomplished not only by having the organs detect the sensorial signals reflecting the constant spatial/temporal motion/change of matter/energy around them but also by possessing the ability for infoautopoiesis or information self-production, so as to homeorhetically inform the necessary syntactical actions to modify matter/energy in a continuous cycle of sensation-information-action starting with semantical interpretation. This chapter uses the process of infoautopoiesis as the foundation to address the mind–body problem.

1 Introduction

The mind–body problem is of fundamental importance in philosophy, yet no definitive solution exists (Horgan, 2018, 2019). Traditionally, the mind–body problem is defined in terms of two separate human entities: the mind and body. The problem is one of resolving their interconnections and dependencies. A more recent nuanced account states, "Adequate explanation in any cognitive science must at some stage address the matter-mind problem, that is, the problem of symbol reference or how the world and our images of the world are coordinated"(Pattee, 2012). In other words, how do human beings internalize their sensory perceptions of the environment around them in a correlated manner to derive meaning from their perceptions, so as to act effectively in the world? This is a process that is difficult to pinpoint but that we hope to uncover. To do so, we rely on Gregory Bateson's characterization of information as "a difference which makes a difference" to an organism (Bateson, 1978). Such a characterization of information evolves from the notion that an organism-in-its-environment is subjected to a noisy environment where matter and/or energy are in constant motion. If the organism is to make sense of its environment, it needs to be selective of the sensorial signals from the environment that allow it to identify elements in that environment that permit satisfaction of its physiological and/or relational needs. In short, we anticipate that information is an important link.

But the concept of information for many scientists appears to elude definition and explanation. In spite of this, the consensus is to postulate information as one of three fundamental elements in the universe, besides matter and/or energy (Barbieri, 2012, 2013; Battail, 2009, 2013; Brier, 1999, 2008; Burgin, 2010; Floridi, 2010, 2011; Hidalgo, 2015; Lloyd, 2006; Pattee, 2013; Stonier, 1997; Umpleby, 2007; Vedral, 2010; Wiener, 1948; Yockey, 2005; Rovelli, 2016; Chalmers, 1995, 2020; Bayne and Chalmers, 2003; Clark and Chalmers, 1998; Koch, 2009, 2019). This widely held but unfounded perspective can be summarized as "it from bit"(Wheeler, 1991).

This chapter seeks to find the connection between information and the mind–body problem under the light of infoautopoiesis (info = information; auto = self; poiesis = creation/production) or "bit from it". As suggested above, the process of infoautopoiesis relies on Gregory Bateson's definition of information as "a difference which makes a difference" to a living being (Bateson, 1978). Infoautopoiesis is the individuated sensory commensurable, self-referential, recursive, interactive homeorhetic feedback process relevant to all living beings in their endless quest to satisfy physiological and/or relational needs (Cárdenas-García, 2020). It relies on the factual observation that self-produced information is not a fundamental quantity of the universe but the result of the detection and self-referential assessment of sensorial signals which reflect the environment and play an important role in making the external environment meaningful.

To discuss the process of infoautopoiesis and its implications for the mind–body problem, this chapter is divided into several sections. First, the concept of information is naturalized by examining its etymological origins as well as Gregory Bateson's characterization as "a difference which makes a difference". Next, the process of infoautopoiesis is shown as central to the self-production of information and is used to present and frame the body–mind problem. This includes consideration of the human organism-in-its-environment, and the self-generation, processing, and transmission of information.

2 What is Information?

The concept of information, even from a cursory examination of the literature, has great diversity. Due to its broad applicability in many fields of knowledge, a wide range of criteria seem to apply in its study (Bateson, 1978; Barbieri, 2012, 2013; Pattee, 2013; Bawden and Robinson, 2022; Burgin and Hofkirchner, 2017; Capurro and Hjørland, 2003; Shannon *et al.*, 1993; Shannon, 1948; Crick, 1970; Sharov, 2009, 2016; Heras-Escribano and de Jesus, 2018). While the kinds of information are diverse, what we want to identify are general characteristics of information that naturalize its meaning. The etymology of the word information yields its Latin roots in the word *informatio*, which comes from the verb *informare* (to inform) in the sense of giving shape to something material, as well as the act of communicating knowledge to another person (Capurro and Hjørland, 2003; Capurro, 2009; Díaz Nafría, 2010; Peters, 1988). Parallel to its etymological origins, Bateson defines information as "a difference which makes a difference" (Bateson, 1978). These two approaches to information imply interactivity. Its etymology accentuates an organism's capacity for recursive interaction with matter/energy in its environment to achieve its ends. The Batesonian approach manifests the human ability for sensorially commensurable spatial/temporal comparison (between two instances) in pursuit of the satisfaction of its physiological (internal/external) and/or relational needs. The first, objective "difference" needs to be sufficiently large in magnitude yet suitably slow to be detectable by human sensory organs. The second, subjective "difference which makes a difference" is assessed by the human organism as to whether it is able to satisfy its physiological (internal/external) and/or relational needs (Cárdenas-García, 2020; Cárdenas-García and Ireland, 2019; Burgin and Cárdenas-García, 2020). These "differences" reflect the dynamic self-referential, interactive, and recursive acts of the human organism-in-its-environment in an ascending virtuous cyclical spiral of sensation-information-action. In summary, information is not a fundamental quantity of the universe but rather the most important element created by human organisms-in-their-environment to discover themselves and their impact on their environment.

Another important aspect of Batesonian information is that "difference" and "idea" are found to be synonymous and homeorhetic (Bateson, 1978). The endless cyclic, self-referential, recursive, and interactive process of sensation-information-action related to Bateson's "difference which makes a difference" is illustrated by the actions of a woodcutter on a tree:

> Consider a tree and a man and an axe. We observe that the axe flies through the air and makes certain sorts of gashes in a pre-existing cut in the side of the tree. If now we want to explain this set of phenomena, we shall be concerned with differences in the cut face of the tree, differences in the retina of the man, differences in his central nervous system, differences in his efferent neural messages, differences in the behavior of his muscles, differences in how the axe flies, to the differences which the axe then makes on the face of the tree. Our explanation (for certain

purposes) will go round and round that circuit. In principle, if you want to explain or understand anything in human behavior, you are always dealing with total circuits, completed circuits. This is the elementary cybernetic thought (Bateson, 1978).

The woodcutter performs self-referencing, interactive, and recursive labor to harvest wood and is applicable to many human job tasks. The activities of the woodcutter are homeorhetic as they converge toward a dynamic trajectory where the activity of the woodcutter follows a moving and developing target during the activity, i.e., shaping the surface of the tree to make it fall. Repetition of an activity such as cutting down trees generally leads to skill improvement and efficiency.

In short, this approach to information naturalizes information as something palpable in our daily lives. Information is something that we internalize as we actively connect with our environment and also impart to matter and other living beings in the dynamic engagement of promoting the ever-expanding cycle of sensation-information-action.

3 Infoautopoiesis and the Mind–Body Problem

The process of infoautopoiesis encapsulates the cyclic process of sensation-information-action immanent to Bateson's *difference which makes a difference*, that engages all humans in their efforts to satisfy their physiological (internal/external) and relational needs (Cárdenas-García, 2020). This self-referential, sensory commensurable, recursive, and interactive individuated homeorhetic feedback process illustrates how the human organism interacts with its environment and how these interactions are constitutive of information generation, information exchange, information relations, and life.

The representation in Fig. 1 shows a human organism-in-its-environment engaged in the process of infoautopoiesis. It is an illustration of the cyclic, self-referential, interactive, and recursive process of sensation-information-action between the organism and the environment. The direction of the arrowheads in Fig. 1 shows that the depicted flow begins as environmental noise; it then transforms into

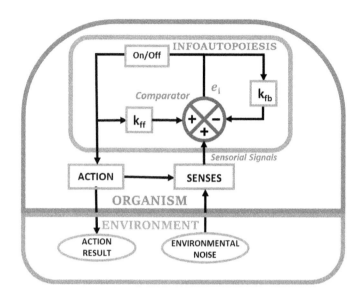

Figure 1. The human organism-in-its-environment and infoautopoiesis.

sensorial signals after detection and transduction by the senses, which are then converted, in the box identified as infoautopoiesis, into useful/meaningful information for the organism-in-its-environment, and, as a result, elicits an action that is exerted on the environment and identified as an action result (Cárdenas-García, 2020, 2022; Cárdenas-García and Ireland, 2019; Burgin and Cárdenas-García, 2020).

An internal and external circuit are identified: the internal homeorhetic cybernetic self-referential circuit is the one that makes effective Bateson's (1978) definition of information as "a difference that makes a difference" and is represented in Fig. 1 as a box labeled as infoautopoiesis. The external circuit allows the organism to influence its environment in a self-referential, interactive, and recursive way in line with the internal circuit. This external circuit is defined to begin from the environmental noise, which is admitted and processed by the organism in the infoautopoiesis box in response to its physiological and/or relational needs which results in an action that impacts the environment as an action result, as well as the senses of the human organism. The feedback of the action into the human organism is by way of the senses and the action result that cuts through the environmental noise since it is an expected signal by the human organism-in-its-environment.

In Fig. 1, after the sensorial signals reach the box labeled infoautopoiesis, a meaning-making box, they are processed and are converted into internalized semantic information. It is here that after an accumulation of internalized semantic information the organism deploys its capacity for action that is specific to the action result that is sought in its environment to pursue satisfaction of physiological (internal/external) and/or relational needs. This action leading to an action result may be characterized as externalized syntactic information which is a consequence of the externalization of the internalized semantic information. In other words, an externalized syntactic informational homeorhetic action occurs depending on the internalized semantic information generated by the human organism-in-its-environment due to the acquired sensorial signals identified by the self-referenced needs of the organism. This is a never-ending self-referential, interactive, and recursive sensation-information-action cycle. This arrangement of components serves to illustrate, in a rudimentary way, how the matter–mind problem might be addressed since it shows ... *how the world and our images of the world are coordinated* (Pattee, 2012, 1982). Another implication is that the mind–body problem is a misnomer since the body of the organism-in-its-environment is an integrated, indivisible whole.

Recent research serves to illustrate this type of behavior. Researchers discovered that bacterial spores *Bacillius subtitilis*, which are partially dehydrated cells, can analyze their environment, despite being in a lethargic state and considered as physiologically dead for years, to survive disadvantageous environmental conditions. However, they continue to generate information from short-lived environmental signals, leaving their dormant state after accumulating sufficient sensorial signals to confirm that they can activate again and return to life under now more favorable environmental conditions (Kikuchi *et al.*, 2022; Lombardino and Burton, 2022). This is precisely what we want to represent as occurring within the infoautopoiesis block of Fig. 1.

Since internalized semantic information leads to externalized syntactic information in an ever-evolving cycle, the implication is that there is no information in the environment, except for that due to externalized syntactic information. Our senses are incapable of identifying information in the environment. We are only capable of capturing sensorial signals that need interpretation to infoautopoietically create internalized semantic information. This is in contradiction to the postulate of many scientists who believe that information exists in the environment, as noted above.

Another approach to explaining the misnamed mind–matter problem is to rely on Charles Sanders Peirce when he describes semiosis as a process and states, "A sign is anything which determines something else (its interpretant) to refer to an object to which [it] itself refers (its object) in the same way, the interpretant becoming in turn a sign, and so on ad infinitum" (CP 2.303) [This means that the quote comes from The Collected Papers of Charles Sanders Peirce, Section 2, Paragraph 303] (Peirce, n.d., 1958). To illustrate semiosis, we use the organism-in-its-environment shown in Fig. 1; the "sensorial signals" from the environment (the object) result in the internalized semantic information (the interpretant) which yields externalized syntactic information (a sign) and so on *ad infinitum*.

What we intended to show is that the infoautopoietic process is able to address the mind–body problem, i.e., *the problem of symbol reference or how the world and our images of the world are coordinated* (Pattee, 2012, 1982). In other words, human beings can internalize their sensory perceptions of the environment around them in a correlated manner to derive meaning from their perceptions, as well as use that internalized meaning to produce externalized syntactic information to achieve satisfaction of physiological and/or relational needs.

In summary, infoautopoiesis, the sensory commensurable, self-referential, recursive, interactive homeorhetic feedback process of information self-production, is at the center of the capacity of a human organism-in-its-environment to mold its surroundings in its own self-image.

4 Discussion

There is a tendency by most researchers to postulate the existence of information as a mysterious quantity that is to be found everywhere in our environment, except that it is difficult to describe, and no identifiable sense organs seem to detect it. Yet, the notion that information can be identified gives credence to the colloquial expression "I know it when I see it".

Norbert Wiener attempted to be more specific when he stated that "Information is information, not matter or energy. No materialism which does not admit this can survive at the present day" (Wiener, 1948). While describing information in terms of itself, this points to a general belief among many scientists that information is a third quantity of the universe besides matter and/or energy. Something that is wholly dependent on a postulate that has no basis in fact.

The approach promoted in this chapter is that information is paramount to the functioning of the human organism-in-its-environment. An etymological perspective ties information to giving shape to matter and using communication as the means to shape the minds of other individuals. While also using Gregory Bateson's definition of information as "a difference which makes a difference", it makes for the possibility that information may be identified and analyzed. Indeed, both perspectives coincide in promoting a naturalized and dynamic view of information. They promote the view that information is a means to describe change in matter and/or energy and that humans have an individuated role to play in acting to promote and observe that change. This approach ties the finding of an answer to phenomena that may be observed in the daily lives of humans and how they interact with their environment.

The result is a new paradigm to study information, that of infoautopoiesis, or the self-referential, sensory commensurable, recursive, and interactive homeorhetic feedback process immanent to Bateson's "difference which makes a difference" by which a human organism-in-its-environment pursues satisfaction of its physiological (internal/external) and/or social needs. This yields the discovery that the information process is a never-ending sensation-information-action cycle that allows us to discover and act on our environment.

The process of infoautopoiesis transforms the "sensorial signals" of the noisy environment in which all living beings live, through their motivated efforts to satisfy their physiological and/or relational needs to improve their ability to engage in their ever-changing environment. The human organism-in-its-environment, through a triadic process involving personal-subjective-relative, impersonal-objective-absolute, and Shannon-distilled information (Cárdenas-García, 2020), can internally generate semantic information that it can then externalize as syntactic information. Our syntactic creations are all the artificial creations that we have created and surround us, some being very rudimentary, but others of great sophistication and technological scope.

Previously, it has been possible to make the argument that there exists a connection between the mind–body problem, information, and meaning. Figure 1 is elucidatory of the possibility of resolving the mind–body problem by illustrating how errors/differences/information/ideas are self-produced from "sensorial signals". These errors/differences/information/ideas, as images of the world, reflect the meaning that helps the human being coordinate its actions in the external environment (Pattee, 1982, 2012). Generalizing, the self-produced errors/differences/information/ideas are always meaningful for the organism, whether or not they are the result of voluntary or involuntary infoautopoiesis. Determining differences is the source for satisfaction of our most basic physiological needs, such as breathing and eating, changing our surroundings by acting on our environment, and when engaged in discussions with others.

5 Conclusions

Infoautopoiesis is a new paradigm for understanding information in the context of all living beings-in-their-environment. Infoautopoiesis is the process of self-production of information, an individuated sensory commensurable, self-referential, recursive, interactive homeorhetic feedback process immanent to Bateson's "difference which makes a difference". A basic premise to infoautopoiesis is that information is not a fundamental quantity of the universe, yet its importance cannot be underestimated for human organisms-in-their-environment. Humans self-produce information to discover the bountifulness of matter and/or energy as expressions of their environmental spatial/temporal motion/change, as information or "differences which make a difference", to satisfy their physiological (internal/external) and relational needs.

Infoautopoiesis defines a connection between the mind–body problem, information, meaning, and life and results in the generation of internalized and externalized information relevant to human organisms-in-their-environment. The self-production of inaccessible internalized or semantic information makes the external environment meaningful. Semantic information is made accessible by communicating through externalized syntactic expressions using language, gestures, pictographs, musical instruments, sculptures, writing, coding, etc. We live in and are surrounded by our artificial creations. This means that all externalized expressions of human-created knowledge are syntactic in nature and require re-interpretation by other peers through "sensorial signals". Syntactic artificial creations surround us and make us believe that information exists in the environment, yet there is no information in the environment or in the universe independent of living beings. Information cannot be the primary element that allows living beings their unique existence.

Acknowledgment

This chapter is dedicated to the memory of Mark Burgin. Writing it brings fond memories of our interactions and collaboration. Mark was someone whose friendly and humble approach allowed our collaboration to occur, resulting in a dialogic publication about information. Mark's friendly but incisive questioning helped me immensely in thinking more fundamentally and thoroughly about infoautopoiesis. For this, I will always remember him as a warm soul that sought to bring the best intellectual outcome out of our interactions. Mark is affectionately remembered and lives forever in his works and in his deeds.

References

Barbieri, M. (2012). What is information? *Biosemi.* 5(2), 147–152.
Barbieri, M. (2013). The paradigms of biology. *Biosemi.* 6(1), 33–59.
Bateson, G. (1978). *Steps to an ecology of mind; collected essays in anthropology, psychiatry, evolution, and epistemology.* New York: Ballantine Books, xxviii, p. 545.
Battail, G. (2009). Applying semiotics and information theory to biology: A critical comparison. *Biosemi.* 2(3), 303.
Battail, G. (2013). Biology needs information theory. *Biosemi.* 6(1), 77–103.
Bawden, D. and Robinson, L. (2022). *Introduction to information science,* 2nd edn. London: Facet Publishing.
Bayne, T. and Chalmers, D. J. (2003). *What is the Unity of Consciousness? The Unity of Consciousness: Binding, Integration, and Dissociation.* New York, NY, US: Oxford University Press. pp. 23–58.
Brier, S. (1999). Biosemiotics and the foundation of cybersemiotics: Reconceptualizing the insights of ethology, second-order cybernetics, and Peirce's semiotics in biosemiotics to create a non-Cartesian information science. *Semi.* 127(1–4), 169–198.
Brier, S. (2008). *Cybersemiotics: Why Information is not Enough!* Toronto; Buffalo: University of Toronto Press. xx, p. 477.
Burgin, M. (2010). *Theory of Information — Fundamentality, Diversity and Unification.* Singapore: World Scientific Publishing Co. Pte. Ltd.
Burgin, M. and Cárdenas-García, J. F. (2020). A dialogue concerning the essence and role of information in the world system. *Inform.* 11(9), 406.
Burgin, M. and Hofkirchner, W. (eds.). (2017). *Information Studies and the Quest for Transdisciplinarity — Unity Through Diversity.* New Jersey: World Scientific Publishing Company.
Capurro, R. and Hjørland, B. (2003). The concept of information. *Ann. Rev. Inform. Sci. Techno.* 37(1), 343–411.
Capurro, R. (2009). Past, present, and future of the concept of information. *TripleC.* 7(2), 125–141.
Cárdenas-García, J. F. and Ireland, T. (2019). The fundamental problem of the science of information. *Biosemi.* 12(2), 213–244.
Cárdenas-García, J. F. (2020). The process of info-autopoiesis — The source of all information. *Biosemiotics.* 13(2), 199–221.
Cárdenas-García, J. F. (2022). The central dogma of information. *Inform.* 13(8), 365.

Chalmers, D. (1995). Facing up to the problem of consciousness. *J. Conscious. Stud.* 2(3), 200–219.
Chalmers, D. (2020). Is the hard problem of consciousness universal? *J. Conscious. Stud.* 27(5-6), 227–257.
Clark, A. and Chalmers, D. (1998). The extended mind. *Anal.* 58(1), 7–19.
Crick, F. (1970). Central dogma of molecular biology. *Nat.* 227(5258), 561–563.
Díaz Nafría, J. M. (2010). What is information? A multidimensional concern. *TripleC.* 8(1), 77–108.
Floridi, L. (2010). *Information: A very short introduction.* Oxford University Press.
Floridi, L. (2011). *The Philosophy of Information.* Oxford New York: Oxford University Press, xviii, p. 405.
Heras-Escribano, M. and de Jesus, P. (2018). Biosemiotics, the extended synthesis, and ecological information: Making sense of the organism-environment relation at the cognitive level. *Biosemi.* 11(2), 245–262.
Hidalgo, C. A. (2015). *Why Information Grows: The Evolution of Order, from Atoms to Economies.* New York: Basic Books, xxi, p. 32.
Horgan, J. (2018). *Who Invented the Mind–Body Problem?* Scientific American.
Horgan, J. (2019). Mind-body problems: Science, subjectivity & who we really are. https://johnhorgan.org/books/mind-body-problems.
Kikuchi, K., Galera-Laporta, L., Weatherwax, C., Lam, J.Y., Moon, E. C., and Theodorakis, E. A., et al. (2022) Electrochemical potential enables dormant spores to integrate environmental signals. *Sci.* 378(6615), 43–49.
Koch, C. A. (2009). Theory of consciousness. *Scienti. Amer. Mind.* 20(4), 16–19.
Koch, C. (2019). *The Feeling of Life Itself: Why Consciousness is Widespread but can't be Computed.* Cambridge, MA: The MIT Press.
Lloyd, S. (2006). *Programming the Universe.* New York, NY: Alfred A. Knopf.
Lombardino, J. and Burton, B. M. (2022). An electric alarm clock for spores. *Sci.* 378(6615) 25–26.
Pattee, H. H. (1982). Cell psychology: An evolutionary approach to the symbol-matter problem. *Cogn. Brain The.* 5(4), 325–341.
Pattee, H. H. (2012). Cell psychology: An evolutionary approach to the symbol-matter problem. In H. H. Pattee and J. Rączaszek-Leonardi (eds). *Laws, Language and Life — Howard Pattee's Classic Papers on the Physics of Symbols with Contemporary Commentary.* Dordrecht: Springer, pp. 165–179.
Pattee, H. H. (2013). Epistemic, evolutionary, and physical conditions for biological information. *Biosemi.* 6(1), 9–31.
Peirce CS. *The Collected Papers. Volumes 1–6.* In C. Hartshorne and P. Weiss (eds.), Cambridge, MA: Harvard University Press, pp. 1931–1936.
Peirce CS. (1958). *The Collected Papers. Volumes 7 & 8.* In A. Burks (ed.), Cambridge, MA: Harvard University Press.
Peters, J. D. (1988). Information: notes toward a critical history. *J. Commun. Inq.* 12, 9–23.
Rovelli, C. (2016). Meaning = Information + Evolution.

Shannon, C. E., Sloane, N. J. A., and Wyner, A. D. (eds.) (1993). Claude Elwood Shannon: Collected Papers.
Shannon, C. E. (1948). A mathematical theory of communication. *The Bell Syst. Techn. J.* 27(379–423), 623–656.
Sharov, A. A. (2009). Coenzyme autocatalytic network on the surface of oil microspheres as a model for the origin of life. *Inter. J. Mol. Sci.* 10(4), 1838–1852.
Sharov, A. A. (2016). Evolution of natural agents: Preservation, advance, and emergence of functional information. *Biosemi.* 9(1), 103–120.
Stonier, T. (1997). *Information and Meaning — an Evolutionary Perspective.* Berlin Heidelberg, New York: Springer-Verlag.
Umpleby, S. A. (2007). Physical relationships among matter, energy and information. *Syst. Resear. Behavior. Sci.* 24(3), 369–372.
Vedral, V. (2010). *Decoding Reality — The Universe as Quantum Information.* Oxford, UK: Oxford University Press.
Wheeler, J. A. (1991). Proceedings of the first international. In A. D. Sakharov (ed.). *Memorial Conference on Physics*, Moscow, USSR: Nova Science Publishers, Commack, NY. pp. 27–31.
Wiener, N. (1948). *Cybernetics: or Control and Communication in the Animal and the Machine.* New York: John Wiley.
Yockey, H. P. (2005). *Information Theory, Evolution, and the Origin of Life.* Cambridge. UK: Cambridge University Press.

© 2025 World Scientific Publishing Company
https://doi.org/10.1142/9789811294921_0013

Chapter 13

Why Artificial Intelligence Alone Will Not Save Mankind: Digital Imperatives for Planetarity and Conviviality

Wolfgang Hofkirchner[*,†,‡,||] and Hans-Jörg Kreowski[†,‡,§,¶,**]

[*]Human Computer Interaction, TU Wien, Vienna, Austria
[†]The Institute for a Global Sustainable Information Society (GSIS), Vienna, Austria
[‡]Leibniz-Sozietät der Wissenschaften zu Berlin e.V., Berlin, Germany
[§]Department of Computer Science, University of Bremen, Bremen, Germany
[¶]Forum Computer Professionals for Peace and Social Responsibility (FIfF), Bremen, Germany
[||]wolfgang.hofkirchner@gsis.at
[**]kreo@uni-bremen.de

Today, to continue social evolution on Earth, objective and subjective requirements must be met to master the apparent poly-crisis. They include interventions in social, ecological, and technological affairs to change the logics of development on a planetary scale as well as adaptations of human cognition, communication, and cooperation capabilities to make them fit for a new level of conviviality. The interventions and adaptations are imperative. They form planetary imperatives and convivial imperatives, respectively. The change of development logics and the adaptation of social actors' information capabilities can be bolstered or blocked by the interference of information technologies. Information technologies can secure or undermine the replacement of development logics. They can fit the information requirements for adapting to the transformation of world society or make these adaptations impracticable. Information technologically supported the advancement of dismantling the poly-crisis and enhancing social information capabilities is imperative. Such an advancement can be achieved when based on digital humanism. Digital humanism can show the way to form so-called digital imperatives that support planetary as well as convivial imperatives. In this respect, applications of artificial intelligence (AI) may become even more important in the near future. However, AI needs human guidance to provide a positive impact on anthroposociogenesis.

1 Introduction

There continues to be a big hype about artificial intelligence (AI) based on brilliant achievements and even more phenomenal promises. One encounters quite some spectacular successes of AI in complex games like chess and Go as well as in more practical applications like language and image processing, on one hand, and robotics, on the other hand. Moreover, some AI experts and many AI non-experts promise further breakthroughs:

- For decades, the development of autonomous systems has been one of the major AI applications in progress. If it will work eventually along a broad front, then it may contribute to better working conditions avoiding inhumane working practices in factories or long-distance transportation. But millions of taxi and lorry drivers, factory workers, and many others will lose their jobs.
- The military counterpart of autonomous weapons is considered as a means to transform the conduct of war. But the "new war" will be as inhumane as ever (cf., e.g., Williams and Scharre, 2015).
- There are some researchers in AI who propagate the idea of a superintelligence, meaning program systems that can "think" as good or even better than humans. Such notions of artificial general intelligence are being fueled right now by applications of large language models, such as ChatGPT, which are reminiscent of ELIZA — the program Joseph Weizenbaum developed in the 1960s that simulated human psychotherapists. This is a bestselling

topic, however, without any technological indication of realization (cf., e.g., Hofkirchner and Kreowski, 2021).

These developments trigger the expectation in politics and economy that AI will become the key technology of future innovation (see, e.g., European Commission, 2021). Therefore, many states have national AI strategies and are going to spend a lot of money on the development of AI. But can a technological push, in particular in AI, help solve the global challenges? The truth is that a significant part of AI developments and applications so far is not at all compatible with the aims of a proper handling of the climate crisis, a sustainable economy or peaceful global politics.

Global challenges that have been besetting (human) social evolution in the so-called Anthropocene are existential threats that are mediated by technology but not the root cause that is social.

- The Atomic Age was ushered in by the atomic bombs, but it is social factors, namely political decisions, that determined their use and still threatened their use.
- The degradation of the global environment became manifest in the 1960s when technically produced herbicides and pesticides and other substances nature cannot bear began to be carried out, but it is social factors, namely economical rationality, that have been and still are responsible for it.
- Poverty distributed across the world and within states is conveyed by technologies that cannot provide a decent life, but it is social factors, namely a kind of war pitting the rich against the poor, that characterize the actual mode of living.

These threats have been intermingling and signify an open-ended poly-crisis: the space of possible trajectories shows a hitherto unprecedented bifurcation between an integration of the diversity of interdependent social systems into a single meta-/supra-system unity that allows coping with the increased complexity of the challenges, on the one hand, and a disintegration of social systems falling back to barbarism, collapse, or extinction, on the other hand (Hofkirchner, 2023).

The reorganization of world society is task number one, prior to all other tasks on the agenda of humanity, if breakdown shall be avoided. All resources available need to be pooled to enable a breakthrough to a new step in anthropogenesis. Technology development, including AI, needs to be subject to this reorientation, enabling the greatest effort of humanity by boosting that transformation.

2 Objective Requirements for Artificial Intelligence in the Anthropocene

This new step requires an in-depth analysis of the root causes of the poly-crisis characteristic of the Anthropocene in order to remove them. Since the root causes are anthropogenic, the focus is on the structure of social systems, the framework of social relations, that incentivizes which kinds of actions are carried out in which kind of areas.

The kinds of actions depend on the type of social relations. There are three types (Hofkirchner, 2023, pp. 67–70):

- **Antagonisms:** They demand uniformity, that is, unity without diversity. This type inheres to conflictive positions that form mutually exclusive oppositions; they are contradictory.
- **Agonisms:** They demand a mere plurality, that is, diversity without unity. Such positions co-exist by being indifferent to their differences.
- **Synergisms:** They demand unity through diversity. Positions complement each other; they are mutually supportive for the sake of sociality.

Social relations extend to different kinds of areas. There are another three of them:

- social relations among the actors in social systems,
- social relations among the actors with integrated parts of natural systems in so-called eco-social systems, and
- social relations among the actors with integrated parts of natural systems, mediated by artificially transformed integrated parts of

natural systems into means towards social ends in so-called techno-eco-social systems.

Since the latter relations are extensions of the former relations, the type of social relations enacted in the first area concatenates to the second and the third areas.

It is a matter of fact that the social evolution under capitalist and neoliberal supremacy has become dominated by antagonistic relations. Such relations exclude more and more actors from sharing the common good and from being enabled to live a good life. Once the surface of Earth has objectively become the shared habitat of all living humans, the provision of systemic synergy effects cannot be sufficiently afforded through compartmentalized humanity. Antagonistic relations in all areas have become outdated as follows:

- A logic of self-centeredness governing the development of social systems allows individual and collective actors to pursue particularistic interests at the expense of other actors, of their own social system as well as of a world society to-be.
- A logic of anthropocentrism governing the development of eco-social systems allows, in addition, to subjugate the natural environment and the human body to particularistic interests.
- A logic of megalomania and almightiness governing the development of techno-eco-social systems allows, in the end, the misuse of productive forces as destructive forces due to their broad impact and depth of penetration, side effects, long-distance effects, and long-term effects that affect the interests of others.

2.1 Planetary imperatives

Thus, the implementation of new logics that is compatible with successfully governing social evolution in the Anthropocene is imperative (Hofkirchner and Kreowski, 2023):

- Human–human relations must abandon the self-centric logic and take over the inclusive logic of pan-humanism enveloping all of humanity (Morin, 2021). Pan-humanism is apt to become the logic of a united humanity building a single whole in which all actors are entitled to share the benefits with each other. The planetary imperative of pan-humanism can be formulated as follows: "Act in such a way that the effects actualize the objective potential to build or maintain a social meta- or suprasystem which, by replacing any logic of self-centeredness in social relations with a logic of self-limitation of the actors, ensures unity through diversity for all of humanity!" This is imperative for globality as to the spatio-temporal dimension of the new society (Hofkirchner, 2017).

- Human–nature relations must leave behind the exclusive logic of anthropocentricity and adopt so-called anthropo-relationalism (Deutsches Referenzzentrum, n.d.; Barthlott *et al.*, 2009). This is a logic that — while doing justice to the unique place of human systems in evolution — does also justice to the diverse places natural systems have been taking in evolution, allowing an alliance with nature (Bloch, 1986). The planetary imperative of anthropo-relational humanism can run as follows: "Act in such a way that the effects actualize the objective potential to build or maintain an eco-social meta- or suprasystem which, by replacing any logic of self-exaltation in the social relations of nature with a logic of self-limitation of the actors, ensures an alliance with the agents of natural self-organising systems!" This is imperative for sustainability including the social sphere as to the level of organization of the new global and sustainable society (Hofkirchner, 2017).
- Human–technology relations, factoring pan-humanism and anthropo-relational humanism, need to replace the logic of power-centricity, the illusion of having superpowers that are not designed to take sufficient care of natural systems as well as human systems. Such a logic has been incrementally hypostatising the effectivity of technology beyond any rational measure. So-called digital humanism must become the logic of civilizational self-limitation — a term coined by Austrian-born writer Ivan Illich in his book *Tools for Conviviality* (Illich, 1973). This can here be explicated as a limitation of the technological tools to their role of serving an alliance-with-nature and pan-human purposes for the common good. In detail, the planetary imperative of digital humanism is

as follows: "Act in such a way that the effects actualise the objective potential to build or maintain a techno-eco-social meta- or suprasystem which, by replacing any logic of self-overestimation in social technological relations with a logic of self-limitation of the actors, ensures an appropriate use of artificial, hetero-organised artefacts!" This is imperative for informedness as a state of intelligence of the new global sustainable information society including the natural and the social sphere (Hofkirchner, 2017).

2.2 *Digital imperatives for planetarity*

Altogether, digital humanist, anthropo-relational humanist, and pan-humanist social relations form the objective requirements of the techno-eco-social transformation. The design and use of AI need to be shaped to provide information-technological means for this transformation. AI needs to work as part of the solution and shall not become part of the problem. It shall build back existing technology not apt to the existing challenges and adapt it accordingly, making digitalization fruitful in an overall humane way.

The role of digital technology, of IT, and of AI can be characterized by the fact that it can support these three planetary imperatives in terms of information technology. Digital imperatives regarding planetarity are formulated as shown in the following:

- The digital imperative to support pan-humanism is as follows: "Act in such a way that the effects contribute to the production and usage of digital technologies that support the construction or maintenance of a social meta- or supra-system of all humanity in its unity-through-diversity!"
- The digital imperative to support anthropo-relational humanism is as follows: "Act in such a way that the effects contribute to the production and usage of digital technologies that support the construction or maintenance of an eco-social meta- or supra-system in alliance with the agents of natural self-organising systems!"
- And the digital imperative to support digital humanism is as follows: "Act in such a way that the effects contribute to the production and usage of digital technologies that support the

construction or maintenance of a techno-eco-social meta- or supra-system with the appropriate applications of artificial, hetero-organised artefacts of whatever kind!"

3 Subjective Requirements for AI in the Anthropocene

One subtask of IT is to re-establish social relations to technology, with nature and among society with the help of new material artifacts and newly materialized procedures that take care of the special situation of humanity as an objective community of destiny. But humanity has yet to become a subject in its own right to come to terms with its destiny. Since IT deals with the support of information processes in the social realm, another subtask of IT is to re-shape social information in order to support the emergence of new ideations. Generating information is a usual event in the evolution of systems of any kind in the universe, as they are at the same time informational agents. Complex, self-organizing systems have the capacity to generate requisite information to overcome crises that are due to complexity gaps. A complexity gap occurs when challenges are more complex than the capacity of the systems to pass them. Requisite information is just that information that allows increasing the complexity of the systems to match or exceed the complexity of the challenges. It is tantamount to just that re-organization of the systems that capacitate them to pass the challenges. Creating the necessary mindfulness about the human–human, the human–nature, and the human–technology relations on a global scale is what will enable social systems today to close the gap and master this crucial transformation. Mindfulness is subjective, morally imperative social information. It is imperative if, again, the continuation of social evolution is deemed desirable.

Social information, that is, information in the social realm, appears — similar to information in the realm of non-human systems — in three different contexts:

- in the context of human cognition,
- in the context of human communication, and
- in the context of human cooperation.

These contexts build an evolutionary hierarchy: human cooperation is emerging from communicative actors to direct their communication, human communication is emerging from cognitive actors to direct their cognition, and human cognition is emerging from lower-level systemic agents (biota) to direct their cooperation in service of an individual actor, while co-operating actors can be viewed as higher-level cognitive actors (social groups) (Hofkirchner, 2013).

What needs to be done to fulfill the subjective requirements is to open up the human capabilities of cognition, communication, and cooperation beyond the frontiers reached so far by anthroposociogenesis (Tomasello, 2014, 2016) to the planetary level.

3.1 *Convivial imperatives*

The first important philosophy that responded to the global challenges is the *Imperative of Responsibility*, formulated in *In Search of an Ethics of the Technological Age* by philosopher (Hans Jonas, 1984), five years after the original German edition. Updating Kant's Categorical Imperative "Act so that you can will that the maxim of your action be made the principle of a universal law", the new imperative runs as follows: "Act so that the effects of your action are compatible with the permanence of genuine human life", which is tantamount to the negative expressions "Act so that the effects of your action are not destructive of the future possibility of such life" or "Do not compromise the conditions for an indefinite continuation of humanity on earth" (Jonas, 1984, p. 11).

Convivialism is a philosophy/social idea/art. Two manifestos have been published: "Convivialist Manifesto: a declaration of interdependence" (*The Convivialists*, 2014) and "The Second Convivialist Manifesto: towards a post-neoliberal world" (*The Convivialist International*, 2020), initiated by mostly French intellectuals, among them philosophers and sociologists Edgar Morin, Alain Caillé, or Serge Latouche, now signed by about 300 international first signatories. They state the following: "A different kind of world is not just possible; it is a crucial and urgent necessity. But where do we start when it comes to envisaging the shape it should take and working out how to bring it about? The Convivialist Manifesto seeks to highlight the similarities between the many initiatives already engaged in building that world and to draw out the common political philosophy that underlies them" (*The Convivialists*, 2014). The idea makes use of the term "conviviality" in the sense of Illich (1973). The term has Latin origins and means the quality of living together in the manner of dining together (convivor) of hosts (convivatores) and guests (convivae) at joint feasts (convivia). The manifestos introduced initially four and later five principles and one imperative as follows:

- **The principle of common humanity:** "There is only one humanity" that "must be respected in the person of each of its members".
- **The principle of common naturality:** Humans "have a responsibility to take care of" nature.
- **The principle of common sociality:** "The greatest wealth is the richness of concrete relationships" that human beings as social beings maintain among themselves.
- **The principle of legitimate individuation:** "Legitimate is the policy that allows each individual to develop their individuality to the fullest by developing his or her capacities, power to be and act, without harming that of others, with a view toward equal freedom".
- **The principle of creative opposition:** "It is normal for humans to be in opposition with each other [...] as long as this does not endanger the framework of common humanity, common sociality, and common naturality that makes rivalry fertile and not destructive".
- **The imperative of hubris control:** "The first condition for rivalry to serve the common good is that it be devoid of desire for omnipotence, excess, hubris (and a fortiori pleonexia, the desire to possess ever more). On this condition, it becomes rivalry to cooperate better" (*The Convivialist International*, 2020).

Starting from these formulations, we coin three convivial imperatives that:

- address individual and collective actors,
- do not focus on the mere survival of mankind but demand a developmental thrust geared toward a socially desirable future destination that allows for the continuation of social evolution,
- determine this destination as a real, concrete utopia (Wright, 2010; Bloch, 1986) of a unique meta-/supra-system, comprising pan-human societal relations among actors, anthroporelational ones with regard to natural agents, and digital humanist ones with regard to technology,
- determine the path to this destination as techno-eco-social transformation, and
- determine the subjective preconditions of this path as the ability of actors to create the requisite consciousness to anticipate the path and the destination through univerzalizing of social cognition, communication, and cooperation.

The convivial imperatives break down the issues of planetarity into three social information contexts:

- The first convivial imperative universalizes cooperation and runs as follows: "As collective entity of actors, act in such a way that the overall effect of your action can expand the subjective capacity for co-operation already achieved towards a planetary practice that realises the objective requirement of a next step of social evolution so that universal wisdom can be achieved that morally anticipates the value of a unity-through-diversity organisation of humanity, in particular, of pan-humanism in societal relations, of anthroporelational humanism in ecological societal relations and of digital humanism in technological ecological societal relations!" This is the imperative of a planetary ethos for global governance.
- The second convivial imperative universalizes communication and runs as follows: "As diverse actors, act in such a way that the effects of your many actions can extend the subjective communication skills already achieved to a mutual understanding of

the objective requirements of a next step of social evolution so that knowledge can be created and shared on a planetary scale, anticipating in conciliatory discourses the tasks of a techno-eco-social transformation of all humanity into a common system!" This is the imperative of planetary conciliatoriness for global dialog.

- The third convivial imperative universalizes cognition and runs as follows: "As single actor, act in such a way that the effect of your action can enlarge the subjective cognitive abilities already achieved by insights into the objective requirements of a next step of social evolution so that facts and figures can be collected and interpreted that mentally anticipate the meaning of operations according to your own positioning in the overall planetary context as well as the positioning of other actors and the positioning of parts that are not actors at all!" This is the imperative of planetary mindsets for global citizenship.

3.2 *Digital imperatives for conviviality*

Digital humanism goes back to philosopher Julian Nida-Rümelin, who published, together with cultural scientist Nathalie Weidenfeld, a book titled in German *Digitaler Humanismus* (Nida-Rümelin and Weidenfeld, 2018) (in English 2022). In 2019, the Vienna Manifesto on Digital Humanism was elaborated (Vienna Manifesto, n.d.): "We must shape technologies in accordance with human values and needs, instead of allowing technologies to shape humans. Our task is not only to rein in the downsides of information and communication technologies, but to encourage human-centered innovation. We call for a Digital Humanism that describes, analyzes, and, most importantly, influences the complex interplay of technology and humankind, for a better society and life, fully respecting universal human rights." Digital Humanism can be defined as humanism dealing with digitalization. It promotes ethically aligned design of human-centered technology.

In particular, the manifesto proclaims eleven "core principles:

- Digital technologies should be designed to promote democracy and inclusion. This will require special efforts to overcome current

inequalities and to use the emancipatory potential of digital technologies to make our societies more inclusive.
- Privacy and freedom of speech are essential values for democracy and should be at the center of our activities. Therefore, artifacts such as social media or online platforms need to be altered to better safeguard the free expression of opinion, the dissemination of information, and the protection of privacy.
- Effective regulations, rules, and laws, based on a broad public discourse, must be established. They should ensure prediction accuracy, fairness and equality, accountability, and transparency of software programs and algorithms.
- Regulators need to intervene with tech monopolies. It is necessary to restore market competitiveness as tech monopolies concentrate market power and stifle innovation. Governments should not leave all decisions to markets.
- Decisions with consequences that have the potential to affect individual or collective human rights must continue to be made by humans. Decision-makers must be responsible and accountable for their decisions. Automated decision-making systems should only support human decision-making, not replace it.
- Scientific approaches crossing different disciplines are a prerequisite for tackling the challenges ahead. Technological disciplines such as computer science/informatics must collaborate with social sciences, humanities, and other sciences, breaking disciplinary silos.
- Universities are the place where new knowledge is produced and critical thought is cultivated. Hence, they have a special responsibility and have to be aware of that.
- Academic and industrial researchers must engage openly with wider society and reflect upon their approaches. This needs to be embedded in the practice of producing new knowledge and technologies, while at the same time defending the freedom of thought and science.
- Practitioners everywhere ought to acknowledge their shared responsibility for the impact of information technologies. They need to understand that no technology is neutral and be sensitized to see both potential benefits and possible downsides.

- A vision is needed for new educational curricula, combining knowledge from the humanities, the social sciences, and engineering studies. In the age of automated decision-making and AI, creativity and attention to human aspects are crucial to the education of future engineers and technologists.
- Education on computer science/informatics and its societal impact must start as early as possible. Students should learn to combine information-technology skills with awareness of the ethical and societal issues at stake" (accentuations deleted).

We take up the intentions of digital humanism which originally did not yet explicitly include the issue of the poly-crisis due to global threats, extend them, and formulate digital(-humanist) imperatives that are compatible with the convivial imperatives described above (Hofkirchner and Kreowski, 2022a, 2022b). Digital imperatives are to complement the convivial ones in that they design and propagate digital solutions that boost the convivial aims. By boosting social aims, the social aims integrate technology and become techno-social aims — social systems integrate technological artifacts and become techno-social systems in which technical functions are subservient to social functions.

The digital imperatives build a three-leveled hierarchy as do the convivial imperatives. They demand any actor, whether collective or single, be they producers or consumers, to contribute to the IT support of the universalization of cooperation, communication, and cognition capabilities:

- The highest level is built by tools for convivial governance that technologize the constitution and institution of consensualized transformative goals. This technologization is value-based such that the planetary ethos is informationalized into the techno-social entity of digital conscience. No technology will replace the human ethos. But digital-humanist informatization can help the human ethos improve by becoming a digitally supported human ethos — in short, a digital(ized) conscience of humanity. The first digital imperative for conviviality is to support the planetary ethos and runs as follows: "Act so as to contribute to the production and use

of digital technologies that support the emergence and sustenance of a planetary ethos as digital conscience!"
- The next lower level is the level of tools for a convivial dialog. The consilient designing and assigning of transformative tasks is supported by human-centered technologies so as to yield digital intelligence as a new techno-social quality of planetary discourses. Intelligence is first and foremost a human feature and this human intelligence, when enacting a conciliatory dialog, becomes a digital(ized) intelligence for humans, when supported by digital tools. The second digital imperative for conviviality is to support planetary conciliatoriness and runs as follows: "Act so as to contribute to the production and use of digital technologies that support the emergence and sustenance of planetary conciliatoriness as digital intelligence!"
- The bottom level encompasses tools for convivial citizenship transformed into netizenship. The technologisation of devising and supervising transformative operations conceptually integrates an assessment and design cycle for continual adjustment of digital ingenuity as a transformed planetary mindset. Ingenuity is not a feature of technology; it is a feature of humans, and this feature is supported by digital technology — thus, a digital(ized) ingenuity of netizens. The third digital imperative for conviviality is to support planetary mindsets and runs as follows: "Act so as to contribute to the production and use of digital technologies that support the emergence and sustenance of planetary mindsets as digital netizens ingenuity".

4 The Needful Shaping of AI

According to these digital imperatives for planetarity and conviviality, AI is called to bolster the planetary imperatives and the convivial imperatives to overcome the poly-crisis. AI applications conforming to the digital imperatives may even turn out to be indispensable to gain the requisite information for transforming our societal systems and support conscience, intelligence, and ingenuity of humans on a planetary scale. However, there are ambivalences and AI technology can quantitatively reinforce existing, or qualitatively spawn new, dysfunctions of cognitive, communicative, or cooperative social information and hamper the aim of contributing to the techno-eco-social transformation as well as quantitatively reinforce existing, or qualitatively innovate new, transformative social information functions. Thus, deliberate design is mandatory.

Design should be aware that the term intelligence cannot signify the property of machine processes or a machine itself because machinic entities are not informational agents, not self-organising systems, and work along hetero-organized determinacies, as one of us stated in 2011 (Hofkirchner, 2011). This idea has been taken up in a publication of the IEEE Global Initiative on Ethics of Autonomous and Intelligent Systems (A/IS) on ethically aligned design. It states, with reference to Hofkirchner and philosopher Rafael Capurro (2012) whose assignment of the distinction between agents and patients is helpful to understand the different roles that humans and technology play, "Of concern for understanding the relationship between human beings and A/IS is the uncritically applied anthropomorphistic approach toward A/IS that many industry and policymakers are using today. This approach erroneously blurs the distinction between moral agents and moral patients (i.e., subjects), otherwise understood as a distinction between 'natural' self-organizing systems and artificial, non-self-organizing devices" (IEEE, 195). This is consequential for the issue of autonomy. "[...] A/IS devices cannot, by definition, become autonomous in the sense that humans or living beings are autonomous" (IEEE, 195). In particular, attempts "to implant true morality and emotions, and thus accountability (i.e., autonomy) into A/IS is both dangerous and misleading in that it encourages anthropomorphistic expectations of machines by human beings when designing and interacting with A/IS" (IEEE, 196).

In the same vein, a German philosopher of technology, Klaus Kornwachs, insists that ethics should not be designed for robots but for people who build and use robots. Such an ethics must guarantee that the conditions for the possibility of responsible action are maintained for all humans involved (Kornwachs, 2023, p. 288]. Thus, he formulates rules for building and using machines. Machines that cannot be turned off must not be allowed (Kornwachs, 2021, p. 42). "Do not fake a machine as a human subject" [43]. It must always be clear to all people involved in so-called human–machine "communication" that the "partner" is a machine.

Having said this, it is clear that AI *per se* will not be able to execute a unity-through-diversity solution for mankind's poly-crisis — an idea that, among others, author Roberto Simanowski (2020) promoted. No automaton can really reach a decision, since a decision is a judgment, an emergent on the deliberation on grounds, whereas machines perform merely on the basis of mechanical determinacy devoid of any deliberation of a self (Hofkirchner, 2022). Faking decisions would be anti-humanism.

What, in principle, AI is missing is humanness, as the Austrian writer Franzobel put it in a nutshell (2023). But with the roll-out of ChatGPT, promises of a future more human than ever are booming again. Instead of admitting that large language models are prone to errors in that the output of the algorithms based upon them may be completely manufactured and describe things that do not exist at all, architects and boosters of generative AI call this phenomenon "hallucinations" and feed thereby "the sector's most cherished mythology: that by building these large language models, and training them on everything that we humans have written, said and represented visually, they are in the process of birthing an animate intelligence on the cusp of sparking an evolutionary leap for our species", as Naomi Klein (2023) criticizes. According to her, it is not AI that is having hallucinations, "it's the tech CEOs who unleashed them along with a phalanx of their fans". She uncovers these hallucinations as ideologemes of profiteers with vested interests in what (Zuboff, 2019) coined Surveillance Capitalism 2019:

- AI will solve the climate crisis. That's wrong because there are already sufficiently robust data sets, but the governments are not listening.
- AI will deliver wise governance. That's wrong because politicians don't suffer from a lack of evidence but are dependent on lobbying campaigns.

- Tech giants can be trusted not to break the world. That's wrong because these companies do "not make decisions based on what's best for the world" but on what's best for their shareholders.
- AI will liberate us from drudgery. That's wrong because the working people set free by technologies will not become philosophers or artists but "will find themselves staring into the abyss" (Klein, 2023).

As justified as these objections to the uncritical reliance on AI promises are, they do not mean that AI has no potential to help solve the global problems. There is a potential (see, e.g., Gomes, 2023). But the technological potential must be deployed under friendly social conditions for it to be truly effective. AI must be deliberately centered on the social solution: the cultural sphere needs to provide digital imperatives, politics needs to regulate compliance with digital imperatives, and the economy needs to align the design and usage of AI tools with digital imperatives.

In sum, AI alone will not save mankind. In fact, there is no pure technological solution. AI can even disrupt the autonomy of humans, if it is allowed to, and interfere with social evolution causing negative impacts. A comprehensive social solution is the necessary condition for the emergence of techno-social tools. All that is needed is the implementation of the appropriate design and assessment of AI applications to make it serve the most meaningful purpose in human history.

5 Conclusion

Homo sapiens has reached a point in the evolution of social systems where a continuation in the same direction would lead to a catastrophe that will cause the impoverishment of a large part of the population and may even endanger the existence of humankind. Digitalisation and AI in particular can only be part of the solution if they are integrated into a planetary perspective and a novel convivialist mode of living with worldwide freedom and peace and in harmony with nature. Humankind must be aware that it forms a community of destiny and must now act accordingly, that is, act as global netizens, engage in a global dialog to find agreements on how to reach a world worth living for all people, and build a global governance to organize a planetary politics and sustainable economy. AI tools can become an indispensable means on that path if digital imperatives for planetarity and conviviality are followed.

Acknowledgment

We dedicate this contribution to Mark Burgin. It was me, Hofkirchner, who came in touch with Mark in 2008 when being invited by Mark to contribute to a volume entitled "Information and Computation" to be edited by Mark and Gordana Dodig-Crnkovic. In the contribution "Does Computing Embrace Self-Organisation?", I laid down my position on the distinction between mechanism and emergentism as a core tenet of a paradigm shift in science, insinuating that computers as long as they compute according to the mechanistic paradigm cannot bring about new information. Mark involved me then in a deep discussion which was helpful in clarifying my standpoint. In the end, Mark agreed to publish my contribution. Shortly after that, he also promoted my book project on "Emergent Information: A Unified Theory of Information Framework" (2013) as he did with my latest book project that rounded off my view of the new paradigm with an elaboration of digital humanism concerning AI — "The Logic of the Third: A Paradigm Shift to a Shared Future for Humanity" (2023). Mark had asked for membership in the Unified Theory of Information Research Group I had founded in 2003 and became the Information Officer of The Institute for a Global Sustainable Information Society into which the group had merged in 2019. On Mark's initiative, he started with me a project for a Single Integrated Science of Information. I will miss his expertise, his humility, and his open-mindedness.

References

Barthlott, W., Linsenmair, K. E., and Porembski, S. (eds.) (2009). *Biodiversity: Structure and Function*, vol. II. Oxford, UK: EOLSS.
Bloch, E. (1986). *The Principle of Hope*. Cambridge, USA: MIT Press.

Capurro, R. (2012). Toward a comparative theory of agents. *AI & Society* 27(4), 479–488.

Deutsches Referenzzentrum für Ethik in den Biowissenschaften (ed.) (n.d.). *Anthroporelational*, https://www.drze.de/de/forschung-publikationen/im-blickpunkt/biodiversitaet/module/anthroporelational (retrieved 31 December 2023).

European Commission (2021). *Coordinated Plan on Artificial Intelligence 2021 Review*. https://digital-strategy.ec.europa.eu/en/library/coordinated-plan-artificial-intelligence-2021-review (retrieved 21 January 2023).

Franzobel (2023). Ist menschengefertigte Kunst obsolet?, *Der Standard*, 27 May 2023, Album, https://www.derstandard.at/story/3000000139466/ist-menschengefertigte-ku.

Gomes, C. P. (2023). AI for Scientific Discovery and a Sustainable Future. Vienna Gödel Lecture 2023, TU Wien Informatics, 5 June 2023.

Hofkirchner, W. (2011). Does computing embrace self-organisation? In G. Dodig-Crnkovic and M. Burgin (eds.). *Information and Computation*, Singapore: World Scientific, pp. 185–202.

Hofkirchner, W. (2013). *Emergent Information: A Unified Theory of Information Framework*. Singapore: World Scientific.

Hofkirchner, W. (2017). Information for a global sustainable information society. In W. Hofkirchner and M. Burgin (eds.). *The Future Information Society: Social and Technological Problems*, Singapore: World Scientific, pp. 11–33.

Hofkirchner, W. (2022). Artificial intelligence: Machines of loving grace or tools for conviviality, *Signifikant* 4/5, 35–48.

Hofkirchner, W. (2023). *The Logic of the Third: A Paradigm Shift to a Shared Future for Humanity*. Singapore: World Scientific. https://doi.org/10.1142/12985.

Hofkirchner, W. and Kreowski, H.-J. (eds.) (2021). *Transhumanism: The Proper Guide to a Posthuman Condition or A Dangerous Idea?* Switzerland: Springer, Cham. https://doi.org/10.1007/978-3-030-56546-6.

Hofkirchner, W. and Kreowski, H.-J. (2022a). Digital Humanism: How to shape digitalisation in the age of global challenges? *Proceed.* 81(1), https://www.mdpi.com/2504-3900/81/1/4.

Hofkirchner, W. and Kreowski, H.-J. (eds.) (2022b). Digital humanism and the future of humanity. *New Explorations*, 2(3), https://jps.library.utoronto.ca/index.php/nexj/issue/view/2600.

Hofkirchner, W. and Kreowski, H.-J. (2023). Digital humanism and AI. In S. Lindgren (ed.). *Handbook of Critical Studies of Artificial Intelligence*, Cheltenham, UK: Edward Elgar, pp. 152–162.

IEEE (n.d.). *Ethically Aligned Design: A Vision for Prioritizing Human Well-Being with Autonomous And Intelligent Systems*, version 2, https:// standards.ieee.org/industry-connections/ec/ead-v1/ (retrieved 4 June 2023).

Illich, I. (1973). *Tools for Conviviality*. London, UK: Marion Boyars.

Jonas, H. (1984). *The Imperative of Responsibility: In Search of an Ethics of the Technological Age*. Chicago, USA: University of Chicago.

Klein, N. (2023). AI machines aren't "hallucinating": But their makers are, *The Guardian*, 8 May 2023, https://www.theguardian.com/commentisfree/2023/may/08/ai-machines-hallucinating-naomi-klein.

Kornwachs, K. (2021). Transhumanism as a derailed anthropology. In W. Hofkirchner and H.-J. Kreowski (eds.), *Transhumanism: the Proper Guide to a Posthuman Condition Or A Dangerous Idea?* Cham, Switzerland: Springer, pp. 21–47.

Kornwachs, K. (2023). *KI und die Disruption der Arbeit, Tätig jenseits von Job und Routine*. München: Hanser.

Morin, E. (2021). Abenteuer Mensch, *Freitag*, 28. https://www.freitag.de/autoren/the-guardian/abenteuer-mensch.

Nida-Rümelin, J. and Weidenfeld, N. (2018). *Digitaler Humanismus: eine Ethik für das Zeitalter der Künstlichen Intelligenz*. München: Piper.

Nida-Rümelin, J. and Weidenfeld, N. (2022). *Digital Humanism: For a Humane Transformation of Democracy, Economy and Culture in The Digital Age*. Cham: Springer. https://link.springer.com/book/10.1007/978-3-031-12482-2.

Simanowski, R. (2020). *Todesalgorithmus: das Dilemma der künstlichen Intelligenz*. Wien: Passagen.

The Convivialist International (2020). The second convivialist manifesto: Towards a post-neoliberal world. *Civic Sociology*. https://doi.org/10.1525/001c.12721.

The Convivialists (2014). Convivialist manifesto: A declaration of interdependence. *Global Dialogues*. https://www.academia.edu/45647218/_Convivialist_Manifesto_Declaration_of_Inteabdrdependence_Global_dialogues_3_Duisburg_Centre_for_Global_Cooperation_Research (retrieved 4 June 2023).

Tomasello, M. (2014). *A Natural History of Human Thinking*. Cambridge, Massachusetts: Harvard University Press.

Tomasello, M. (2016). *A Natural History of Human Morality*. Cambridge, Massachusetts: Harvard University Press.
Vienna Manifesto on Digital Humanism. https://dighum.ec.tuwien.ac.at.
Williams, A. and Scharre, P. (eds.) (2015). *Autonomous Systems: Issues for Defence Policymakers*. NATO Allied Command Transformation. ISBN: 9789284501939.
Wright, E. O. (2010). *Envisioning Real Utopias*. London: Verso.
Zuboff, S. (2019). The age of surveillance capitalism: The fight for a human future at the new frontier of power. *Public Affairs*, New York.

© 2025 World Scientific Publishing Company
https://doi.org/10.1142/9789811294921_0014

Chapter 14

Artificial Text: Prolegomena

Rafal Maciag

*Institute of Information Studies,
Jagiellonian University, Kraków, Poland*
rafal.maciag@uj.edu.pl

The text attempts to create the basis for the analysis of artificial text, i.e., text, which is a product of the large language models technology, developed within the field of natural language processing, which is part of the so-called artificial intelligence. This attempt is based on the assumption that the text is a phenomenon of fundamental importance and has an autonomous character. This situation requires the formulation of basic premises that should underlie any reasoning about an artificial text. This chapter formulates three main premises. The first premise is historical and refers to the human project proposed by René Descartes, which assumed mistakenly the complete uniqueness of human cognitive competencies. The second premise refers to the field of philosophy that has accumulated extensive reflection on language and text, i.e., hermeneutics. It proposes several conclusions to confirm its unique and autonomous character. The third premise analyzes the specific properties of the large language models technology mentioned at the beginning, in particular paying attention to their specific use of natural language resources. The described premises allow for a discussion concluding this chapter regarding the existence of the phenomenon of an intelligible and fully fledged artificial text.

1 Introduction

The text attempts to formulate the research circumstances based on which the analysis of the artificial text, i.e., of the text generated by algorithms based on language models such as the GPT-3 model (Brown et al., 2020) and especially the GPT-4 model (OpenAI, 2023), can begin. We assume that, against the background of the existing research experience, the issues of the text are of a fundamental nature, and hence the aforementioned analysis must be as broad as possible. This is due to the special role historically associated with the phenomenon of the text. There is no reason to ignore this enormous legacy. A small part of it which is recalled here justifies the advisability of the opposite approach. For this purpose, at the beginning, three basic premises of such an analysis will be collected, concerning in turn: (1) the idea of man and the role of language, especially in the context of its machine creation, presented by René Descartes; (2) hermeneutics as a great and old area of reflection linking the text with the basic areas of metaphysical research, i.e., ontology and epistemology; (3) the design and nature of technological inventions which are so-called language models in the area of natural language processing. This chapter ends with a kind of summary and discussion. Although this chapter tries to elaborate on several issues, i.e., proposing a definition of an intelligible text and pointing to the similarity of linguistic models to axiomatic systems, it should be treated as an introduction that tries to show several research circumstances related to artificial text. This chapter omits purely formal issues, for example, in the field of computability.

2 Premise #1: René Descartes and the Uniqueness of Man

The first premise refers to a certain cultural and civilizational context against which the problem of an artificial text should be considered. To understand it, it is necessary to recall the firm position of the philosopher who was one of the most important fathers of the European project of human subjectivity and shaped the basis for perceiving man in the context of his presence in the world. We are talking about René Descartes, a French thinker who lived in the years 1596–1650. It is his words that are used here as a starting point of reasoning. Ernest Sosa emphasizes the unusual and usually over-simplified role of Descartes. He is commonly called the father of modern philosophy, and this judgment is based on the epistemological foundations of his thought. Sosa emphasizes, however, that this opinion does not take into account the much broader project of Descartes, which resulted, among other things, from his diverse knowledge, including those fields that today belong to science. Descartes writes Sosa, "is not primarily concerned with the criteria for knowledge claims, or with definitions of the epistemic concepts involved; his aim, rather, is to provide a unified framework for understanding the universe. In place of the fragmented scholastic world of separate disciplines, each with its methods and standards of precision, he aimed to construct a coherent theory of the world and man's place within it" (Dancy et al., 2010, p. 307). His reflections on the epistemic situation of man, considered fundamental today, were the basis of such a holistic approach.

The significance of these foundations is described by Woleński, who points out that Descartes' key problem in the construction of his universal system was to find "the fundamental starting point, obvious and free of any doubt" (Woleński, 2004, p. 15). Descartes sought that point in man himself and his predispositions. The strategy of methodological doubt, based on strict rules, ultimately breaking down at the place of the very act of thinking, the denial of which led to a logical contradiction, allowed one to indicate this point. The two sentences that ultimately contain this reasoning are strikingly obvious: "from the very fact that I thought of doubting the truth of other things, it followed incontrovertibly and certainly that I myself existed, whereas, if I had merely ceased thinking, I would have no reason to believe that I existed, even if everything else I had ever imagined had been true. I thereby concluded that I was a substance whose whole essence or nature resides only in thinking, and which, to exist, has no need of place and is not dependent on any material thing" (Descartes, 2006, p. 29).

Although, as Woleński writes, this belief was not entirely new, it led to new consequences. In combination with an analytical methodology, strongly inspired by mathematics, it served as the basis for true knowledge about the world. The content of this knowledge was, as a rule, supposed to be as complete as possible and in this sense go beyond contemporary philosophy and include not only theology but also medicine or mechanics. Out of necessity, this knowledge also became a story about the special status of man in which the two most important substances that create the world meet: things (res extensa) and the mind (res cogitans). They were realizations of fundamental ideas, respectively, of the body and the soul, the set of which was ultimately closed by the infinite, and in this sense remaining outside the world of God. Man was a specific and unique place where the world of things met and penetrated with the world of the mind, constituting opposition to the infinite God.

In the same work in which Descartes formulated his famous thought, in the fifth part, he dealt with the description of the way man functions as a certain mechanical system. He focused on the circulatory system, analyzing at the level of anatomical knowledge of the time the principles of blood circulation, and then went on to briefly comment on the structure of the brain. At this point, however, Descartes abandoned the anatomical description and moved on to the extraordinary abilities of man, related to his cognitive skills, which he presented as unique, especially in comparison with animals or machines. To justify his thesis, he gave two skills that cannot be artificially imitated in any way, and thus they become evidence of a special role played by man, and so they were the keystones of Descartes' entire epistemological system. The first was concerned with the use of language, and the second was with intelligent action. Descartes described the first skill relevant here, pointing to the limitations of potential machines that would try to effectively imitate the human ability to speak: "they would never be able to use words or other signs by composing them as we do to declare our thoughts to others. For we can well conceive of a machine made in such a way that it emits words, and even utters them about bodily actions which bring about some corresponding change in its organs (...); but it is not conceivable that it should put these words in different orders to correspond to the meaning of things said in its presence, as even the most dull-witted of men can do" (Descartes, 2006, p. 46).

The opinion quoted is not just a curiosity. It belongs to a broader system of reasoning concerning the role and status of man, which has become the foundation of the modern humanistic project, setting the framework of the modern world. Its axis was the establishment of a special role of man, perceived as, on the one hand, a kind of worldview revolution and, on the other, as the beginning of a new civilizational project, which determined the future of Europe's social transformations. Tzvetan Todorov wrote about him as follows: "It was revolutionary to claim that the best justification of an act, one that makes it most legitimate, issues from the man himself: from his will, from his reason, from his feelings. The center of gravity shifts, here, from cosmos to anthropos, from the objective world to subjective will; the human being no longer bows to an order that is external to him, but wishes to establish this order himself. The movement is therefore double: a disenchantment of the world and a sacralization of man; values, removed from one, will be entrusted to the other. The new principle, whose consequences may still affect us, is responsible for the present face of our politics and our law, our arts and our sciences. This principle also presides over the modern nation-states, and if we accept them, we cannot deny the principle without becoming incoherent" (Todorov, 2002, p. 9/10).

In this context, the issue of the uniqueness of man and his capabilities becomes a serious anthropological and social problem, in the sense that the humanistic appreciation of man has become the constructional basis for a system of values, and then culture, and finally a social organization, materializing as a system of laws or a political system. A specific anthropological project was created, covering all these spheres, lasting until today and constituting the basis for contemporary axiological and legal-constitutional constructions. All these circumstances reflect, on the one hand, the proper background of Descartes' system and his enunciation of exceptional human skills, including, in particular, the ability to formulate linguistic statements. However, they cover all the issues one has to deal with when dealing with an artificial text.

3 Premise #2: Hermeneutics

The second premise is based on a centuries-old tradition of how a text is treated from a philosophical and social perspective. As one can see, this problem is huge. Fortunately, we are in a partially favorable position, because there is a field of analysis, which is hermeneutics, which has set itself the goal of interpreting the text. The partiality is because hermeneutics itself has developed into a great field of study and reflection. However, it allows us to formulate the main reasons that justify the significance of the text.

The textbooks are fairly consistent on the basic but also historically earliest subject of hermeneutics. Here are some examples: "It is commonplace to refer to hermeneutics as 'the art or science of interpretation', or sometimes as the 'theory of interpretation'. In this sense, hermeneutics refers to any systematic approach to the questions of interpretation as those questions might arise in some particular domain — so one can speak of Talmudic or Biblical hermeneutics or the hermeneutics of literature or the hermeneutics of social discourse" (Malpas, 2015, p. 1). "In its most basic sense hermeneutics refers to the many ways in which we may theorize about the nature of human interpretation, whether that means understanding books, works of art, architecture, verbal communication, or even nonverbal bodily gestures" (Porter and Robinson, 2011, p. 1). "In its original sense hermeneutics is the theory of interpretation and understanding. Hermeneutical questions (What is human meaning? How do we understand others? What happens in textual interpretation and how is it best done? What about understanding discourse? Art? How can we facilitate understanding between cultures and across time periods?) are at the heart of academic, aesthetic, political, legal, and religious practices" (Forster and Gjesdal, 2019, p. 1).

Forster and Gjesdal remind us that the forms of hermeneutic thinking are very old. They already appeared in the reflections of Protagoras, Plato, Aristotle, and the Stoics, but the modern development of hermeneutics began thanks to the Protestant Reformation, which transferred the task of interpreting the Bible from the institution of the Church to the person. The sacred text is also constructed, especially the phenomenon of the author. Bringing this aspect closer, Porter and Robinson refer to the figure of Hermes, whose name corresponds to the Greek meaning of the word *hermeneuein*, which means "to interpret" or "to translate" (Porter and Robinson, 2011, p. 2). They write that "He was a medial figure that worked in the 'in-between' as an interpreter of the gods, communicating a message from Olympus so humans might understand the meaning" (Porter and Robinson, 2011, p. 3). The "in-between" position is a metaphorical term for a certain mobile condition of the content, which cannot be precisely determined, understood as meaning or message. For they remain in constant motion between their ultimate instances: utterance and understanding. It is this vague and key place, personified by Hermes, that is at the same time the source of meaning and interpretation.

Forster and Gjesdal write that 18th century Protestant theorists have paid attention to the role of context, i.e., continuing the metaphor of Porter and Robinson, the numerous complications and obstacles that the message must overcome between the source and the destination. At the same time, they pointed out the quite secular character of these circumstances: "historical and cultural distance, the importance of paying close attention to word usage to discover a text's distinctive meanings, the need to take into account a broader social context" (Forster and Gjesdal, 2019, p. 2). In this way, they opened up a field far beyond that of theology and made possible a hermeneutical analysis of the text as that such as of art. In the 19th century, thinkers such as Johann Gottfried Herder and Friedrich Schleiermacher, as well as Wilhelm Dilthey and Friedrich Nietzsche, entered this field along with the progressive, profound transformation of hermeneutics toward a fundamental generalization of its problems.

Bleicher gives a brief description of the source and process of this transformation: "The realization that human expressions contain a meaningful component, which has to be recognized as such by a subject and transposed into his system of values and meanings, has given rise to the 'problem of hermeneutics': how this process is possible and how to render accounts of subjectively intended meaning objective in the face of the fact that they are mediated by the interpreter's subjectivity" (Bleicher, 1980, p. 1). This description contains at least two thoughts: first, it points to the fundamental role of the act of understanding as an activity that is ultimately mediated at the same time and explodes with its impact on every instance of such activity, regardless of what it concerns and in what environment of thinking it takes place. Second, it sets the right perspective for perceiving the text whose universality in terms of the problems posed by this phenomenon remains similarly unlimited. The scope of this explosion is summarized by Gianni Vattimo, for whom the question of understanding ultimately leads to the question of existence, because the experience of existence is always based on epistemological foundations, on the process of perceiving oneself and everything else. Therefore, he writes, "the question concerning a rationally grounded understanding of texts has progressively tended towards the thinking of a general ontology" (Vattimo, 2015, p. 721).

The nature of this explosion is of extraordinary scope and complexity, as evidenced by Martin Heidegger's reflection. He provides the most advanced reflection of this kind, in which being itself becomes a kind of interpretation made from a higher level of viewing. Such a reflection tries to rise above this instance, which is the deepest subject of philosophical considerations to date and sees in it the possibility of defining the proper, forgotten role of man. Language plays a special role in this process, allowing one to perceive and shape this possibility. Heidegger calls language the "house of Being", explaining the special situation elevated above ordinary communication experiences of everyday life. He writes, "But man is not only a living creature who possesses language along with other capacities. Rather, language is the house of Being in which man ek-sists by dwelling, in that he belongs to the truth of Being, guarding it" (Heidegger, 1993, p. 237).

In his essay on the future of hermeneutics, Vattimo emphasizes its indefinite and complex nature, which is understandable even against the background of Heidegger's reflections. For it is at the same time a certain branch of philosophy and, on the other hand, a philosophical school, similar to positivism or historicism. He sees its development as a convergence of both spaces that combines a general philosophical perspective with a certain discipline. It is a convergence of two different levels of understanding, showing the proper importance of this phenomenon, covering on the one hand a certain, relatively well-defined fragment of reality being the subject of reflection with the general mode of reasoning, which is simply a kind of philosophizing.

Bleicher, trying to capture this complexity from a slightly different perspective, proposes to consider three strands of contemporary hermeneutics: hermeneutical theory, hermeneutic philosophy, and critical hermeneutics. The first, as he writes, "focuses on the problematic of a general theory of interpretation as the methodology for the human sciences (or *Geisteswissenschaften*, which include the social sciences)" (Bleicher, 1980, p. 1). The text is a natural object of analysis within it. The second strand turns to the researcher, who remains entangled in the various contexts that determine his views and ultimately himself. Knowledge (meaning) is produced "through the dialogical dialectical mediation of subject and object" (Bleicher, 1980, p. 3). In such a process, circumstances such as temporality and historicity become essential. To grasp them deeply enough, Bleicher refers to Heidegger's *Dasein*, a very advanced way of describing man's being in the world, which goes far beyond ordinary existence. Such *Dasein* creates the appropriate, i.e., appropriately extensive and fundamental context of the aforementioned process, in fact, the deepest one that has been able to be formulated. In such circumstances, text "can consequently no longer be the objective re-cognition of the author's intended meaning, but the emergence of practically relevant knowledge in which the subject himself is changed by being made aware of new possibilities of existence and his responsibility for his future" (Bleicher, 1980, p. 3). At this point, the text acquires a special and completely autonomous significance. The author's instance is rejected, and the text becomes a rightful "participant" in the epistemic situation, equating itself with a human being. The last and third strand Bleicher describes primarily in the context of the contribution of Jürgen Habermas, who "challenges the idealist assumptions underlying both hermeneutical theory and hermeneutic philosophy: the neglect to consider extxa-Iinguistic factors which also help to constitute the context of thought and action, i.e., work and domination" (Bleicher, 1980, p. 3). This approach updates the ideological and political background of the formation of knowledge (meanings), paying attention to the materialistic (in the Marxian sense) circumstances of their existence.

Vattimo points to Hans Georg Gadamer as the one whose reflection unequivocally concretizes the ontological culmination of hermeneutics. It takes place, which is extremely important, through the text, which is an instance of language. Gadamer in his 1960 book, *Truth and Method* (Gadamer, 1960), regarded as one of the greatest philosophical achievements of the 20th century, returns to the phenomenon of the text, which has historically been the first area of interest for hermeneutics. The text understood as the fullest instance of language marks the only possible field for the existence of the world, reduced to a constant and insurmountable interpretation, which takes place in the process of interaction between the object and its observer/interpreter. This process, as Vattimo writes, takes the shape of an ongoing conversation and — as he writes — "this conversation is what we all are" (Vattimo, 2015, p. 723).

For Bleicher, Gadamer's postulate is an innovative step that sets a new way of understanding human existence: "In Gadamer's work, the problem of hermeneutics is given a language-philosophical turn: as the 'hermeneutic problem' it is concerned with achieving an agreement with somebody else about our shared 'world'. This communication takes the form of a dialogue that results in the 'fusion of horizons'" (Bleicher, 1980, p. 3). Hermeneutics, thanks to Heidegger and Gadamer, funds an extensive and multithreaded movement of language analysis and its role, in which there are outstanding thinkers representing continental Europe, such as Jacques Derrida, Roland Barthes, Michel Foucault, and Gilles Deleuze, English analytical philosophy, e.g., Bertrand Russell, Gilbert Ryle, and John Longshaw Austin, or America like Richard Rorty and Paul de Man. This extremely rich movement developed in the second half of the 20th century is largely inspired by the aforementioned book by Gadamer. It also refers to the research undertaken within the so-called Vienna circle, with its most famous representative, Ludwik Wittgenstein, and its sources also go back to Friedrich Nietzsche.

Vattimo recalls the famous sentence in which Gadamer succinctly summarizes the fundamental role of language: "Being that can be understood is language" (Gadamer, 2004, p. 470). From our point of view, the questions about the metaphysical consequences of this thought are not important, but the establishment of the absolute primacy of language as the basis of existence, immersed and destined for dialog, of this incessant conversation that produces a game between man and the world. Gadamer also does not doubt that language reveals itself as text and that text encompasses all that is written: "Everything written is, in fact, the paradigmatic object of hermeneutics. What we found in the extreme case of a foreign language and in the problems of translation is confirmed here by the autonomy of reading: understanding is not a psychic transposition. The horizon of understanding cannot be limited either by what the writer originally had in mind or by the horizon of the person to whom the text was originally addressed" (Gadamer, 2004, p. 396).

Probably one of the most advanced reflections on the role of language and text is contained in the publications of Jacques Derrida, in particular the book *De la Grammatologie* (Derrida, 1967). It is an in-depth and extensive project to reconstruct the interdependencies and roles played by language, utterance, and writing, also realized as a text. It tells of a slow and very long process, lasting, according to Derrida, about 20 centuries, in which an extraordinary reversal of the interdependence of these phenomena takes place. This process comes down to capturing the main and superior role by writing. This thought is also a polemic with the common belief about the primacy of language, which somehow produces speech and writing from itself, and about the primacy of speech, which seems to be inherent in man. According to Derrida, language gradually took shape and exists as writing, which also found itself above speech. Writing ceased to play only the role of a shell, an impermanent duplicate, and a material carrier of meaning, but it enabled both the possession of meaning by language and the formation of this meaning over time. The text can be understood as a technical implementation of writing, reduced to the role of a tissue of signs and drawing strength from the constructions provided by writing. Thus, not only is it secondary to writing, but when it appears as an artifact, for example, as a book, it limits it in a way. The book, as Derrida notes, is based on the idea that it is possible to construct a certain whole, covering a selected horizon of content. Thus, through this choice, she takes away from writing its constant and mobile ambiguity.

In this context, Derrida formulates one of the most important circumstances of the existence and development of writing: it is the basis and condition of science. He writes, "Writing is not only an auxiliary means in the service of science — and possibly its object — but first, as Husserl, in particular, pointed out in *The Origin of Geometry*, the condition of the possibility of ideal objects and therefore of scientific objectivity" (Derrida, 1997, p. 27). In the civilizational and philosophical project of the West, as Derrida claims, these two phenomena appeared together and in combination. Mathematics is a kind of challenge for writing because, in principle, it is not dependent on its historical context of any kind, while writing remains entangled in the goal it sets for itself each time. Writing therefore in a constant need to justify its meaning, which turns into a kind of game that enables the continuous emergence and formation of meanings. It is this game that enables the movement of thought needed by science and thus shows the successive steps of reasoning in the form of notation.

Walter Ong, another classical researcher of these problems, although polemicizing with Derrida, also gives writing a fundamental role in shaping the reality of meanings: "Without writing, the literate mind would not and could not think as it does, not only when engaged in writing but normally even when it is composing its thoughts in oral form. More than any other single invention, writing has transformed human consciousness" (Ong, 2002, p. 77). Ong recalls the classic text of Phaedrus, in which Plato, through the mouth of Socrates, criticizes the strangeness, stability, and silence of writing, incapable of responding to the surroundings and the recipients. For Derrida, it is a kind of usurpation through which the writing reveals the existence of this environment. In this way, writing paradoxically gains autonomy, which enables it to assume an active role in constructing meanings. On the other hand, according to Ong's idea, it can also influence and shape a person. In this multilevel and complex game, the reality of the written text is created, the uniqueness of which Gadamer summed up simply by writing that "nothing is so strange, and at the same time so demanding, as the written word" (Gadamer, 2004, p. 156).

A short story about hermeneutics is intended to present, on the one hand, the depth of the issues related to the text, which, on the other hand, builds its unique and autonomous character. The text, although it arises as an act, is never reduced to this act but establishes its reality. This reality includes all the forms and instances of sense that arise in every place of the game that the text plays with the circumstances of its reception, which is the second opposite act that determines its existence. The text seems to be stretched between these acts, but, as the authors discussed here show, at the moment of its production, it gains autonomy, which means that the act of reception engages the entire environment of the text, which may depend on almost any circumstances: historical events, a person encountered, current culture, metapsychological factors, etc. The analysis of this baggage, colloquially called interpretation, extends to the farthest areas of reasoning, including the conditions and ways of existence of the world, which serves as the environment.

4 Premise #3: Language Models

The most important achievement in artificial text, which also became the cause of this text, is to overcome the barrier of effective digitization of text semantics. The measure of this effectiveness is the spectacular results achieved by solutions such as GPT-3 (Brown *et al.*, 2020) or BERT (Devlin *et al.*, 2018), and especially GPT-4 (OpenAI, 2023) with its popular implementation ChatGPT, in generating artificial texts or other text-related tasks. An extensive description of the background to these achievements is provided by Jurafsky and Martin in their continuously improved natural language processing (NLP) textbook, the latest draft of which was published in January 2023 (Jurafsky and Martin, 2023). It is without a doubt the most comprehensive and competent source on NLP today and will serve as your primary guide here. In this book, as no other, you can also find extensive literature on the subject which for this reason does not need to be repeated here. The common feature of these solutions is the construction of a language model based on the use of vector semantics techniques and the so-called embeddings. Due to this, it was possible to interpret the semantic dependencies that appear in sentences, i.e., in practical constructions of language, in a previously unattainable way. This achievement made it possible to apply more sophisticated solutions to analyze such a model, such as the technique of attention (Vaswani *et al.*, 2017).

The language model allows one to assign a probability to a sentence or other sequence of words. The first ideas for carrying out such an operation are quite old. Jurafsky and Martin go back to Markov's proposal from the beginning of the 20th century and also recall, among others, influential papers by Noam Chomsky from the 1950s and 1960s in the context of the simplest, basic models of this kind, the so-called n-grams (Jurafsky and Martin, 2023, p. 55). The description of the probability distribution of words in sentences made it possible to structure the linguistic resource and give it a formal (mathematical) shape, which could then be used in such practical tasks as speech recognition or machine translation. This description proposed its version of the implementation of the key and canonical step from the point of view of the foundations of computing ontology, which was the recognition of a certain physical reality in the form of a mathematical model. The n-gram models took this step earlier and were the first attempt to reflect the internal structure of a real language. However, it is debatable to what extent these first models constituted a formalization of the syntax and to what extent the semantics of the language, although they certainly also entered the latter area. In this context, the publications of Chomsky provided particular theoretical support. Therefore, to develop the possibility of formalizing semantics, a slightly different interpretation of this structure was needed, based on a new formal idea. Such a solution turned out to be the so-called vector semantics and the so-called embedding.

As Jurafsky and Martin write, "The idea of vector semantics arose out of research in the 1950s in three distinct fields: linguistics, psychology, and computer science, each of which contributed a fundamental aspect of the model" (Jurafsky and Martin, 2023, p. 131).

This diversity indicates a certain specificity of the concept, which requires a new look at the semantics of language and the use of new analytical methods. The deep foundations of this concept are philosophical and refer to the reflections of Descartes or Leibniz. Descartes perceived the world as a set of relationships that can be expressed in the form of measurable (quantitative) dependencies, which, according to him, would be the basis of the universal scientific approach of the so-called *mathesis universalis*. Leibniz developed this thought in the idea of a universal, formal language, *characteristica universalis*, with which such a world could be captured. Both of these ideas introduced mathematics and its structures as a basic and sufficient means of scientific insight into reality.

A different type of philosophical approach to language was represented by Ludwig Wittgenstein, whose books became the basis for a very broad reflection on language. It was heading toward the analysis of language as a certain activity that fulfills many practical roles going beyond the mere exchange of information and thus, above all, representing reality. It turned out that the process of this representation is extremely complex, variable, and dependent on various contexts. This reflection is thus significantly different from the contributions of Descartes and Leibniz, who postulated a complete formalization of language and saw it as possible and necessary. Their idea was later continued; it is worth mentioning in this context the idea of Frege (1879) and also, in a sense, Boole (1854), to which we return, and it was finally abandoned at the beginning of the 20th century. Wittgenstein, on the other hand, drew attention to the complexity of a system such as language and its existence "to imagine a language means to imagine a form of life" (Wittgenstein, 1958, p. 8). In this sense, Wittgenstein gave the ultimate idea of the phenomenon of language, opposite to Descartes and Leibniz, which could form the conceptual basis of the idea of vectorization.

Suggestions for the practical implementation of Wittgenstein's vision came from linguistics in the form of the so-called distributional hypothesis, whose proponents were published in the 1950s. These included Joos (1950), Harris (1954), and Firth (1957). Jurafsky and

Firth also mention Thomas Sebeok in this context (Jurafsky and Martin, 2023, p. 131). Harris wrote, "All elements in a language can be grouped into classes whose relative occurrence can be stated exactly [...] it is possible to state the occurrence of any element relative to any other element" (Harris, 1954, p. 146). Firth wrote, "You shall know a word by the company it keeps" (Firth, 1957, p. 11). In this way, a technical basis appeared for the search for semantic dependencies that could be formalized. Juraffsky and Martin describe a whole series of studies that shaped the idea of capturing these relationships in a vectorized, abstract, and multidimensional space, starting with the ideas of Osgood *et al.* (1957).

Individual words are single points in this space, and their coordinates are treated as vectors, which is the basis of a technique called embeddings. Juraffsky and Martin looked for the origins of this technique in the 1990s of the 20th century, but its intensive development took place at the beginning of the 21st century. The symbolic text that opened the era of advanced currently used language models based on the idea of vector space, used to represent words, is considered to be the text of Mikolov *et al.* (2013). They used the technique of dense vectors, which consisted of reducing the number of space dimensions to an arbitrarily selected, relatively small number. Dimensions were assigned to words, and then to smaller units, derived from their division, the so-called tokens by analyzing their locations in a real set of texts. This process was called training. The set mentioned above became more and more extensive. In the case of the aforementioned GPT-3 model, it had 499 billion tokens and was based on various sources, which in total made up about 700 GB of plain text (Brown *et al.*, 2020).

The location of words or their parts in the space prepared in this way reflected their semantic relations by giving them numerical values (vectors). The effectiveness of this procedure turned out to be extremely high. It also allowed mathematical operations on those vectors that gave correct results in the semantic relations mapped by these vectors. In addition, the embedding building of the space took place inside the text analysis procedure without human participation, implementing the principle of unsupervised learning. In this way, the subjective factor of analysis was eliminated and the original corpus of texts remained the only and final source of semantic relations. However, for this purpose, it had to undergo cleaning procedures, the so-called normalization consisting of simplifying grammatical forms (lemmatization), sentence segmentation, and finally tokenization. Jurafsky and Martin consider the text solution as "implicitly supervised training data" a revolutionary intuition (Jurafsky and Martin, 2023, p. 120). It seemed like a completely objective and impartial insight into the semantic structures of language.[a] Such technical data are not published for GPT-4.

However, it should be noted that words and other semantic units of language, according to Wittgenstein's approach, are strongly contextualized. The next stages of artificial text development can be interpreted as taking this fact into account. First, it was possible to capture this context by introducing the so-called positional embedding, which consisted of combining a vector calculated for a given word following its semantic environment and a vector calculated based on its physical position in the text, which allowed us to distinguish words with the same sound but different meaning. A breakthrough approach in this regard was provided by Vaswani *et al.* (2017). It implemented, above all, the idea of calculating the contextual semantic weight of a word in a sentence, introducing the so-called self-attention technique based on text-only calculations. It also used whole sentences, incorporating memory mechanisms into the calculations, taking into account previously made calculations. The improved results were the basis for creating a new class of artificial neural network algorithms called transformers, which have now become state-of-the-art NLP. The GPT-3 and GPT-4 models also belong to this class. A relatively up-to-date comparative review of these solutions is provided by Zhao *et al.* (2023).

According to the review, it is clear that we are dealing with spectacular progress in building language models, also known as

---

[a]This is a certain simplification because human participation appears earlier in the process of selecting texts and preparing them for training. Also in the case of ChatGPT, people were used to evaluate query results in the Reinforcement Learning from Human Feedback (RLHF) process (Ouyang *et al.*, 2022).

pretrained language models. It concerns not only their size measured by the number of parameters, which can be roughly understood as the number of weights resulting from the complexity of artificial neural networks that are their basis or the volume of language corpora on which they are trained. First of all, the effects of their use available in the case of GPT-3 and GPT-4, for example, on the OpenAI Company website, are spectacular. The development leading to transformer technology is based on the disclosure of increasingly complex semantic inferences inside the text corpus, referring to the semantic elements of the set, most often tokens. These inferences are implicitly stored in a large artificial neural network. Thus, there is no direct access to the semantic structure, but the mathematical computational model of the language can be used for several different procedures that, implicitly mapped to the semantic structure inside the model, allow us to generate meaningful semantic constructions, i.e., texts with a high degree of discursive complexity: stories, essays, essay, description, etc. The implementation of the GPT-4 model, which is ChatGPT, has become particularly popular, both as a widely available semantic tool and as a subject of scientific analysis. As summarized in (Liu *et al.*, 2023), "The findings of this study reveal that the interest in these models is growing rapidly, and they have shown considerable potential for application across a wide range of domains" (Liu *et al.*, 2023, p. 26).

There is a certain similarity between the linguistic model and the axiomatic system, which bases its sense only on defined inferences. Kline defines it as follows: "The essence of an axiomatic development (...) is to start with undefined terms whose properties are specified by the axioms; the goal of the work is to derive the consequences of the axioms. In addition, the independence, consistency, and categoricalness of the axioms (...) are to be established for each system"[b] (Kline, 1972, p. 1026). This understanding became the basis

---

[b]Keeping the principle of categoricalness, however, would be difficult in the context of its meaning described by Kline (1972, p. 1014) due to the implicit nature of the axioms at the current level of analysis. This does not preclude better elaboration of this problem in the future.

of Giuseppe Peano's approach (Peano, 1889) and David Hilbert, who was based on it (Hilbert, 1899), concerning the so-called axiomatic systems. The linguistic model can be roughly interpreted as a kind of axiomatic system whose axioms determine the internal logical structure of this model. The primary semantic structure, recorded in the language model, is treated as the basis of a certain (logical) order, which is a set of inferences that make up the language model. This order is implicit in the linguistic model and is ultimately unknown, making language and its products difficult or impossible to describe precisely.

The axioms, in this case, would be hidden inferences, implicitly stored in the language model, present in it as semantic dependencies, formulated thanks to the analysis of the corpus of real texts. Attention should be paid to the strongly pragmatic nature of such an axiomatic system, which would describe real and strictly defined linguistic instances of an empirical nature, i.e., texts of a real and spontaneous nature, collected by the authors of language models in the corpora on which the models are trained. Such an axiomatic system can generate any number of logical sentences that take the form of a text, i.e., linguistic sentences of an artificial text. The specificity of such an axiomatic system lies in the fact that the axioms are implicit, although they certainly exist and are implicitly recorded in the structure of the neural network. Such a record, although incomprehensible to a human being, is nevertheless strict and does not preclude a better interpretation in the future. It is also characterized by a high degree of complexity. It is also not the result of an individual decision but a socially shaped structure in the course of constant evolution. On the other hand, because a linguistic model depends on a set of selected linguistic instances, different language models are possible, and correspondingly different axiomatic systems are hidden in them, which, however, is not the subject of research and reflection. Finally, there is no doubt that an axiomatic system is essentially a structure of knowledge. From this point of view, the linguistic model is also a knowledge structure, which is confirmed by other textual research approaches, in particular, discourse theory.

5 Discussion: Intelligible Artificial Text

At the beginning of the second decade of the 21st century, the phenomenon of artificial text appeared, the degree of sophistication of which is unprecedented. It is a text generated by a specialized computer algorithm and achieves a level of meaningfulness equal to that of human texts. The latter property is particularly difficult to define precisely because it is ultimately at the disposal of man as the sole arbitrator. This type of problem is well known in the field of computer science as the Turing test. In 1950, Turing proposed a test, called the imitation game (Turing, 1950), which was supposed to answer the question "Can machines think?". This test was based on the question-and-answer method, which "seems to be suitable for introducing almost any one of the fields of human endeavor that we wish to include" (Turing, 1950, p. 435). Three people were to participate in the experiment based on this method: a woman, a man, and an unspecified interrogator, designated by Turing A, B, and C, respectively. The interrogator was supposed to ask questions to the other people using a communication device that masked voices and finally assigned the roles of women and men to their interrogators. Turing replaced the original question with another: "What will happen when a machine takes the part of A in this game? Will the interrogator decide wrongly as often when the game is played like this as he does when the game is played between a man and a woman?" (Turing, 1950, p. 434). Furthermore, Turing explicitly pointed to the digital computer as the machine used in the test and stated: "I believe that in about fifty years it will be possible to program computers, with a storage capacity of about 10^9, to make them play the imitation game so well that an average interrogator will not have more than 70 percent, chance of making the right identification after five minutes of questioning" (Turing, 1950, p. 442).

Oppy and Dowe in the context of the description of the Turing test resemble another experiment proposed by John Searle known as The Chinese Room (Oppy and Dowe, 2021). This experiment contradicted Turing's vision and pointed to a kind of delusion that we can succumb to when we observe the manipulation of text by a computer (Searle, 1980). It consisted of imagining a situation in which an operator, actually Searl himself, who does not know Chinese, can successfully translate texts given to him in this language, having the appropriate device which only manipulates symbols and numerals. Such a translation turns out to be only a mindless implementation of procedures and is not based on cognitive competence. His reflection became the beginning of an intense discussion regarding the fundamental property of artificial intelligence, which is mastering the human reasoning process.

In both cases, the Turing test and the Chinese Room experiment, their authors tackle the problem of intelligent machine behavior and in both cases use examples related to text manipulation, respectively, dialogue and translation. This is a very important shift in the issue, which takes into account the human competence formulated by Descartes, described earlier. It allows one to avoid questions of similar breadth and complexity as the question of the ability to think. At the same time, it refers to the artifact of the text, which has a material representation, well-established empirically. This makes it easier to analyze and describe. However, this is not true, given the existential basis of the text, which is language. There are fundamental problems with this artifact.

For example, the question of what kind of text can be considered full-fledged text arises. For this purpose, the concept of intelligibility can be used as a feature to conditioning this full-fledgedness. Furthermore, this term can become a descriptive collection of properties of such a text. Intelligibility as defined by Merriam-Webster means "1. capable of being understood or comprehended, or 2. apprehensible by the intellect only" (Merriam-Webster.com). The sense of this definition finds support in the rich hermeneutical reflection described earlier. However, it is not clear where the line lies between an intelligible text and a text that does not have this feature. It is not determined by formal correctness, e.g., grammatical. Examples of such far-reaching intelligibility are literary texts such as Ulysses by James Joys or poetry, which commonly violates grammatical and logical rules or uses a word-formation technique. The meaning of the text can be hidden and indirect, emerging as a result of complex associations and unexpected combinations of words, remaining outside the rules of syntax and punctuation. The only depositary of meaningfulness turns out to be the recipient, who ultimately decides about the intelligibility of the text and does so as a result of an individual act of understanding, which does not have to be shared by others. However, this act of understanding can be described, according to a hermeneutical approach, as an interpretive act of making sense, which in turn can be seen as knowledge.

The most advanced reflection on these processes, referring to the basic ontological acts, was presented by Heidegger, who pointed to the special role of poetry. It allowed us to realize human existence in its fullness, which Heidegger calls the truth of beings. It is the implementation of a sublime idea concerning the very essence of the human being, which rises above the experience of existence, as well as above traditional ontology, allowing it to be interpreted on a new, higher level. The key area in which this process takes place is language, and the only possessor of the latter is man. Heidegger also does not doubt that recognizing this situation and completing the task of reaching the truth of human beings remains to be done: "Language still denies us its essence: that it is the house of the truth of Being. Instead, language surrenders itself to our mere willing and trafficking as an instrument of domination over beings. Beings themselves appear as actualities in the interaction of cause and effect" (Heidegger, 1993, p. 223).

In the 20th century, language became the subject of very intensive research. One of its most important theoreticians was Ludwig Wittgenstein, the author of the two books mentioned above, which became the beginning of a wave of reflection on language: *Tractatus logico-philosophicus* (Wittgenstein, 1922) and *Philosophical Investigations* (Wittgenstein, 1958). The first tried to justify the formal and logical coherence of language, which is an image of reality while formulating its fundamental role. The most famous thesis attributed to Wittgenstein comes from this work, which at the same time ends the entire text: "What we cannot speak about we must pass over in silence" (Wittgenstein, 1974, p. 89). It entered the legacy of the philosophy of language as the first and final thesis establishing language as a condition of cognition. *Philosophical investigations* abandoned the style of logical precision of reasoning in favor of poetic metaphors that showed language as a complex entity in which utterances are produced according to the logic of the game. Language is presented as the result of a certain practice that Wittgenstein explicitly calls life, he writes, "to imagine a language means to imagine a form of life" (Wittgenstein, 1958, p. 8). Such a form is the old town, which develops spontaneously and abounds in complicated and illogical structures. This is another of his language metaphors. With these books, Wittgenstein decrees, on the one hand, the fundamental nature of language and, on the other hand, its inherent complexity, precluding formal unification. Later linguistic reflection moves only within the limits of these two observations, which remain valid.

According to the concept presented by Wittgenstein, the limit of textual analysis is language, whose complex nature makes it impossible to formally unify its analysis and to reduce it to the form of a coherent system. For this reason, formalizing the language semantics turns out to be extremely difficult. On the other hand, there is also no doubt that the way to a deep analysis of the text leads through language, which in turn becomes a condition for access to reality at every level. Both Heidegger and Wittgenstein, limit names to the most important, present language as the sole and ultimate epistemic tool. On the one hand, it defines the available world, and, on the other hand, it allows for an extremely sublime interpretation of the essence of its presence. These two, in a way complementary, reflections determine the complete cognitive competence of a human being.

Language in the first solutions proposed in the field of artificial intelligence in the area of natural language processing was treated in this way as a knowledge container. Deng and Liu proposed a three-wave description of NLP development. The first, the earliest, dominant from the 1960s to the 1980s, was based on the belief that it is possible to construct open formal systems that would allow the generation of meaningful sentences. This belief was based on the well-known theory of Noam Chomsky, which linked grammar and linguistic semantics into one system, e.g., (Chomsky, 1957). Various solutions inspired by this approach have emerged, e.g., Schank's conceptual dependency theory (Schank and Tesler, 1969), semantic networks by Collins and Quillian (1969), frames (frame-based systems) by Minsky (1974), and scripts by Schank and Abelson (1975). The next two waves were based on an empirical approach, observation, and data corpora analysis. It has become clear, as Deng and Liu write, that "learning and perception capabilities are crucial for complex artificial intelligence systems but missing in the expert systems popular in the previous wave" (Deng and Liu, 2018, p. 4). It was a fundamental change in the way language was treated as a container of knowledge, which eventually became realized as a deep learning technology. In phase two, as they write, "the associated shallow models lack the representation power and hence the ability to form levels of decomposable abstractions that would automatically disentangle complex factors in shaping the observed language data" (Deng and Liu, 2018, p. 6). The third phase, which we are currently observing, is characterized by the development of advanced forms of language semantics using deep learning technology.

However, knowledge remains the key area, but the nature of its representation is changing. In the early solutions of artificial intelligence and, accordingly, NLP solutions, knowledge appeared in a symbolic way, i.e., it maintains interpretable compliance with empirical data, i.e., language and its elements. Flasiński calls this variant "symbolic Artificial Intelligence" and gives the fundamental principles on which it is based: "a model representing an intelligent system can be defined explicitly, knowledge in such a model is represented symbolically, mental/cognitive operations can be described as formal operations over symbolic expressions and structures which belong to a knowledge model" (Flasiński, 2016, p. 15). Advanced deep learning solutions do not meet these assumptions, and knowledge is hidden in numerical values that cannot be interpreted. Flasinski calls this variant "Computational Intelligence": "numeric information is basic in a knowledge representation, knowledge processing is based mainly on numeric computation, usually knowledge is not represented in an explicit way" (Flasiński, 2016, p. 23).

The transition from the first to the third wave, described by Deng and Liu, as well as the transition from symbolic artificial intelligence to Computational Intelligence, allows one to define the essential premises for understanding the essence of artificial text as it appears today. First of all, knowledge remains the key context, although the way it manifests is hidden, implicit, and not subject to direct interpretation, i.e., regarding input linguistic data. This character of the approach points to an advanced and complex way of dealing with knowledge. Second, the foundation in the sense of an empirical resource and the source of ordering structures remains practical language, i.e., what de Saussure calls *parole* — speech, the act of practical use of the language system — *langue* (Saussure, 1959) (1916). Third, the barrier of digitization of language semantics is overcome, which is based on the formalization of deep inferences extracted from the repositories of practical language, large corpora, used to train language models. Among the aforementioned circumstances, the first one also allows us to specify the meaning of the concept of an intelligible text, as a text that contains/transfers a certain amount of knowledge.

The link between the issues of knowledge and language was noticed relatively long ago, in the 1950s, so at the time when the other foundations of NLP, mentioned earlier, were being shaped. One of the most important fields in which a powerful and multi-threaded reflection on this topic has developed is discourse analysis. Paltridge describes it as follows: "It looks at patterns of language across texts and considers the relationship between language and the social and cultural contexts in which it is used. Discourse analysis also considers how the use of language presents different views of the world and different understandings. It examines how the use of language is influenced by relationships between participants as well as the effects the use of language has on social identities and relations. It also considers how views of the world, and identities, are constructed through the use of discourse" (Paltridge, 2006, p. 2). This opinion clearly shows that the subject of the study is the world that appears in a certain way to its observers, producing knowledge about it.

Paltridge, going back to the beginnings of discourse analysis, points to Zellig Harris, already quoted here, who in 1952 presented the first basic principles of discourse analysis (Harris, 1952). Although he declared a formal approach, he also took into account social circumstances, which allows Patrigde to indicate two basic and related paths that this analysis follows: formal and social. The former is well defined by Kamath *et al.*: "Discourse analysis is the study of the structure, relations, and meaning in units of text that are longer than a single sentence. More specifically, it investigates the flow of information and meaning by a collection of sentences taken as a whole" (Kamath *et al.*, 2019, p. 105). The second turns out to be extremely diverse. Fairclough, seeking a good description, proposes an explicitly epistemological approach: "I see discourses as ways of representing aspects of the world — the processes, relations, and structures of the material world, the 'mental world' of thoughts, feelings, beliefs and so forth, and the social world. Particular aspects of the world may be represented differently, so we are generally in the position of having to consider the relationship between different discourses" (Fairclough, 2003, p. 124). Jørgensen and Phillips interpret these differences literally, as a problem of different instances of knowledge: "the struggle between different knowledge claims could be understood and empirically explored as a struggle between different discourses which represent different ways of understanding aspects of the world and construct different identities for speakers" (Jørgensen and Phillips, 2002, p. 2).

At the end of this thread, it is worth recalling the earliest idea of discourse, which at the same time opened the widest research perspective, ultimately with an epistemological focus. It comes from a philosopher and mathematician whose name has already been mentioned here: George Boole. He proposed already around the middle of the 19th century a descriptive construction — the universe of discourse, developing the earlier idea of Augustus de Morgan, formulated as a logical problem, but based on principles given in a general and descriptive way. Boole writes, "In every discourse, whether of the mind conversing with its thoughts or of the individual in his intercourse with others, there is an assumed or expressed limit within which the subjects of its operation are confined. (...) Now, whatever may be the extent of the field within which all the objects of our discourse are found, that field may properly be termed the universe of discourse" (Boole, 1854, p. 42). These principles include, first, the idea of separating certain ontological entities from the real world based on the perception of the fundamental epistemic limitations of man and science. However, Boole places no restrictions on these parts, leaving this act at the disposal of the researcher. This is the second extremely important step, consisting of transferring the epistemic initiative to man and his decisions. Keynes writes, "No general criterion can be laid down for determining what is the universe of discourse in any particular case. It may, however, be said that knowledge as to what is the universe referred to is involved in understanding the meaning of any given proposition; and cases in which there can be any practical doubt are exceptional" (Keynes, 1990, p. 213). Allowing for the possibility of an infinite dispersion of parallel discourse universes ultimately also opens up the possibility of the existence of different instances of the universe of discourse. Third, Boole establishes a close connection between three key areas: the world, that is, the area of described entities, thought, that is, cognitive competencies, and language, as the area of naming: "The office of any name or descriptive term employed under the limitations supposed is not to raise in the mind the conception of all the beings or objects to which that name or description is applicable, but only of those which exist within the supposed universe of discourse" (Boole, 1854, p. 42). Each of these conditions develops into a separate and broad research reflection, which I have tried to describe elsewhere (Maciąg, 2022). From the point of view of this text, the issue of cognitive dispositions assumed by Boole, which can be understood as dispositions in the field of knowledge, is of key importance. Boole, by introducing a descriptive cognitive construction, enters the area of knowledge. Although he does not deal directly with this issue, he emphasizes the fundamental nature of his treatise, "which is the investigation of scientific truths and laws" (Boole, 1854, p. 23).

6 Conclusions

The text tries to present and justify the importance of the historical fact of the emergence of artificial text by recalling the roles of text and language formulated in historical research on them. A special role in this process was played by the field of philosophical and linguistic research, which is hermeneutics. Inspired by her approach, the analysis contained in the text tries to indicate the phenomenon of knowledge, which should be considered in the context of artificial texts. At the same time, this phenomenon can become one of the types of specialist insights into the phenomenon of artificial text, opening up various ways of its interpretation, e.g., as an axiomatic system. Ultimately, it also determines the essence of an artificial text that can be considered intelligible.

References

Bleicher, J. (1980). *Contemporary Hermeneutics: Hermeneutics as Method, Philosophy and Critique*. Reprint, 1982 edition. London; Boston: Routledge & Kegan Paul.

Boole, G. (1854). *An Investigation of the Laws of Thought: On Which are Founded the Mathematical Theories of Logic and Probabilities*. London: Walton and Maberly. http://archive.org/details/investigationofl00boolrich [Accessed 2 August 2018].

Brown, T. B., Mann, B., Ryder, N., Subbiah, M., Kaplan, J., Dhariwal, P., et al. (2020). Language Models are Few-Shot Learners. *arXiv:2005.14165 [cs]*; http://arxiv.org/abs/2005.14165.

Chomsky, N. (1957). *Syntactic Structures*. The Hague: Mouton.

Collins, A. M., Quillian, M. R. (1969). Retrieval time from semantic memory. *Journal of Verbal Learning and Verbal Behavior*. 8(2), 240–247. https://doi.org/10.1016/S0022-5371(69)80069-1.

Dancy, J., Sosa, E., Steup, M., (eds.). (2010). *A Companion to Epistemology*. Malden: Wiley-Blackwell.

Deng, L., Liu, Y., (eds.). (2018). *Deep Learning in Natural Language Processing*. Singapore: Springer Singapore.

Derrida, J. (1967). *De la grammatologie.* Paris: Les Éditions de Minuit.
Derrida, J. (1997) *Of Grammatology.* Baltimore, London: The Johns Hopkins University Press.
Descartes, R. (2006). *A Discourse on the Method.* Oxford, UK: Oxford University Press.
Devlin, J., Chang, M. W., Lee, K., Toutanova, K. (2018). BERT: Pre-training of deep bidirectional transformers for language understanding. *CoRR.* abs/1810.04805. http://arxiv.org/abs/1810.04805.
Fairclough, N. (2003). *Analysing Discourse: Textual Analysis for Social Research.* London: Routledge.
Firth, J. R. (1957). A synopsis of linguistic theory 1930–55. *Studies in Linguistic Analysis (special volume of the Philological Society).* 1952–59, 1–32.
Flasiński, M. (2016). *Introduction to Artificial Intelligence.* Cham, Switzerland: Springer.
Forster, M. N., Gjesdal, K., (eds.). (2019). *The Cambridge Companion to Hermeneutics.* Cambridge; New York: Cambridge University Press.
Frege, G. (1879). *Begriffsschrift, eine der arithmetischen nachgebildete Formelsprache des reinen Denkens.* Halle: Verlag von Luis Nebert.
Gadamer, H. G. (1960). *Wahrheit und Methode: Grundzüge einer philosophischen Hermeneutik.* Tübingen: J.C.B. Mohr (Paul Siebeck).
Gadamer, H. G. (2004). *Truth and Method,* revised edn. London; New York: Bloomsbury Academic.
Harris, Z. S. (1954). Distributional structure. *WORD.* 10(2–3), 146–162. https://doi.org/10.1080/00437956.1954.11659520.
Heidegger, M. (1993). Letter on humanism. In Krell DF (ed.) *Basic Writings. from Being and Time (1927) to The Task of Thinking (1964).* San Francisco: HarperSanFrancisco pp. 213–265.
Hilbert, D. (1899). *Grundlagen der Geometrie.* Leipzig: Verlag von B.G. Teubner http://archive.org/details/grundlagendergeo00hilb [Accessed 14 May 2019].
Joos, M. (1950). Description of language design. *The Journal of the Acoustical Society of America.* 22(6), 701–707. https://doi.org/10.1121/1.1906674.
Jørgensen, M., Phillips, L. (2002). *Discourse analysis as theory and method.* London; Thousand Oaks, California: Sage Publications.
Jurafsky, D., Martin, J. H. (2023). *Speech and Language Processing. An Introduction to Natural Language Processing, Computational Linguistics, and Speech Recognition. Third Edition draft.* Stanford, CA. https://web.stanford.edu/~jurafsky/slp3/ed3book_jan72023.pdf [Accessed 11 January 2023].
Kamath, U., Liu, J., Whitaker, J. (2019). *Deep Learning for NLP and Speech Recognition.* 1st edn. Springer.
Keynes, J. N. (1900). *Studies and exercises in Formal logic,* 4th edn. Macmillan And Company Limited.; http://archive.org/details/studiesandexerci02942 7mbp [Accessed 12 April 2021].
Kline, M. (1972). *Mathematical Thought from Ancient to Modern Times,* 1st edn. New York: Oxford University Press.

Liu, Y., Han, T., Ma, S., Zhang, J., Yang, Y., Tian, J., et al. (2023). *Summary of ChatGPT/GPT-4 Research and Perspective Towards the Future of Large Language Models.* https://doi.org/10.48550/arXiv.2304.01852.

Maciąg, R. *Wiedza jako opowieść. Przestrzeń dyskursywna.* Kraków: TAiWPN Universitas (2022).

Malpas, J. (2015). Introduction. In Malpas, J., Gander, H.H. (eds.) *The Routledge Companion to Hermeneutics.* Abingdon, Oxon: Routledge pp. 1–9.

Merriam-Webster.com. *Definition of INTELLIGIBLE.* https://www.merriam-webster.com/dictionary/intelligible [Accessed 25 October 2021].

Mikolov, T., Sutskever, I., Chen, K., Corrado, G., Dean, J. (2013). Distributed Representations of Words and Phrases and their Compositionality. *arXiv:1310.4546.* http://arxiv.org/abs/1310.4546.

Minsky, M. (1974). *A Framework for Representing Knowledge.*

Ong, W. (2002). *Orality and Literacy. The Technologizing of the Word.* London: New York: Routledge, Taylor & Francis Group.

OpenAI. (2023). *GPT-4 Technical Report.* https://doi.org/10.48550/arXiv.2303.08774.

Oppy, G., Dowe, D. (2021). The Turing Test. In Zalta E.N. (ed.) *The Stanford Encyclopedia of Philosophy.* Winter 2021. Metaphysics Research Lab, Stanford University.

Osgood, C. E., Suci, G. J., Tannenbaum, P. H. (1957). *The Measurement of Meaning.* Oxford, England: Univer. Illinois Press p. 342.

Ouyang, L., Wu, J., Jiang, X., Almeida, D., Wainright, C. L., Mishkin, P., et al. (2022). *Training Language Models to Follow Instructions with Human Feedback.* https://doi.org/10.48550/arXiv.2203.02155.

Paltridge, B. (2006). *Discourse Analysis. An Introduction.* London; New York: Continuum.

Peano, G. (1889). *Arithmetices Principia: Nova Methodo.* Romae, Florentiae: Fratres Bocca.

Porter, S. E., Robinson, J. C. (2011). *Hermeneutics. An Introduction to Interpretative Theory.* Grand Rapids, Michigan, Cambridge, U.K.: William B. Erdmans Publishing Company.

Saussure, F. de. (1959). *Course in General Linguistics.* New York: Philosophical Library.

Schank, R. C., Tesler, L. (1969). A conceptual dependency parser for natural language. In: *Proceedings of the 1969 Conference on Computational linguistics.* Sweden: Association for Computational Linguistics, pp. 1–3.

Schank, R. C., Abelson, R. P. (1975). Scripts, Plans, and Knowledge. In: *Proceedings of the 4th International Joint Conference on Artificial Intelligence — Volume 1.* San Francisco, CA, USA: Morgan Kaufmann Publishers Inc, pp. 151–157.

Searle, J. R. (1980). Minds, brains, and programs. *Behavioral and Brain Sciences.* 2010/02/04 ed. 3(3), 417–424. https://doi.org/10.1017/S0140525X00005756.

Todorov, T. (2002). *Imperfect Garden: The Legacy of Humanism.* Princeton, Oxford: Princeton University Press.

Turing, A. M. (1950). Computing machinery and intelligence. *Mind*. 236(14), 433–460.

Vaswani, A., Shazeer, N., Parmar, N., Uszkoreit, J., Jones, L., Gomez, A. N., *et al.* (2017). Attention is all you need. *arXiv:1706.03762 [cs]*. http://arxiv.org/abs/1706.03762.

Vattimo, G. (2015). Conclusion: the Future of Hermeneutics. In Malpas J., Gander H.H. (eds.) *The Routledge Companion to Hermeneutics*. Abingdon, Oxon: Routledge p. 721–728.

Wittgenstein, L. (1922). *Tractatus logico-philosophicus*. London: Kegan Paul, Trench, Trubner & Co.

Wittgenstein, L. (1958). *Philosophical Investigations*, 2nd edn. Oxford: Basil Blackwell.

Wittgenstein, L. (1974). *Tractatus Logico-Philosophicus*, 2nd edn. London: New York: Routledge.

Woleński, J. (2004). The History of Epistemology. In: Niiniluoto, I., Sintonen, M., Woleński, J. (eds.) *Handbook of Epistemology*. Dordrecht: Springer Netherlands, pp. 3–54.

Zhao, W. X., Zhou, K., Li, J., Tang, T., Wang, X., Hou, Y., *et al.* (2023). *A Survey of Large Language Models*. https://doi.org/10.48550/arXiv.2303.18223.

© 2025 World Scientific Publishing Company
https://doi.org/10.1142/9789811294921_0015

Chapter 15

Human Intelligence and the Phenomenon of Information: Open Science, Human Competencies, and Higher Education

Yurii Mielkov*, Iryna Drach, and Olha Petroye

Institute of Higher Education of the National Academy of Educational Sciences of Ukraine, 9 Bastionna Street, Kyiv, 01014, Ukraine
**uka7777@gmail.com*

This chapter deals with the problems of human intelligence and especially knowledge formation in the situation of open access to a huge amount of information in the era of widespread digital technologies. Information plays a vital role in today's society, but philosophical reflection draws a strict distinction between information and knowledge: information appears as an alienated knowledge deprived of its subject carrier, while knowledge is the information made personal and processed through the whole complex of human intelligence including not just rational mind but also will and emotions. Thus, the major problem is that the abundance of information and the proliferation of information technologies coincide with the deficit of human knowledge and human competencies. It is argued that the movement of Open Science could present a solution for humans to deal with the vast volumes of available information by both developing personal competencies and digital literacy required to be able to handle information in the digital age and reconstruct science as a public enterprise. And that in turn implies new understanding of the higher education process aimed at the formation of the said competencies and at providing a person with the ability to create one's own knowledge.

1 Introduction

The present-day social life of humankind is characterized by deep penetration of information technologies into many, if not most, spheres of activity so that today's society itself is often being called "the information society". The proliferation of computer devices and the spreading of Internet technologies have transformed our lifestyle and the way we are accustomed to performing both our jobs and everyday chores rather radically, especially during the last few years under the situation of the COVID-19 pandemic. The changes in question are even more evident in the sphere of education and scientific investigations that face rapid digitalization while presenting both challenges and opportunities for the development of human cognition and higher education in the 21st century. And in order to realize those opportunities, it is not enough just to follow the general trends of digitalization and computerization in a passive and non-reflexive way, but it is of vital necessity to achieve the due and adequate understanding of the processes in question, to review their meanings and their features, to analyze the phenomenon of information and its role in today's world in a philosophical way, particularly in order to overcome the possible misbalance in the development of human society and human intelligence, when that development is being reduced to mostly just dissemination of machines and technologies.

The authors of this chapter intend to review the phenomenon of human intelligence by considering the meaning of information in its relation to human persons and their knowledge, as well as to try to provide philosophical consideration of the impact of common trends of the development of the ICT technologies on today's science, education, and society in general. In Section 2, the theoretical grounds are laid out, providing some insight into the phenomena of information and knowledge, their correlation, and problems that arise out of the imbalance in their comprehension. Section 3 is dedicated to the main principles of Open Science that outline the grounds for the creation of human knowledge in the 21st century. In Section 4, personal and educational aspects of intelligence are discussed. Section 5 deals with specific competencies — especially digital literacy — that should be the results of lifelong learning in order to enable humans to deal with the challenges of the information age.

2 The Phenomenon of Information and its Relation to Human Knowledge

Despite its widespread usage, the concept of information remains one of the most controversial in science, and the term can have different meanings in different disciplines and areas of human activity. In the middle of the 20th century, scholars in cybernetics and philosophy from many countries of the world have spent a lot of effort trying to lay out the grounds for the information theory (Mindell et al., 2003). However, for our current investigation, it is not sufficient to have information defined in its specific cybernetic or physical meaning, as a decrease in uncertainty (negentropy), which increases the probability of a certain choice for the recipient of information (Claude Shannon), or as diversity and a kind of inhomogeneity in the distribution of matter in the universe (Viktor Glushkov), or as an entity brought by a signal to a receptor, not in terms of supports necessary to exist physically, but regarding different configuration, which is to be identified (Dominique Dubarle). It would be quite interesting to note though that academics who tried to provide a rather common and broad understanding of the phenomenon of information had in fact acknowledged it being virtually impossible to arrive at a comprehensible definition. According to the already famous quotation by Norbert Wiener: "Information is information, neither matter nor energy" (Wiener, 1961, p. 132). On the other hand, Vasiliy Nalimov has pointed out the polymorphism of the semantic meaning of the term, coming to the conclusion that "[w]e cannot define what 'information' is and we shall consider it to be a complicated concept whose meaning is revealed by its context" (Nalimov, 1981, p. 24).

The context of this chapter is related to the broad comprehension of the phenomenon of information that is required for the evaluation of the impact of the current development of information technologies on human society and for a better understanding of human intelligence, so we can try to develop a certain universal interpretation of what information is. Indeed, all the mentioned different approaches and definitions have something in common: they refer to information as some contents of a message that could be received by a recipient enabling the latter to make certain choices based on what was received, i.e., enhancing his or her knowledge. "Some contents" could be then described as *data*: in fact, for example, the Merriam–Webster English dictionary defines information mostly as "knowledge obtained from investigation, study, or instruction", "intelligence", "data", and "a signal or character (as in a communication system or computer) representing data".

Still, it could be argued that information is in no way just identical in its meaning to "knowledge" and "intelligence". The correlation between information and knowledge is what seems to be especially important in the context of our investigation (see Mielkov, 2006). Information could then be defined as the contents of a message transmitted from a communicant to a recipient, but the data being transmitted is not exactly knowledge — it is a more generic concept in relation to knowledge. Knowledge could be in turn described as personalized information, as information that belongs to a certain human subject: it can't be separated from its carrier who has assimilated some information in order to transform it into his or her own knowledge. Correspondingly, information could be described as a special kind of knowledge — as knowledge taken in the process of its transmission, in isolation from its subject, whether it be the subject that transmits this knowledge (the communicant) or the subject that perceives it (the recipient). In other words, information is knowledge that is extremely impersonal; it is knowledge deprived of its human subject carrier (see Mielkov, 2022).

Information is thus alienated knowledge — knowledge that could be measured, codified, transferred, or even sold. The problem is that in today's society information easily becomes the standard representation for knowledge and even intelligence itself. With the usage of the latest models of mobile gadgets, a human person finds himself or herself in a virtually continuous, unstoppable process of communication, i.e., in an eternal state of "receiving" information. As a result, it is not difficult for a person to get lost in the vast amount of data available — and in the virtual space that the information constitutes. For example, Bernard Stiegler argues that global tracking systems based on satellite observation, like GPS, although having been designed to make it easier for a person to navigate in an unfamiliar area, have led to the fact that a person acquires this orientation only in a virtual information space — this means "that the user also becomes data, travelling through 'data landscapes' — that is, through electronic data that is physically located and situated on the interfaces simulating territorial space" (Stiegler, 2003).

On the other hand, the availability of such vast amounts of information thanks to Internet technologies leads to another problem — the illusion of ready-available knowledge. The disastrous effect of such an illusion could be seen, for example, in catastrophic levels of violation of academic integrity. Why study, why compose one's own papers, if "everything" is virtually available? According to the empirical data obtained as a result of a survey conducted by sociologists from Kharkiv, Ukraine, more than 90% of Ukrainian university students use plagiarism in one form or another. In particular, 37% of students acknowledged downloading essays and other papers from free sites and using them as their own, while 49% confessed to rewriting text from a source in their own words without references.[1]

Thus, it is important to draw a distinction between information and knowledge — as the latter could not be obtained in a technological way. It is impossible to share intelligence or to transfer knowledge — it is only alienated and depersonalized knowledge that could be transferred in the form of information, and it requires a lot of personal effort for a human person to convert the information he

---

[1] *Academic culture of Ukrainian students: main factors of formation and development* (In Ukrainian). (2015). https://surgery-four.pdmu.edu.ua/storage/common/docs/R1vyfIum4hNfJntUrILfR6KRh1FhLdBWxXuD6Amv.pdf, pp. 40–41.

or she receives back into knowledge. That is, knowledge is information processed through the whole complex of human intelligence that has to assert that information as being true. Literally, no one could bear the responsibility of creating knowledge except the very subject of knowledge in question.

In other words, intelligence in its philosophical comprehension is always *human* intelligence; it can't belong to a machine or any other kind of entity besides human beings. Any human can send and receive information, and any material carrier (from paper to computers) can serve as a bearer of information, but no machine can be a bearer of knowledge, i.e., no machine can be *intelligent*. One can even relate this argument to the known discussions on whether a machine could "think", which were quite popular since the middle of the 20th century. Even in our days, just like in the 1960s, philosophers find it necessary to repeat the idea that a computer can never "think" in the exact way humans do — as a machine is by definition unable to duplicate human intelligence. As argued by Adriana Braga and Robert Logan (Braga and Logan, 2021, p. 133):

> The notion of intelligence that advocates of the technological singularity promote does not take into account the full dimension of human intelligence. Human intelligence... is not based solely on logical operations and computation, but rather includes a long list of other characteristics that are unique to humans... ...no computer can ever duplicate the intelligence of a human being because of the many dimensions of human intelligence that involve characteristics that we believe cannot be duplicated by silicon-based forms of intelligence because machines lack a number of essential properties that only a flesh and blood living organism, especially a human, can possess.

The proposed distinction between information and knowledge, with the human dimension of the latter being stressed, and the noted problems of plagiarism related to incorrect understanding of knowledge make it important to review the situation in the sphere of education, as it is the activity at least partly dedicated to the transfer of knowledge, and even science in general. The changes brought to the sphere of education, and especially higher education, by the proliferation of information and Internet technologies are indeed quite radical. Just a few decades ago, university students used to study almost in the same way their grandparents were doing it in the previous centuries, especially in the sphere of humanities — searching for sources in traditional libraries, reading books, and sometimes facing difficulties in obtaining newer publications, first of all, those published in foreign languages and in other countries. The ways today's students study are indeed quite different — as well as the problems they face: finding access to the vast amounts of available sources is no longer an issue, but the temptation of plagiarism certainly is. And the task of distinguishing primary and reliable sources from secondary and dubious ones is now much more difficult for a non-prepared person who is not a specialist in the field of the knowledge in question, especially taking into account that many reliable sources, different from less credible ones could be not accessible openly. One of the undesirable consequences of this situation could be different forms of plagiarism and other violations of academic integrity in science and education — while the movement of *Open Science* that becomes more and more influential in today's European educational and scientific community can be presented as a solution for the current problems of information management and knowledge exchange.

3 Open Science and its Values and Principles

What is Open Science then? Based on high standards of transparency, cooperation, and communication, it is a priority for policy development in the European Research Area (ERA) and the European Higher Education Area (EHEA) (Drach, 2022, p. 90). The most complete definition of Open Science is given by the General Conference of UNESCO on a meeting in Paris in November 2021 (UNESCO, 2021):

> ...**open science** is defined as an inclusive construct that combines various movements and practices aiming to make multilingual scientific knowledge openly available, accessible and reusable for everyone, to increase scientific collaborations and sharing of information for the benefits of science and society, and to open the processes of scientific knowledge creation, evaluation and communication to societal actors beyond the traditional scientific community. It comprises all scientific disciplines and aspects of scholarly practices, including basic and applied

sciences, natural and social sciences and the humanities, and it builds on the following key pillars: open scientific knowledge, open science infrastructures, science communication, open engagement of societal actors and open dialogue with other knowledge systems.

Open Science is thus a system change that allows science to be improved through an open and collaborative way to produce and share knowledge and data as early as possible in the research process, as well as to communicate and share the results. This new approach affects research institutions and scientific practices, creating new ways of funding, evaluating, and rewarding researchers. Open science increases the quality and impact of science by promoting reproducibility and interdisciplinarity. This makes science more efficient through a better sharing of resources, making it more reliable and more responsive to the needs of society.

Out of the noted key pillars, *open scientific knowledge* refers to open access to scientific publications, research data, metadata, open educational resources, software, and source code and hardware that are publicly available or protected by copyright. This also applies to the possibility of open research methodologies and evaluation processes of their results.

Open science infrastructures are shared research infrastructures: virtual or physical, including basic scientific equipment or toolkits, knowledge-based resources such as collections, journals and open access publishing platforms, repositories, archives and scientific data, current research information systems, open bibliometric and scientometric systems for evaluation and analysis of scientific areas, open computing, and service infrastructures of data manipulation that provide collaborative and interdisciplinary data analysis and digital infrastructures. Open Science infrastructures must be non-commercial and guarantee permanent and unlimited access for the entire public to the greatest extent possible (see UNESCO, 2021).

Open engagement with society means increased collaboration between scientists and society beyond the closed academic community. The scientific process is becoming more inclusive and accessible to wider selections of interested social groups based on such new forms of collaboration and work as crowdfunding, crowdsourcing, and scientific volunteering. In addition, citizen science and citizen participation have evolved as models of scientific research conducted by non-professional academics following evidence-based methodologies and often carried out in conjunction with formal scientific programs or professional scientists using web-based platforms and social media, as well as open-source hardware and software.

Open dialogue with other knowledge systems means communication between different knowledge carriers that recognize the richness of diverse knowledge systems and epistemologies, as well as the diversity of knowledge producers, according to the 2001 UNESCO Universal Declaration on Cultural Diversity. It aims to strengthen the interrelationships and complementarities between different epistemologies, respect for international human rights norms and standards, respect for the sovereignty of knowledge and governance, and recognition of the rights of knowledge holders to receive a fair and equitable share of the benefits that may arise from the use of their knowledge. In particular, it is necessary to establish connections with the knowledge systems of the indigenous population (see UNESCO, 2021).

But Open Science is not just a series of static problems but a complex combination of topics that have yet to be defined. Open Science that makes research accessible to everyone today steadily becomes a standard way of producing knowledge. Within the paradigm of Open Science, we see the growth of the role of universities, which consists in the fact that the community of students, scientists, and professionals, as well as graduates and a wide circle of partners and citizens linked with institutions by networks of local, national, and international levels, build bridges between countries, cultures, and sectors, thus demonstrating peaceful and constructive European and international cooperation for high-quality research and innovation, as well as for learning and teaching. European universities will operate equally on global and domestic levels, ensuring the exchange of knowledge and data and maximizing the benefits of the free mobility of knowledge, researchers, and learners. As seen in footnote 2, Open Science

---

[2] Open Science and its role in universities: A roadmap for cultural change (2018). League of European Research Universities. https://www.leru.org/files/LERU-AP24-Open-Science-full-paper.pdf (accessed June 5, 2023).

could thus be defined as one of the fundamental freedoms of an open world, alongside the freedom of movement of persons, goods, services, and capital.

At the same time, while different aspects of Open Science are often discussed with an emphasis on research practices, the impact of Open Science on teaching and learning in higher education is an important issue as well. For example, Hecka *et al.* (2020) by analyzing the results of an online survey of 210 participants from higher education institutions in Germany note that 60% of respondents do not use open educational resources. At the same time, researchers argue that the use of Open Education by students should increase their awareness of the future goals of Open Science and teach them the skills needed to achieve these goals. Thus, Open Science should indeed be supported by appropriate educational practices.

4 Higher Education and the Human Dimension of Knowledge

We already mentioned that the changes in the way information is now being transferred lead to radical transformations in the field of education and especially higher education — and such transformations relate not only to the formation of Open Science as a way to create knowledge but also to methods of teaching the creation of knowledge and to the role of a university professor. In the contemporary world, there could not be any ready-made knowledge regardless of how much information could be available to anyone thanks to the Internet and information technologies. Let us again remind you of the distinction between information and knowledge and of the amount of effort required to convert the former into the latter: without such efforts, a person could only be a carrier of information — just like a computer or any other unintelligent medium is. For example, we can read a book on, say, quantum mechanics without understanding any word of it, we can even memorize the texts — but the information contained in that book will never become our knowledge and will remain unintelligible to us unless we comprehend the terms, acknowledge the arguments and the ideas of the author of the book.

That's why knowledge is in fact creative knowledge, as it is being produced out of information by a cognizing person using not only his or her rational mind but the whole sphere of human intelligence. Already in 1958, the personal nature of human knowledge and the deep interconnection of knowledge with passion and conviction have been noted by Michael Polanyi (2005, p. 319):

> I can speak of facts, knowledge, proof, reality, etc., within my commitment situation, for it is constituted by my search for facts, knowledge, proof, reality, etc., as binding on me. These are proper designations for commitment targets which apply so long as I am committed to them; but they cannot be referred to non-committally. You cannot speak without self-contradiction of knowledge you do not believe...

The idea of personal, humane knowledge has been reflected in the conception of post-non-classical science proposed by Stepin (2005) in order to provide a special philosophical definition for the contemporary new type of scientific rationality. According to this view, currently, we experience a global scientific revolution leading to the formation of a new type of rationality that features the introduction of human cultural values into the very core of scientific knowledge as science turns its attention toward complex objects that are found to be *human-commensurable*. In other words, it becomes apparent that science is in fact grounded on ethics and on human values — instead of pretending to be value-neutral. It is no coincidence that the major principles of Open Science well correspond to the main points of the scientific ethos singled out by Robert Merton already in 1943. Namely, the very openness of scientific enterprise and the free availability of the research results correspond to the principles of universalism and communism, which oppose any form of particularism in science and which state that findings of science constitute a common heritage of the community and could not be a private property (Merton, 1973, pp. 270–275).

Of course, it should be stressed that the introduction of the value dimension into science does not lead to it abandoning the quest for the objective impersonal truth — quite the contrary: acknowledging the presence of values in scientific activity allows us to comprehend the nature of those values and their impact on the methods and results obtained according to the norms of critical rationality. Some researchers note that the idea of "openness" is closely related to "smartness", which is also in line with such trends as Smart Society, Smart Government, Smart Cities, etc. (see Nikiforova, 2021). The concept of Smart Society is in turn related to the initiative called *Society 5.0* — it was proposed by the Japanese government in 2016 in order to create a sustainable society and solve social issues by creating a "superintelligent" society based on technologies, particularly based on the Internet of Things and artificial intelligence (AI), facilitating digital and physical infrastructures for human beings (Narvaez Rojas *et al.*, 2021). However, we would argue that basing society on technologies alone with human values being a kind of addendum to industrial progress is not enough to solve the problems of not only sustainability but of technologies themselves as well. The human dimension of sustainability is often being neglected serving just as a means for economical development. As stated by French philosopher André Gorz, the current crisis requires a transition from the society of production and the society of labor toward the society of culture that in fact opposes the market economy — and requires even not sustainability, but "degrowth" (*la décroissance*), meaning another lifestyle, other economics, and other social relations (Gorz, 2008, p. 29).

In relation to the structure of intelligence, this leads us to comprehend an opposite type of knowledge-information, which could be presented as *knowledge-wisdom*. The whole hierarchy of knowledge thus presents itself as quite a complex phenomenon and even process. Robert Logan outlines it in the following way: first, it is *data*, as simple unorganized bits of information; second, it is *information per se* as structured data with meaning and context added; third, it is *knowledge* defined as the ability to use information strategically to achieve one's objectives; and finally, *wisdom* as the capacity to choose objectives consistent with one's values (Logan and Stokes, 2004, pp. 38–39). We would add that this hierarchy follows the path of humanization, going from impersonal information (a set of organized data) to personal knowledge (which is non-transferrable and presents a result of elaborating, mastering, and assimilating the information received) to even more humane wisdom (the complex knowledge dealing not only with the rational comprehension of information but with values, life experience, ethics, will, and imagination as well).

In fact, it is the lack of will and personal commitment that hinders the development of human knowledge and intelligence even in the situation of vastly available information technologies. For example, according to the empirical data obtained as a result of a survey conducted by the authors and their colleagues from the Institute of Higher Education of the National Academy of Educational Sciences of Ukraine in 2019, in response to questions about the factors hindering the improvement of the quality of higher education, representatives of all groups of the academic community demonstrated complete unanimity, giving priority to the "motivational sphere", namely "lack of motivation of university staff" was chosen by 50% of university managers, 53% of teachers/researchers and 30% of students; and "lack of motivation of students" was chosen by 47% of managers, 49% of teachers/researchers and 47% of students. At the same time, the lack of modern information resources was designated as a problem by only 13–16% of management and teachers as opposed to 28% of students (Kalashnikova, 2019, pp. 190–191).

Wisdom as the form of intelligence dealing not only with knowledge but with values, wills, and emotions as well prevails in philosophy, but we would argue that knowledge of science indeed becomes now more close to that of humanities and philosophy. And that in turn leads to some core changes in the methodology of scientific investigations and in the strategies of contemporary higher education. Particularly, we now have to aim at a holistic personality endowed with certain human qualities, certain values, and modes of existence that would allow a person to deal with the challenges of the new, constantly changing circumstances in the world of complexity and uncertainty — in case there would be some motivation and some competencies for doing that. In other words, mastering information and creating new knowledge requires now a different set of human competencies, namely digital and informational competencies.

5 Human Competencies in the Era of Information

5.1 *Digital competencies: Content and models*

The comprehension of digital competencies of all participants in educational and research activities is anticipated by both the introduction of Open Science and by the requirements for information literacy, which have increased as a factor of its effective use and formation of new knowledge, which, in turn, is becoming more specialized and expanding exponentially. The term "information literacy" often reflects the processes in the structure of human competencies that characterize its ability to obtain specific information to meet a wide range of personal and business needs, in particular, to recognize information needs, determine what information is needed, find information, evaluate information, systematize information, use information effectively, etc.[3]

Under the current conditions of rapid development of the digital environment and the global network, which covered almost all spheres of life, the information space is undergoing radical changes. We are witnessing the proliferation of the Fourth Industrial Revolution (Industry 4.0), an era of innovation with cutting-edge technologies (cloud technologies, development of big data collection and analysis tools, crowdsourcing, biotechnology, drones, 3D printing, bitcoin cryptocurrencies, blockchain technology, artificial intelligence, etc.) that radically change entire sectors of the economy and society as a whole. There is already a completely new type of industrial production, which is based on big data and their analysis, full automation of production, and technologies of augmented reality.

The use of these and other new technologies opens up new opportunities for economic prosperity, social integration, and environmental sustainability and, consequently, leads to new and increased demands on human competencies in the context of digital literacy. Being competent in the digital field is no longer a choice, but a

---

[3] *Presidential Committee on Information Literacy: Final Report*, (1989). Washington, p. 10. https://www.ala.org/acrl/publications/whitepapers/presidential.

must for 21st century citizens. After all, the development of digital competencies is becoming one of the most important conditions not only for personal development but also for the development of the digital market, which is directly or indirectly related to all areas of the economy and society as a whole.

Existing approaches to the interpretation of digital competencies highlight the following basic elements and characteristics of this phenomenon: the set of knowledge, abilities, peculiarities of character, and behavior necessary for a person to use information and communication and digital technologies to achieve goals in personal or professional life[4]; confident, critical and responsible use of digital technologies for interaction, learning, work and participation in public life (Carretero et al., 2017); combination of knowledge, skills, and attitudes on the effective, safe, critical, creative, ethical use of technologies to perform tasks, solve problems, communicate, manage information, collaborate, and create and share content (Skov, 2016).

Over the last decade, digital competencies, their content, and structure have been the subject of research and development by many scholars and experts at various institutional levels. In order to develop the potential for digital transformations, in 2013, the EU developed a reference model Digital Competence Framework for Citizens (DigComp), which was further revised and refined (DigComp 2.0, 2016; DigComp 2.1, 2017; DigComp 2.2, 2022). Today, this model serves as an effective tool to raise the level of digital competence of citizens, to form the policy supporting the development of digital competence, planning of educational and training initiatives and assessment of digital competencies level of specific target groups, etc.

The modern DigComp model identifies key components of digital competence and offers a tool to improve them in five areas: (1) information and digital literacy; (2) communication and cooperation with the help of digital technologies; (3) creation of digital content; (4) security (including protection of personal data in the digital

---

[4]*DigComp 2.0: The Digital Competence Framework for Citizens*, https://publications.jrc.ec.europa.eu/repository/bitstream/JRC101254/jrc101254_digcomp%202.0%20the%20digital%20competence%20framework%20for%20citizens.%20update%20phase%201.pdf (2016).

environment and cybersecurity); (5) solving of various problems and lifelong learning in the digital society (Vuorikari et al., 2022).

According to DigComp content, digital competencies are based on information and digital literacy, which is the ability to identify information needs, to search and retrieve digital data, information, and content, to assess the relevance of sources and their content, to store and manage digital data, information, and content (Vuorikari et al., 2022).

At the same time, in the general structure of the competency model proposed by experts from McKinsey & Company, in the segment of digital competencies 13 detailed digital skills are formed (Distinct elements of talent — DELTAS), united into three main groups: (1) digital literacy, digital skills, digital cooperation, and digital ethics; (2) use and development of software (programming literacy, data analysis and statistics, computational and algorithmic thinking); (3) understanding of digital systems (data literacy, system intelligence, cybersecurity literacy, technical translation, and inclusion) (Punie and Redecker, 2017).

As we can see, the DELTAS model of digital competencies lacks a component of information literacy, which, in our opinion, is not entirely correct. We'd argue that information literacy is the key to the formation of both information and digital competencies.

An important aspect in the development of DigComp digital competence models is their flexibility and ability to take into account the specifics of certain types of activities (Carretero et al., 2017) which serves as a basis for the creation of separate reference professional models of digital competencies, such as "The Digital Competence Framework for Educators" (DigCompEdu) (Punie and Redecker, 2017).

Therefore, it can be assumed that the DigComp models offered by the EU experts are more substantiated and concretized for the formation, development, and implementation of systems for the development and assessment of digital competencies, they are more systematic and meaningful, they are constantly refined and improved according to the challenges posed by dynamic information and technological changes.

5.2 Digital literacy and digital competencies under Open Science

In the situation where digital technologies are being constantly transformed and diversified into different sectors and industries, their use depends on digital literacy — awareness, attitude, and ability to use digital tools and means in a proper way: to detect, access, manage, integrate, evaluate, analyze, and synthesize digital resources, to create new knowledge and media images and to communicate with other people for professional activities and specific life situations (Martin, 2016).

The tasks of investigating and developing digital competencies and digital literacy are relevant to communities of different levels. At that, the development of digital competencies becomes one of the most important conditions not only for the development of a personality but also for the development of digital market, which is directly or indirectly related to all spheres of the economy and society as a whole.

The task of mastering digital competencies by participants in educational and research activities as key subjects in the system of formation of new knowledge appears as the most important and relevant. According to the Eurostat data for 2020, enterprises of scientific and technological spheres are among the leading ones in the list of enterprises that are the most employing, recruiting, and having hard-to-fill vacancies for ICT specialists (Fig. 1).

In general, more than 90% of vocational roles in the EU require a basic level of digital competence.[5] The Digital Economy and Society Index (DESI) 2022 report published by the European Commission highlights that Europeans lag behind in their digital skills compared to the overall progress made by EU member states in the field of digital technologies, as 46% of Europeans do not have basic digital skills. There is also a significant gap in the levels of digital skills between various European countries. If in such states as the Netherlands and

---

[5] *Digital skills: Shaping Europe's Digital Future* (2022). https://digital-strategy.ec.europa.eu/en/policies/digital-skills.

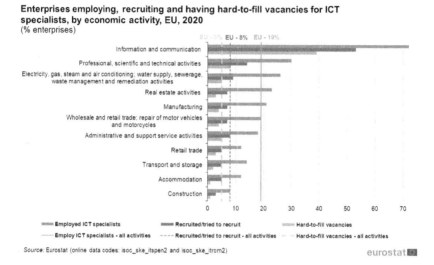

Figure 1. Enterprises employing, recruiting, and having hard-to-fill vacancies for ICT specialists, by economic activity, EU, 2020 (% enterprises) (see Eurostat: ICT specialists — statistics on hard-to-fill vacancies in enterprises, (2021). https://ec.europa.eu/eurostat/statistics-explained/index.php?title=ICT specialists-statistics on hard-to-fill vacancies in enterprises#Employment and recruitment of ICT specialists).

Finland about 79% of people had at least basic digital skills in 2021, then in Romania and Bulgaria the share is only about 30%. Therefore, the task of the next decade defined as a "Digital Decade" for the EU is built around a population that has digital skills. According to this vision of the future, the EU aims to ensure that 80% of Europeans have basic digital skills by 2030.[6]

The "Digital Decade" program also extends to Ukraine as a candidate country for joining the EU. According to the results of the digital skills survey conducted for the first time in Ukraine in 2019 in accordance with the European Commission's methodology for assessing digital competencies, about 47% of Ukrainians had a "basic level" of digital skills. Communication (75.3%) and information (74.4%)

---

[6] Source: Digital Literacy of the Population of Ukraine (2019). https://osvita.diia.gov.ua/uploads/0/585-cifrova_gramotnist_naselenna_ukraini_2019_compressed.pdf.

skills are more developed among Ukrainians. The lowest are problem-solving skills (55.6%) and software skills (28.8%).[7]

The results also proved the importance of higher education in the formation of digital skills (Table 1). As shown in the table, 67.8% of the Ukrainians with higher education have digital skills at the basic level or above the basic level, while in the category of persons with vocational secondary and secondary education, the share is just 31.9% and 26.8% correspondingly.

Several European and national policies have been put in place in order to raise digital literacy. The implementation of the Digital Education Action Plan (2021–2027), which aims to promote the development of a highly productive digital education ecosystem and increase digital skills and competencies for digital transformation, contributes to raising the level of digital literacy and digital competencies in the EU. To increase the digital literacy of the population in Ukraine, there is a National Online Platform for the Development of Digital Literacy "Diia"[8] in use. This platform also provides National testing and monitoring of digital literacy with four types of testing available: for citizens, teachers, medical workers, and civil servants. Also, Ukraine has adapted, in accordance with the national, cultural,

Table 1. General evaluation of digital skills of the Ukrainians in 2019, by type of education level (%).

Education level	No skills	Low skills	Basic skills	Above basic skills
Incomplete/complete secondary	27.9	45.2	14.6	12.2
Vocational secondary	20.7	47.3	19.2	12.7
Incomplete higher/higher	5.2	27.0	26.3	41.8

Note: *Digital literacy of the population of Ukraine* (2019), p. 23. https://osvita.diia.gov.ua/uploads/0/585-cifrova_gramotnist_naselenna_ukraini_2019_compressed.pdf.

---

[7] *Digital Literacy of the Population of Ukraine* (2019). https://osvita.diia.gov.ua/uploads/0/585-cifrova_gramotnist_naselenna_ukraini_2019_compressed.pdf.
[8] *Diia. Digital Education* (2022). https://osvita.diia.gov.ua/en.

educational, and economic characteristics of the country, the conceptual and reference model of the Digital Competence System for EU citizens DigComp 2.1. The Ukrainian model includes a system of knowledge and practical skills that citizens will need to use modern technologies and compete in the Ukrainian and European labor markets, which covers the following areas: basic computer literacy; information literacy and ability to work with data; creation of digital content; communication and interaction in the digital society; security in the digital environment; solving problems in the digital environment; lifelong learning.[9]

Digital space, digital market, and digital society are powerful challenges for researchers, this necessitates the development of a reference professional model of digital competencies as a key component in their overall professional competency structure, which is even more relevant in the conditions of Open Science — DigCompOS. Digital competencies of researchers in the conditions of Open Science are a condition to ensure free, open access to data and scientific information; extension of the principles of openness to the entire research cycle, exchange of data, scientific results, etc.

Given the existing approaches, the basis of the reference model of digital competencies of researchers (DigCompOS) may be the European conceptual and reference model DigComp, supplemented by

- *special* professional digital competencies of researchers, defined in accordance with the special requirements necessary for the successful implementation of their scientific activities with the use of digital technologies in the conditions of Open Science;
- *subject* digital competencies of researchers, determined in accordance with the requirements of digital competencies of researchers in a particular subject area.

Given the national and institutional challenges of today, the reference European DigComp — the framework of digital competencies of researchers can be successfully adapted to the national

---

[9] *Ukraine's Digital Skills Drive: Q&A with Mykhailo Fedorov* (2021). https://www.itu.int/hub/2021/11/ukraines-digital-skills-drive-qa-with-mykhailo-fedorov/.

cultural, educational, and economic characteristics of different countries, research organizations, and universities.

Another factor that must be taken into account in determination of the content and justification of the structure of digital competencies model of the researcher, definition of goals and objectives concerning their formation and development, is that digital competencies should be perceived not only as knowledge, skills, and abilities that are related to technical skills but also as knowledge, largely focused on informational, cognitive, social and emotional aspects of Open Science, research activities in the digital environment.

The proposed topic of developing human competencies, enabling humans to deal with the available amounts of information in today's world and to create their own knowledge, its correlation to the trends of Open Science and its embodiment in the practices and strategies of the higher education development requires more substantial investigation, especially the ethical issues present themselves an urgent topic (Eke *et al.*, 2020) that should a direction for the future research of the problems discussed in this chapter.

6 Conclusions

The decisive role that information plays in today's society should not disclose the profound distinction between information and knowledge: in this chapter, we have argued that information could be presented as alienated knowledge, as knowledge is deprived of its subject carrier and made available for transfer or for sale. In order for information to become knowledge, it has to be "digested", personalized, and adopted by a certain human person — knowledge is thus information processed through the whole complex of human intelligence including not just rational mind that has to assert that information as being true, but also human will and human emotions that evaluate it on the basis of desirability and acceptability.

The abundance of information made available thanks to the proliferation of digital computer technologies coincides with the deficit of human knowledge and human competencies required to transform information into generated knowledge. We think that the movement of Open Science presents itself as a solution for humans to deal with the vast volumes of available information by both developing personal competencies and reconstructing science as a public, citizen's enterprise, as it is Open Science that supports the early exchange of research results in open access modes, enables the participation of representatives of the non-academic sector in the research process and promotes active interaction with the public.

References

Braga, A. and Logan, R. K. (2021). The singularity hoax: Why computers will never be more intelligent than humans. In W. Hofkirchner, H. J. Kreowski (eds.), *Transhumanism: The Proper Guide to a Posthuman Condition or a Dangerous Idea?* Springer, Cham, Switzerland, pp. 133–140.

Carretero, G. S., Vuorikari, R., and Punie, Y. (2017). *DigComp 2.1: The Digital Competence Framework for Citizens with eight proficiency levels and examples of use*. Publications Office of the European Union, Luxembourg.

Digital Education Action Plan (2021–2027), https://education.ec.europa.eu/focus-topics/digital-education/action-plan (2021).

Drach, I. (2022). Open science in universities: Goals and benefits (in Ukrainian). *Scientific Herald of Uzhgorod University* 1(50), 90–94.

Eke, D., Akintoye, S., Knight, W., Ogoh, G., and Stahl, B. (2020). *Ethical Issues of E-Infrastructures: What Are They and How Can They Be Addressed?* https://www.researchgate.net/publication/342833325.

Gorz, A. (2008). *Écologica*. Paris: Galilée.

Hecka, T., Petersb, L., Mazarakisb, A., Scherpc, A., and Blümeld, I. (2020). Open science practices in higher education: Discussion of survey results from research and teaching staff in Germany. *Education for Information* 36, 301–323.

Kalashnikova, S. (ed.) (2019). Results of the national survey "perspectives and needs of Ukrainian universities' development in the context of European integration" (in Ukrainian). *International Scientific Journal of Universities and Leadership* 2(8), 144–220.

Logan, R. and Stokes, L. (2004). *Collaborate to Compete: Driving Profitability in the Knowledge Economy*. Toronto and New York: Wiley.

Martin, A. (2016). A European framework for digital literacy. *Nordic Journal of Digital Literacy* 1, 151–161.

Merton, R. (1973). The Normative Structure of Science. In Merton, R. (ed.). *The Sociology of Science. Theoretical and Empirical Investigations*. London, Chicago: The University of Chicago Press, pp. 267–278.

Mielkov, Yu. (2006). Practical philosophy in the society of information. *Praktychna Filosofiya* 1(19), 3–13.

Mielkov, Yu. (2022). Knowledge in the age of information: Human values in science and higher education. *New Explorations: Studies in Culture and Communication* 2(3), 28–39.
Mindell, D., Segal, J., and Gerovitch, S. (2003). Cybernetics and information theory in the United States, France and the Soviet Union. In Walker, M. (ed.), *Science and Ideology: A Comparative History*, London: Routledge, pp. 66–94.
Nalimov, V. (1981). *In the Labyrinths of Language: A Mathematician's Journey*. Philadelphia, PA: Isi Press.
Narvaez Rojas, C., Alomia Peñafiel, G. A., Loaiza Buitrago, D. F., and C. A. (2021). Tavera Romero, Society 5.0: A Japanese concept for a superintelligent society. *Sustainability* 13(12), 6567.
Nikiforova, A. (2021). Smarter Open Government Data for Society 5.0: Are your open data smart enough? *Sensors* 21(15), 5204.
Polanyi, M. (2005). *Personal Knowledge: Towards a Post-Critical Philosophy.* London: Taylor & Francis.
Punie, Y. and Redecker, C. (eds.) (2017). *European Framework for the Digital Competence of Educators: DigCompEdu.* Luxembourg: Publications Office of the European Union.
Skov, A. (2016). *What is Digital Competence?* https://digital-competence.eu/dc/front/what-is-digital-competence/.
Stepin, V. (2005). *Theoretical Knowledge.* Dordrecht: Springer Verlag.
Stiegler, B. (2003). Our ailing educational institutions/transl. by stefan Herbrechter. *Culture Machine* 5. https://culturemachine.net/the-e-issue/our-ailing-educational-institutions/.
UNESCO (2021). *Recommendation on Open Science,* https://www.unesco.org/en/legal-affairs/recommendation-open-science.
Vuorikari, R., Kluzer, S., and Punie, Y. (2022). *DigComp 2.2: The Digital Competence Framework for Citizens — With new examples of knowledge, skills and stats.* Publications Office of the European Union, Luxembourg.
Wiener, N. (1961). *Cybernetics, or Control and Communication in the Animal and the Machine,* 2nd edn. Cambridge, MA: MIT Press.

Chapter 16

Trend of Increasing Percentages of Mirror-Symmetric Signs in the Cretan Script Family and the Phoenician Alphabet Family*

Peter Z. Revesz

School of Computing
University of Nebraska-Lincoln, Lincoln, NE 68588, USA
peter.revesz@unl.edu

This chapter reports the discovery of the trend of increasing percentages of signs that have mirror symmetry along a vertical line in two major script families. This chapter defines three measures of the direction of change toward less or more mirror symmetry during the adaptation of an old script to a new script: modification bias, addition bias, and overall bias, which each range from −1 to 1. For the Phoenician alphabet family, the average modification bias is 0.17, while for the Cretan script family, the average modification bias is 0.79. These show a strong tendency toward increased mirror symmetry. In the Phoenician alphabet family, the percentage of mirrored signs rises from 40.9% in the Phoenician alphabet to 59.3% in the Euclidean Greek alphabet. In the Cretan script family, the percentage of mirrored signs rises from 28.9 for the Phaistos Disk script to 64.2% for the Carian alphabet. This chapter also identifies possible drivers of the increased use of mirror-symmetric signs, including boustrophedon writing and religious writings with deliberate mirroring as an afterlife symbolism.

---

*A preliminary version of this chapter was presented at the 12th SIS-Symmetry Congress in Porto, Portugal in 2022 (Revesz, 2022b).

1 Introduction

Although the spread of writing and the evolution of scripts are intensely investigated subjects, none of the previous studies has yet identified a curious phenomenon, which is the tendency of gradually increasing percentages of mirror-symmetric signs in many script families. In this chapter, "mirror symmetry" means that a vertical line can be drawn through a sign such that the left- and right-hand sides are reflections of each other. For example, the letters A, U, and X are all mirror symmetric.

The primary goal of this chapter is to statistically analyze and demonstrate the existence of the trend of increasing mirror symmetry within the Cretan Script Family, a script family identified in Revesz (2016), and the well-known Phoenician Alphabet Family (Fisher, 2003). The secondary goal is to identify the possible causes of this trend. We argue that the use of boustrophedon writing was a major reason for the development of mirror symmetry in various script families. However, we point out that apparently other reasons, including certain religious concepts, also played a role in the development of mirror symmetry.

This chapter is organized as follows. Section 2 describes the data sources, which give a comprehensive description of various scripts in the Cretan Script Family and the Phoenician Alphabet Family. Section 3 gives an analysis of the development of mirror symmetry in these two families and demonstrates a trend of general increase in the percentages of mirror-symmetric signs. Section 4 presents a discussion of the results with some ideas about the possible causes of the discovered trend. Finally, Section 5 gives some conclusions and directions for further research.

2 Data Sources

Sections 2.1 and 2.2 review the Phoenician alphabet and the Cretan script families, respectively.

2.1 The phoenician alphabet family

The Phoenician alphabet was the ancestor of many alphabets worldwide. The evolution of the Phoenician Alphabet Family is well understood. Figure 1 shows seven alphabets that belong to the Phoenician Alphabet Family. The first row in Fig. 1 shows the original Phoenician alphabet with 22 letters, which was already used around 1050 BCE (Fisher, 2003). The Phoenician alphabet letters are considered to have depicted various objects. For example, the first letter of the alphabet depicted the head of an ox and was called *alef*, the Phoenician name for "ox". Out of the 22 letters of the Phoenician alphabet, nine letters have vertical mirror symmetry. These nine letters are shaded in gray in Fig. 1.

Figure 1 shows the archaic Western Greek alphabet in the second row and the later Euclidean Greek alphabet in the third row. The archaic Western Greek alphabet was an adaptation of the Phoenician alphabet around 800 BCE. The Euclidean Greek alphabet was a reform of the archaic Western Greek alphabet under the Eucleides, the archon of Athens from 403 until 402 BCE.

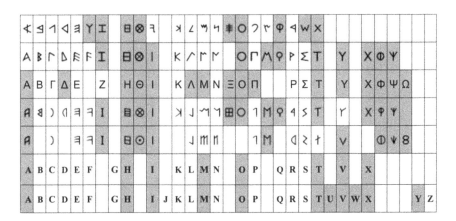

Figure 1. Evolution of the alphabet: Phoenician (1st row), archaic Western Greek (2nd row), Euclidean Greek (3rd row), archaic Etruscan from Marsiliana (4th row), Neo-Etruscan (5th row), Old Latin alphabet (6th row), and Medieval Latin (7th row). The letters with mirror symmetry are shaded in gray. Source of the alphabets: Wikipedia, "Old Italic Scripts" and "Latin Alphabet" entries.

The archaic Western Greek alphabet spread to Italy, where the Etruscans adapted it to their language. This archaic Etruscan alphabet is based on a single example from Marsiliana, which may contain some handwritten variations of ideal letter forms that had mirror symmetry. For example, ᴀ may have been a handwritten variation of a symmetric form of the letter A. Hence, it is indicated in gray. The Neo-Etruscan alphabet was a reformed version of the archaic Etruscan alphabet around 400 BCE.

The Old Latin alphabet derived from the archaic Etruscan alphabet. The Latin alphabet was extended by some letters with respect to the Old Latin alphabet over the centuries up to the Middle Ages. This later alphabet is called the medieval Latin alphabet. The medieval Latin alphabet is also used to write English today. The Latin alphabet letters A, M, V, X, and Y are not perfectly mirror symmetric because there are wider and thinner lines. However, the use of wider and thinner lines is a later stylistic development that the earliest Latin inscriptions did not have. Hence, we consider these letters as mirror symmetric.

2.2 *The cretan script family*

The Cretan Script Family includes the Phaistos Disk script, Cretan Hieroglyphs, Linear A, Linear B, the Carian alphabet, and the Old Hungarian script among other scripts (Revesz, 2016b). In the following subsections, we describe these scripts within the Cretan Script Family.

2.2.1 *The phaistos disk script*

The Phaistos Disk is a unique Minoan artifact that is commonly dated to around 1800 BCE (Evans, 1909; Pernier, 1909). A more recent dating puts the creation of the Phaistos Disk between the Middle Minoan IIB and Middle Minoan IIIA period (Baldacci, 2017). The writing is done in a spiral form on both sides of the clay disk. The reading direction is left-to-right (Revesz, 2022). Although the Phaistos Disk was considered undeciphered since its discovery in

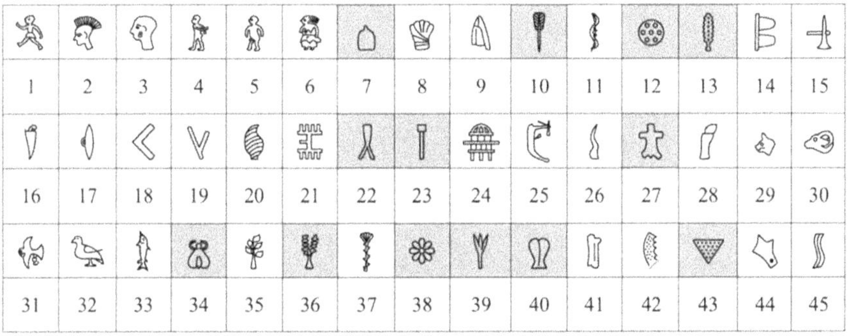

Figure 2. The Phaistos disk signs as enumerated by Evans (1909).

1908, a computer-aided decipherment as a hymn to the sun goddess was recently published (Revesz, 2016). Some earlier decipherment proposals include the proposals of Aartun (1992), Achterberg et al. (2004), Ephron (1962), Faucounau (1999), Fisher (1997), Georgiev (1976), Hempl (1911), Martin (2000), Ohlenroth (1996), and Schwartz (1959), and Stawell (1911).

The Phaistos Disk contains 45 different signs, some of which are repeated to form a text with 241 signs, which is currently the longest known Minoan text. Figure 2 shows an enumeration of the Phaistos Disk signs given by Sir Arthur Evans (1909). The letters with mirror symmetry are shaded in gray.

2.2.2 Cretan hieroglyphs

Figure 3 shows the 96 Cretan Hieroglyphs of Olivier and Godart (1996).

The Minoans used the Cretan Hieroglyphic script between 2100 and 1700 BCE.

2.2.3 The linear A and linear B scripts

The Minoan civilization used the Linear A script between 1800 and 1450 BCE, while the Mycenean civilization used the Linear B script, which was an adaptation of the Linear A script, between 1450 and

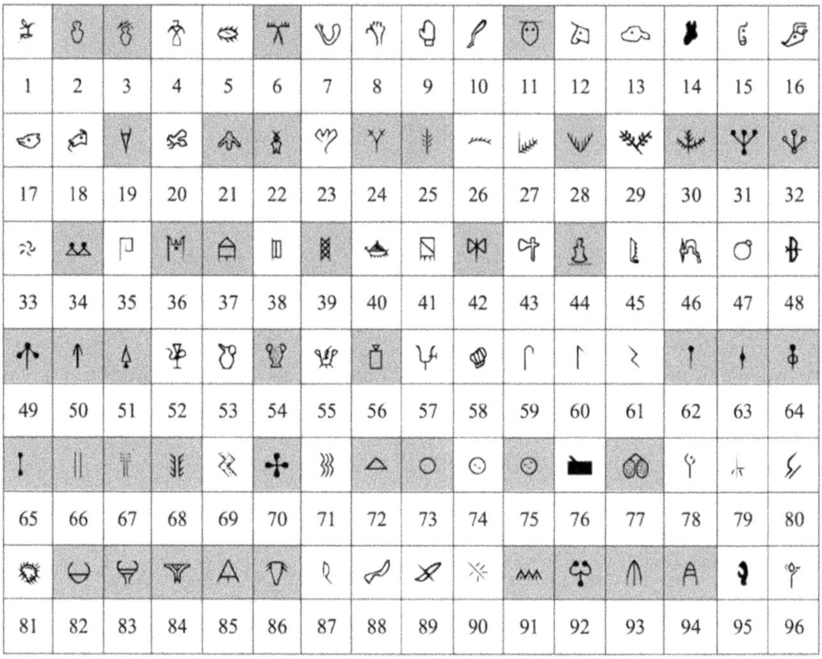

Figure 3. Cretan Hieroglyphs as enumerated by Olivier and Godart (1996). The hieroglyphs with mirror symmetry are shaded in gray.

1200 BCE. The Linear B script is the oldest known script that was used to write the Greek language. Linear B was deciphered by Michael Ventris in 1952 (Ventris and Chadwick, 1973). Scholars have found about 74 signs that are shared by the Linear A and the Linear B scripts. These signs, which are called Linear AB signs, also form the most frequent signs in the two scripts. Hence, in this chapter, we focus on these Linear AB signs. Figure 4 is an enumeration of the Linear AB signs based on Godart and Olivier (1976).

The Linear A script was considered an undeciphered script since its discovery in the early 20th century. Recently, Revesz (2017) proposed a decipherment of twenty-eight inscriptions that identifies the underlying language as belonging to the Ugric branch of the Uralic languages. The AIDA system contains a library of translated Linear A inscriptions (Revesz et al., 2019).

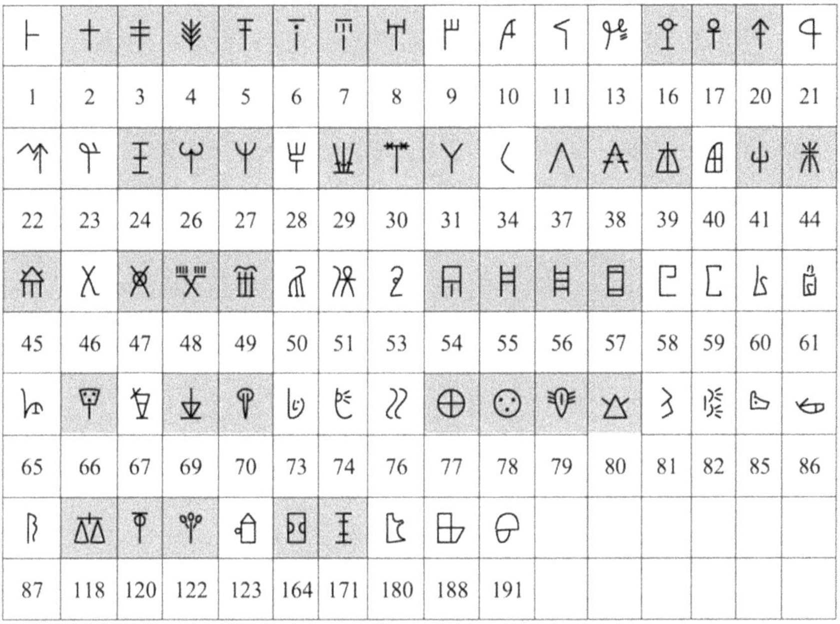

├	╀	╪	⚹	₮	T̄	T̈	T̈	╟	ℱ	ˤ	⅌	⇞	⇞	⇞	⇞
1	2	3	4	5	6	7	8	9	10	11	13	16	17	20	21

↑	ꭍ	⊥	⊻	Ψ	Ψ	Ⱦ	Y	⟨	∧	A	⛉	ᴁ	⇔	⋇
22	23	24	26	27	28	29	30	31	34	37	38	39	40	41	44

𝍨	X	⊠	⁂	⊞	⍉	⋊	⁇	ᖷ	H	ᖯ	⊟	⊑	⊏	⌁	⌁
45	46	47	48	49	50	51	53	54	55	56	57	58	59	60	61

⋎	⚇	⚉	⋤	⚑	⋓	⋛	⟨⟨	⊕	⊙	⚲	⛰	⋑	⋛	⇔	⇐
65	66	67	69	70	73	74	76	77	78	79	80	81	82	85	86

ʖ	⋈	⚇	⚕	⌂	⊠	⊥	⌁	⇄	⊖
87	118	120	122	123	164	171	180	188	191

Figure 4. Linear AB signs as enumerated by Godart and Olivier (1976) with some signs omitted by convention of modern scholars who grouped some signs together. The signs with mirror symmetry are shaded in gray.

2.2.4 The carian alphabet

The Carian alphabet is a strange script because many of its letters look like Greek letters, but their phonetic values are often completely different according to Adiego (2006) and Kloekhorst (2009). Therefore, one can suspect a gradual convergence of the Carian letters to Greek letters in form while keeping their original phonetic values. In addition, in the final period of the Carian alphabet, some Greek letters could have been borrowed with both form and phonetic value. Figure 5 shows a list of the Carian alphabet letters as presented in the Unicode standardization plus four more letters that we added to the list based on Adiego (2006). The additional four letters are considered alternative, more archaic Carian forms of the letters for /a/, /r/, /k^w/, and /j/.

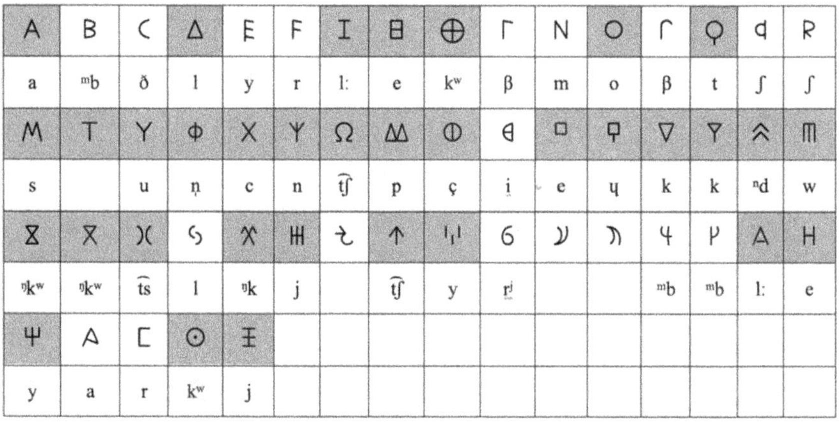

A	B	C	△	E	F	I	日	⊕	Γ	N	O	Γ	Q	d	R
a	ᵐb	ð	l	y	r	l:	e	kʷ	β	m	o	β	t	ʃ	ʃ
M	T	Y	Φ	X	Y	Ω	⋀⋀	⊕	ᵮ	□	ᛂ	▽	Ψ	≈	⋔
s		u	ṇ	c	n	t͡ʃ	p	ç	i	e	ɥ	k	k	ⁿd	w
𐊼	𐊺	)(	⁄	𐊼	H	⋌	↑	¦,¦	6	𝒴	𝒟	4	Þ	△	H
ŋkʷ	ŋkʷ	t͡s	l	ŋk	j		t͡ʃ	y	rʲ			ᵐb	ᵐb	l:	e
Ψ	△	[	⊙	∓											
y	a	r	kʷ	j											

Figure 5. The Carian alphabet. The letters with mirror symmetry are shaded in gray. The International Phonetic Alphabet notation is given below each letter based on Kloekhorst (2009).

Source: Wikipedia, "Carian alphabets" entry.

2.2.5 The old hungarian alphabet

The Old Hungarian script was used by Hungarians before the adaptation of the Latin alphabet, but its origins are disputed with Benkő et al. (2021) advocating Old Turkic script, Hosszú (2013) a Phoenician alphabet, and Revesz (2016b) a Cretan Script Family origin. Figure 6 lists the Old Hungarian script signs based on the Unicode description.

We added two archaic signs at the end of this list. The first archaic sign, which looks like a triangle, represents the phoneme /k/. This sign dropped out of use after the 14th century. The second archaic sign, which is a circle with a dot, represents the phoneme /lj/.

We also replaced the sign that represents /c/ with a mirror-symmetric alternative form that occurs in an Old Hungarian text, which was found in the library of Nikolsburg Castle, Czechia in a book that was printed in 1483 (Hosszú, 2013). This form seems to be the original form of this sign.

Finally, the Old Hungarian signs for /m/ and /x/ are rotated 90 degrees because these were apparently the original forms and were

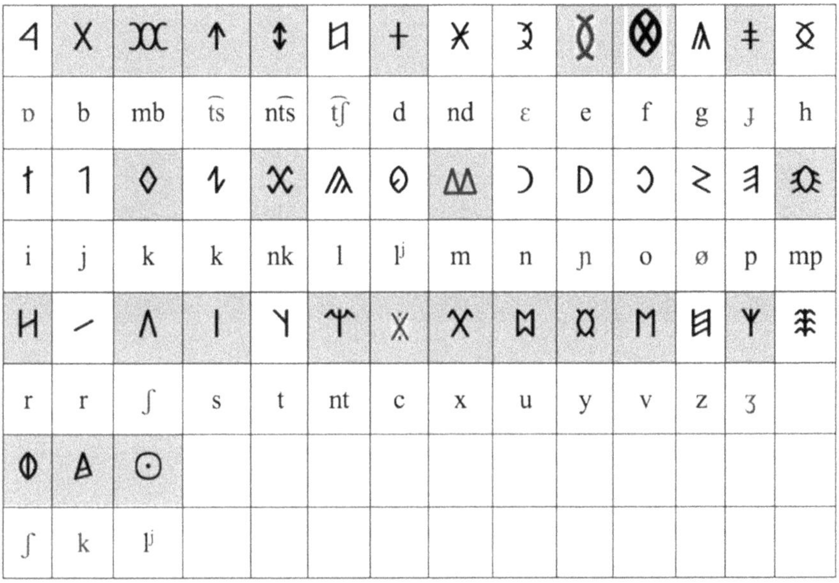

◁	X	↔	↑	↕	⊔	+	✗	Ʒ	◊	⊗	Λ	ǂ	⊗
ɒ	b	mb	t͡s	nts	t͡ʃ	d	nd	ɛ	e	f	g	ɟ	h
↑	1	◊	↙	✕	⋀	⊖	⋙	⊃	D	⊃	⋛	∃	✿
i	j	k	k	nk	l	lʲ	m	n	ɲ	o	ø	p	mp
H	⁃	Λ	I	Ƴ	⋎	X	✕	⋈	Ǫ	M	⊟	Y	⟟
r	r	ʃ	s	t	nt	c	x	u	y	v	z	ʒ	
Φ	A	⊙											
ʃ	k	lʲ											

Figure 6. The Old Hungarian script with mirror-symmetric signs shaded in gray.

rotated later to save horizontal space. The original version of /x/ is found in early medieval inscriptions in the Carpathian Basin (Hosszú, 2013).

Figure 6 omits the long vowels because it is well known that the long vowels were not part of the oldest Old Hungarian script. According to our calculation, there are 45 signs without long vowels. Out of those, 25 have vertical mirror symmetry. Note that the archaic /k/ sign that we added is counted as mirror symmetric. In counting mirror symmetry, we always must take the standard form of the signs. When there are few examples like in this case, then there is a risk that the form that is found on a particular inscription is taken as the standard form.

As an illustration of this risk, the letter A is also not written with perfect mirror symmetry by most people. If we had only one or two slanted handwritten examples of the letter A, then we could wrongly conclude that it needs to be slanted too.

Figure 7. The evolution of the cretan script family: Linear AB sign numbering (1st row), Phaistos disk signs (2nd row), Cretan Hieroglyphs (3rd row), Linear AB signs (4th row), Carian alphabet (5th row), and Old Hungarian script (6th row). The signs with mirror symmetry are shaded in gray.

2.2.6 The evolution of the Cretan Script Family

The evolution of the Cretan Script Family is reconstructed in Fig. 7 based on Revesz (2017) with some extensions.

3 Results and Analysis

Sections 3.1 presents the analysis method, and Sections 3.2 and 3.3 apply it to the Phoenician Alphabet Family and the Cretan Script Family, respectively.

3.1 The analysis method

Each script adaptation is evaluated based on the following four types of changes in mirror symmetry:

(1) *Introduction* occurs when a non-mirror-symmetric sign is modified into a mirror-symmetric one.

(2) *Loss* occurs when a mirror-symmetric sign is modified into a non-mirror-symmetric one.

(3) *Addition* occurs when a new mirror-symmetric sign is added to the alphabet.

(4) *Dropping* occurs when a mirror-symmetric sign is no longer used.

Let I, L, A, and D be the number of mirror-symmetric signs introduced, lost, added, and dropped, respectively. Let A' and D' be the number of non-mirror-symmetric signs added and dropped, respectively. The preference of culture for mirror-symmetric signs can be estimated by various bias measures. We define the *modification bias* B_{mod} as the difference between the number of introduced and lost mirror-symmetric signs divided by the total number of signs whose mirror symmetry status is modified:
$$B_{\mathrm{mod}} = (I - L)/(I + L).$$
Similarly, we define the *addition bias* B_{add} as the difference between the number of added mirror-symmetric and non-mirror-symmetric signs divided by the total number of added signs:
$$B_{\mathrm{add}} = (A - A')/(A + A').$$
Clearly, a high modification bias or addition bias shows a preference for mirror-symmetric signs. In contrast, dropping letters happens because they denote phonemes that are not needed. Whether a phoneme denoted by a sign is needed is independent of the letter having mirror symmetry. That is, no letter is deliberately dropped because it has or lacks mirror symmetry. Hence, no conclusion can be made regarding the preference for mirror symmetry from the dropping of signs. Therefore, D and D' are not considered in the definition of overall bias B, which is the following:
$$B = (I - L + A - A')/(I + L + A + A').$$
In other words, bias B is the net number of deliberate increases in mirror symmetries divided by the total number of modifications that change the mirror symmetry status of a sign and additions of signs. In the subsequent analysis of the Phoenician alphabet and the Cretan script families, we focus on the bias measures, although we also present statistical results for the percentages of mirror-symmetric signs.

3.2 Analysis of the phoenician alphabet family

We collected some statistics about the Phoenician Alphabet Family in Table 1. We give some explanations below for the values in Table 1.

Table 1. Phoenician alphabet family statistics.

Alphabet	Mirror symmetry						Non-mirror symmetry		Bias	
	Intro.	Lose	Add	Drop	Total	%	Add	Total	B_{mod}	B
Phoenician	—	—	9	—	9	40.9	13	13	—	—
Archaic West. Greek	2	2	4	1	12	48.0	0	13	0	0.5
Euclidean Greek	5	1	2	2	16	59.3	0	11	0.67	0.75
Archaic Etruscan	1	1	1	0	13	50.0	0	13	0	0.33
Neo-Etruscan	1	1	1	4	10	50.0	0	10	0	0.33
Old Latin	2	1	0	6	8	38.1	1	13	0.33	0
Medieval Latin	0	0	3	0	11	42.3	2	15	0	0.2
Average (rows 4–9)	1.83	1	1.83	2.17	11.67	48.0	0.5	12.5	0.17	0.35

Within the archaic Western Greek alphabet, mirror symmetry was introduced in two letters, mirror symmetry was lost in two letters, four mirror-symmetric letters were added, and one mirror-symmetric letter was dropped. These developments show a preference for increased mirror symmetry. However, the preference for mirror symmetry was sometimes overridden by other considerations. The mirror symmetry was lost in the letter W by a 90 degree clockwise rotation that yielded Σ. In this case, the preference for mirror symmetry may have been overridden by an even higher preference to save space. The mirror symmetry was also lost in the letter Y which was transformed into the Greek digamma F. In this case, the preference for mirror symmetry may have been overridden by some other reason. This reason may be a desire for a better representation of a hook, which the letter Y, called by the Phoenicians *wāw* "hook", was supposed to depict. The Greek letter Y, which denotes the Greek phoneme /y/, is counted as an addition despite its similarity to the Phoenician *wāw*. The percentage of mirror-symmetric letters increased from 40.9% to 48% from the original Phoenician alphabet to the archaic Western Greek alphabet.

Within the Euclidean Greek alphabet, mirror symmetry was introduced in five letters, mirror symmetry was lost in one letter, two new mirror-symmetric letters were added, and two mirror-symmetric letters were dropped. The percentage of mirror-symmetric letters increased from 48% to 59.3% from the archaic Western Greek alphabet to the Euclidean Greek alphabet.

Within the archaic Etruscan alphabet, mirror symmetry was introduced in the letter **A**, mirror symmetry was lost in one letter, and one new mirror-symmetric letter was added. The percentage of mirror-symmetric letters increased from 48% to 50% from the archaic Western Greek alphabet to the archaic Etruscan alphabet.

Within the Neo-Etruscan alphabet, mirror symmetry was introduced in one letter, mirror symmetry was lost in one letter, and one new mirror-symmetric letter was added. Four mirror-symmetric letters that denoted unused phonemes were also omitted in the Neo-Etruscan alphabet. The percentage of mirror-symmetric letters remained at 50% from the archaic Etruscan alphabet to the Neo-Etruscan alphabet.

Within the Old Latin alphabet, mirror symmetry was introduced in the letters M and V, mirror symmetry was lost in Q, and six mirror-symmetric letters were dropped when the archaic Etruscan alphabet was adapted. Letters U, W, and Y are additional mirror-symmetric letters in the medieval Latin alphabet with respect to the Old Latin alphabet. The following letters have mirror symmetry in the medieval Latin alphabet: A, H, I, M, O, T, U, V, W, X, and Y. That is, a total of $11/26 = 42.3\%$ of the letters of the medieval Latin alphabet have mirror symmetry.

Figure 8 shows the part of the Phoenician Alphabet Family that was considered in this chapter. We see a trend of increasing percentage of mirror-symmetric signs in both the Etruscan and the Greek branches of the tree. The increasing trend also holds for the change from the Old Latin to the medieval Latin alphabet. The only anomaly is the change from the archaic Etruscan to the Old Latin alphabet, where the percentage of mirror-symmetric signs decreases from 50% to 38%. This anomaly is due to the six mirror-symmetric signs that were dropped during the adaptation of the archaic Etruscan alphabet

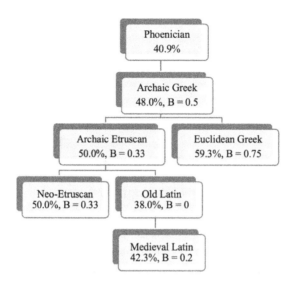

Figure 8. The evolution of the Phoenician script family. The percentage of signs with mirror symmetry and the bias B is shown below each script.

to the Old Latin alphabet. This anomaly is avoided by the bias measure, which is non-negative everywhere in the tree.

3.3 Analysis of the cretan script family

We also collected some statistics about the Cretan Script Family in Table 2. We give some explanations for the values below.

At first, we consider the percentage or mirror-symmetric signs. The Phaistos Disk has 13 different mirror-symmetric signs that are shaded in gray in Fig. 2. Some additional signs such as signs 16 and 24 are near misses because they are just slightly short of a perfect mirror symmetry. One can calculate that $13/45 = 28.9\%$ of the Phaistos Disk signs have mirror symmetry. The Cretan Hieroglyphic script contains 45 mirror-symmetric signs. Therefore, the percentage of mirror syllabic signs is $45/96 = 46.875\%$, which is much higher than in the case of the Phaistos Disk. The Linear AB script contains 41 mirror-symmetric signs out of 74 basic Linear AB signs (some of these signs have variations), that is, $41/74 = 55.4\%$. There are 34 mirror-symmetric Carian letters out of a total of 53, as shown in

Table 2. Cretan script family statistics.

Script	Intro.	Mirror Symmetry					Non-Mirror Symmetry		Bias	
		Lose	Add	Drop	Total	%	Add	Total	B_{mod}	B
Phaistos Disk	—	—	13	—	13	28.9	32	32	—	—
Cretan Hieroglyphs	2	0	40	10	45	46.9	48	51	1	−0.07
Linear AB	3	0	31	0	41	55.4	27	33	1	0.11
Carian Alphabet	3	0	15	0	34	64.2	11	19	1	0.24
Old Hungarian	4	3	5	0	25	55.6	13	20	0.14	−0.28
Average (rows 4–7)	3	0.75	22.75	2.5	36.25	55.5	24.8	30.75	0.79	0

Fig. 5. The Old Hungarian script contains 25 mirror-symmetric signs out of a total of 45 signs as we already mentioned in Section 2.2.5. That means that 55.6% of the Old Hungarian signs have mirror symmetry.

The calculation of I, L, A, D, A', and D' can be done similarly to how they were calculated in the case of the Phoenician Alphabet Family. The A and A' have much higher averages in Table 2 than they had in Table 1. The average value of A changes from 1.83 in Table 1 to 22.75 in Table 2, while the average value of A' changes from 0.5 in Table 1 to 24.8 in Table 2. These high values may be due to not knowing how to match many of the signs. If we could match more of the signs, then the A and A' values would be lower. The problem is that the artificially high A and A' values make B_{add} and B artificially low. The B_{mod} value seems more robust in case of incomplete matchings because modifications can be detected more easily. Figure 9 shows the evolution of the Cretan Script Family with the percentage of the mirror-symmetric signs and the modification bias values.

Figure 9 shows that the percentage of mirror-symmetric signs tends to grow in each branch of the Cretan Script Family. In addition, the modification bias is always non-negative. The general bias value

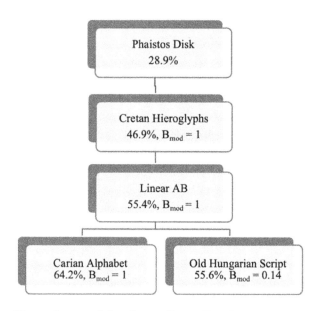

Figure 9. The evolution of the Cretan Script Family. The percentage of signs with mirror symmetry and the modification bias B_{mod} are shown below each script.

also tends to be non-negative, but its calculated value is lower than warranted as we mentioned earlier.

4 Discussion of the Results

While the analysis in Section 3 demonstrated that there is a trend toward an increased percentage of mirror-symmetric signs in both the Phoenician Alphabet Family and the Cretan Script Family, it did not give any explanation for this puzzling phenomenon. In this section, we place this phenomenon in a social context. Sections 4.1 and 4.2 describe two possible drivers of the phenomenon.

4.1 *Boustrophedon writing*

We believe that the unexpectedly high number of symmetric signs is due to a *boustrophedon* type of writing, which was widespread among the scripts that we studied, as shown in Fig. 10.

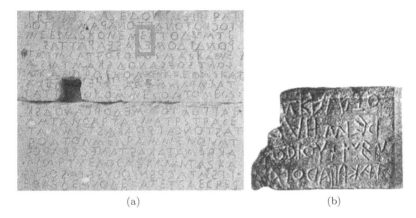

Figure 10. Boustrophedon writings' examples: (a) Gortyn law code, Euclidean Greek alphabet text. Source: Wikimedia, CC BY 4.0, https://en.wikipedia.org/wiki/Boustrophedon. (b) A fragment of the Lapis Niger inscription from Rome, Old Latin alphabet text.
Source: Wikimedia, CC BY 4.0, https://en.wikipedia.org/wiki/Lapis_Niger#The_inscription/. The added boxes indicate mirror images of the letter E.

The name boustrophedon literally means "as the ox goes" and denotes the type of writing where at the end of a line the next line continues right below the end and goes in the opposite direction. Hence, left-to-right and right-to-left lines alternate each other. This is reminiscent of how oxen plow a plot of land.

Figure 10(a) shows a fragment of the Gortyn law code, which was written using the Euclidean Greek alphabet. Figure 10(b) shows a fragment of an Old Latin inscription from Rome. Boustrophedon writing also occurs in the Linear A inscription KN Za 19 from Knossos (Godart and Olivier, 1976, vol. 4, p. 14) and in several Old Hungarian script inscriptions including a medieval calendar (Hosszú, 2013, p. 176).

The main problem with boustrophedon writing is that when we look at a particular line, we do not automatically know which way it should be read, unlike in modern English texts, where every line is read from left to right. As a modern example, suppose we would like to write "GOD" in a row that is to be read from right to left. This looks like an easy task that can be done by simply writing: "DOG". The problem is that the reader may not recognize that the row needs to be read from right-to-left, hence "God" becomes "dog" for the reader. Ancient scribes compensated for this problem by vertically mirroring any non-symmetric letter so that the orientation of the words would indicate the direction. Using this concept, instead of "DOG", they would have written "ꓷOD".

Figure 10(a) shows this mirroring. We can see in the box that in the third line, the letter E is written in the normal form, while in the fourth line, it is written reversed. Similarly, the Old Latin inscription in Fig. 10(b) reverses the letter E on the second line.

Mirroring signs is a difficult mental task that often leads to errors in the case of complex signs. Therefore, to reduce the difficulty of the mental task of mirroring the signs, most signs are changed to mirror-symmetric forms. It is enough to keep only a few signs that are frequent and easily reversed such as the letter E to indicate the reading direction of the lines. In other words, boustrophedon writing leads to mirroring, which is a driver of an increase in the percentages of symmetric signs.

4.2 Afterlife beliefs

While boustrophedon writing with mirroring of asymmetric signs is an attractive explanation, it cannot explain several cases of mirrored writing that occur only on one line. For example, the Linear A inscriptions on PL Zf 1, which is a silver pin, IO Za 9 and VRY Za 1, which are libation tablets, are one-line inscriptions that need to be read right-to-left (Godart and Olivier, 1976). Each of these inscriptions contains a religious text (Revesz, 2017). Apparently, the religious beliefs of the scribes influenced the writing direction. In the following, we propose a possible way that religious belief could influence the writing direction.

Ancient Bronze Age Europeans may have thought that the spirits of deceased people lived on at the bottom of lakes. They also may have considered mirrors magical because they reflected the real world as shown in Fig. 11. Since all objects are reflected in a lake or a mirror with the left and the right sides reversed, in the spirit world, the spirits were thought to be left-handed. This belief led to the Celtic

Figure 11. Illustration of the connection between afterlife beliefs and water reflections.

funeral practice where a dead man's sword or lance was placed at his left side.

The apparent mirroring of all things in the afterlife may have led to the use of mirrored signs on religious inscriptions. This may have been thought to facilitate communication with the dead who wrote using reversed signs. The mirroring needed in religious inscriptions may have become cumbersome and led to the increase in the percentages of mirror-symmetric signs, like how boustrophedon writing may have led to the same.

Mirrored signs can also occur in non-boustrophedon and non-religious writings. These other cases are investigated and listed by Daggumati and Revesz (2021). They focus on the Indus Valley Script, but their observations are applicable to other scripts too. The following are some of the cases that were included in their investigation: (1) when the writer used a mirrored sign to save space, (2) copying errors, and (3) location anomalies.

5 Conclusions and Future Work

The statistical analysis of this chapter demonstrated a trend of increasing percentages of mirror-symmetric signs in the Cretan Script Family and the Phoenician alphabet family. In addition, two drivers of this trend were proposed. We believe that this trend also occurs in other script families where boustrophedon writing or similar beliefs in the afterlife are found. A recent convolutional neural network-based analysis showed that the Sumerian Pictographic script is the ancestor of the Indus Valley Script (Daggumati and Revesz, 2023). Since the Indus Valley Script is known to have examples of boustrophedon writings, the practice of boustrophedon writing may have increased the percentage of mirror-symmetric signs with respect to the Sumerian pictographs. Hence, this chapter can be extended to other script families.

References

Adiego Lajara, I.-J. (2006). *The Carian Language, Volume 86 of Handbook of Oriental Studies*, Leiden: Brill.

Aartun, K. (1992). *Der Diskos von Phaistos; Die beschriftete Bronzeaxt; Die Inschrift der Taragona-tafel in Die minoische Schrift: Sprache und Texte, Vol. 1*, Wiesbaden: Harrassowitz.

Achterberg, W., Best, J., Enzler, K., Rietveld, L., and Woudhuizen, F. (2004). *The Phaistos Disc: A Luwian Letter to Nestor, Henry Frankfort Foundation Publications Series Vol. 13*, Amsterdam: Dutch Archaeological and Historical Society.

Baldacci, G. (2017). Low-relief potters' marks and the Phaistos Disc: A note on the "comb" sign (n. 21). *Annuario della Scuola Archeologica di Atene e delle Missioni Italiane in Oriente*, 95, 65–79.

Benkő, E., Sándor, K., and Vásáry, I. (2021). *A Székely írás emlékei (Historical documents of the Székely script)*. Budapest: Bölcsészettudományi Kutatóközpont.

Daggumati, S. and Revesz, P. Z. (2021). A method of identifying allographs in undeciphered scripts and its application to the Indus Valley Script. *Humanit. Soc. Scien. Commun.* 8(50).

Daggumati, S. and Revesz, P. Z. (2023). Convolutional neural networks analysis reveals three possible sources of Bronze Age writings between Greece and India. *Information* 14(227).

Ephron, H. D. (1962). Hygieia Tharso and Iaon: The Phaistos Disk. *Harvard Stud. Classi. Philol.* 66, 1–91.

Evans, A. J. (1909). *Scripta Minoa, the Written Documents of Minoan Crete, with Special Reference to the Archives of Knossos*, Oxford: Clarendon.

Faucounau, J. (1999). *Le déchiffrement du disque de Phaistos: Preuves et consequences*, Paris: L'Harmattan.

Fisher, S. R. (1997). *Glyph-Breaker*, New York: Springer.

Fischer, S. R. (2003). *History of Writing*, London: Reaktion Books.

Georgiev, V. (1976). Le déchiffrement du texte sur le disque de Phaistos, *Linguisti. Balkani.* 19, 5–47.

Godart, L. and Olivier. J.-P. (1976). *Recueil des inscriptions en Linéaire A. Number 21 in Études Crétoises*, Paris: De Boccard.

Hempl, G. (1911). The solving of an ancient riddle: Ionic Greek before Homer, *Harper's Monthly Magazine* 122(728), 187–198.

Hosszú, G. (2013). *Heritage of Scribes: The Relation of Rovas Scripts to Eurasian Writing Systems*, Budapest: Rovas Foundation.

Kloekhorst, A. (2009). Studies in Lycian and Carian Phonology and Morphology. *Kadmos* 47(1–2), 117–146.

Martin, A. (2000). *Der Diskos von Phaistos — Ein zweisprachiges Dokument geschrieben in einer frühgriechischen Alpha-betschrift.* Ludwig Auer, Germany: Donauwörth.

Ohlenroth, D. (1996). *Das Abaton des lykäischen Zeus und der Hain der Elaia: Zum Diskos von Phaistos und zur frühen griechischen Schriftkultur*, Tübingen: Max Niemeyer.

Olivier, J.-P. and Godart, L. (1996). *Corpus Hieroglyphicarum Inscriptionum Cretae*, Paris: De Boccard.

Revesz, P. Z. (2016). A computer-aided translation of the Phaistos disk. *Inter. J. Comp.* 10, 94–100.

Revesz, P. Z. (2016b). Bioinformatics evolutionary tree algorithms reveal the history of the Cretan Script Family. *Int. J. Appl. Math. Inform.*, 10, 67–76.

Revesz, P. Z. (2017). Establishing the West-Ugric language family with Minoan, Hattic and Hungarian by a decipherment of Linear A. *WSEAS Trans. Inform. Sci. Appl.* 14, 306–335.

Revesz, P. Z. (2022). Experimental evidence for a left-to-right reading direction of the Phaistos Disk. *Mediterran. Archaeol. Archaeom.* 22(1), 79–96.

Revesz, P. Z. (2022b). The development and role of symmetry in ancient scripts, *Symmetry: Art and Science,* (12th SIS-Symmetry Congress), 1–4, pp. 308–315.

Revesz, P. Z., Rashid, M. P., and Tuyishime, Y. (2019). The design and implementation of AIDA: Ancient inscription database and analytics system. In *Proc. 23rd Int. Database Engineering and Appl. Symposium*, ACM Press, pp. 292–297.

Schwartz, B. (1959). The Phaistos disk. *J. Near East. Stud.* 18(2), 105–112.

Stawell, F. M. (1911). An interpretation of the Phaistos disk. *Burlington Magaz. Conno.* 19(97), 23–38.

Ventris, M. and Chadwick, J. (1973). *Documents in Mycenaean Greek*, 2nd edn., Cambridge, UK: Cambridge University Press.

Chapter 17

The Influence of Information Revolution on Human Thinking Paradigm

Liang Wang*, Ziyi Ma[†], and Shengrui Wang[‡]

Center for Science & Technology Innovation and Ethical Governance, College of Marxism, Xi'an Jiaotong University, Xi'an 710049, China
*wangg85@163.com
[†]ma7855@126.com
[‡]Ciri9264@stu.xjtu.edu.cn

The emergence of new disciplines in the contemporary information revolution has challenged human beings' traditional non-complexity thinking paradigm. Edgar Morin proposed a paradigm of complex thinking based on "three theories", "self-organization", and "three principles"; Xiao saw the significant impact of the information revolution on human cognitive ability and proposed three paradoxes related to cognitive ability; Wu was also keenly aware of the change in thinking paradigm caused by the information revolution and proposed a complexity thinking paradigm based on "information thinking". From different perspectives, the three scholars discussed the impact of the information revolution on the human thinking paradigm.

1 Shaking of the Pillars of a Non-Complexity Thinking Paradigm

With the emergence of new clusters of disciplines in the information revolution, there has been a fundamental shift in the paradigm of human thinking from a non-complex to a complex thinking paradigm. Edgar Morin, a leading contemporary French philosopher and sociologist, has profoundly analyzed the realities of this shift and its theoretical progression. First, he argues that the information revolution has powerfully shaken "the three pillars of classical science", "the classical scientific way of thinking was based on the 'ordered', 'segmented' and 'rational' pillars". And now the cornerstone of these pillars has been shaken by the development of science itself, even though modern science was founded on all three pillars at its birth" (Edgar, 1999). The shaking of the three pillars has undoubtedly opened up the possibility of a paradigm shift in thinking, and the shaking of the pillars was not formed spontaneously but driven by the advancement of technology and the development of theory in the information revolution. The development of thermodynamics, chaos physics, and other disciplines pushed "disorder", the opposite of "order", to the forefront of theory (Edgar, 1999). "Disorder" has become a new paradigm for explaining the picture of the world, and it has also become one of the core concepts of complexity paradigm; the development of system science and ecology has made the traditional "segmentation" model helpless, simple segmentation and reduction cannot solve organic complex problems; the formal deductive logic of traditional rationality cannot fully explain the dual nature of some phenomena (Edgar, 1999). In short, non-complexity thinking cannot fundamentally deal with the complex world picture and the emerging discipline groups, and the transformation of thinking paradigm is imperative. Then, Morin proposed three cornerstones of the transformation of the complex thinking paradigm: "three theories", "self-organization", and "three principles" (Edgar, 1999). "The combination of information theory, control theory, and systems theory introduces us to the universe of organized phenomena, where organization arises from both resistance to and dependence on disorder" (Edgar, 1999). The role of "disorder" in the complexity paradigm has been strongly presented in the "three theories". "Three theories" also become the first cornerstone of the complexity thinking paradigm. "Self-organization" is a good explanation of the dialectical relationship between "order" and "disorder" and its important role in the complex world picture, which constitutes the second cornerstone of the complexity thinking paradigm. The "three principles" are the essence of the complexity thinking paradigm, which is the abstract summary and development of the "three theories" and "self-organization", and finally constitutes the third cornerstone of the complexity thinking paradigm. Morin makes a metaphor with them, he thinks the whole complexity thinking paradigm is like a building, and this building is not a single layer but a multilayered building (Edgar, 1999). In terms of this building, information theory, system theory, and control theory have the role of foundation support, which are the cornerstone of the building (Edgar, 1999). Self-organization theory constitutes the second layer of the building. The duality, holographic, and circulation principles reflect the shape of the entire building or the overall framework of the building. It's a kind of overall architectural style, which is also fully reflected through the first and second floors of the building at the same time (Edgar, 1999). Through this metaphor, Morin also pointed out the close relationship between the paradigm of complexity thinking and "three theories", "self-organization", and "three principles", namely the "three principles" is the core embodiment of the complexity thinking paradigm, which itself is derived from the high summary and development of the "three theories" and "self-organization" concepts (Edgar, 1999). Finally, by analyzing traditional scientific theories, Morin summarized the achievements of emerging scientific theories and, on this basis, deeply realized the possibility and inevitability of human thinking paradigm shifting from non-complexity to complexity, and also put forward the basic framework and principles of complexity thinking paradigm.

2 The "Cognitive Ability Paradox" of the Information Revolution

The information revolution has also had a great impact on human cognitive ability. Feng Xiao, a Chinese scholar who has fully studied the philosophy of information technology, once combined the criticism of domestic and foreign scholars on the information revolution and put forward three paradoxes related to cognitive ability. "The data paradox reveals that the information age has seen a surge in the amount of data available to the knower but less insight formed; the knowledge paradox brings to light the tension between popularization and democratization and elitism and authority in the production process of knowledge, as well as the inconsistency between the ephemeral identity and the enduring value of knowledge; and the cognitive capacity paradox demonstrates how modern information technology, while aiding in the increase of one's level of awareness, has also led to a new degradation in the capacity to know" (Feng, 2016). In terms of paradox one, Xiao affirms the positive role of data-based information in broadening human knowledge and thinking, but the "information explosion" caused by the excessive growth of data has made people "slaves to the flood of data" and they lose their ability to judge at a deeper level, which ultimately leads to "more information but less insight" (Feng, 2016). The "information explosion" caused by the excessive growth of data makes people "slaves to the flood of data", and they lose their ability to judge at a deeper level, which ultimately leads to "more information but less insight" (Feng, 2016).

The second paradox concerns the contradiction between the breadth and depth of cognition brought about by the change in how knowledge is created. On the one hand, due to the development of information technology, especially the development of network technology, people can make use of this highly open platform to explore and publish new knowledge, which inevitably makes the acquisition and generation of knowledge easier and more common, expanding the breadth of human cognition. On the other hand, due to the limitations of professions, the phenomenon of being knowledgeable but not specialized often occurs so that the knowledge created by the general public is inevitably superficial (Feng, 2016). On closer analysis, the negative aspects of the information revolution embodied in paradoxes one and two can be reduced to one issue: the quality of the information output. Paradox one reflects the negative effect of the proliferation of the quantity of information output; paradox two reflects the negative effect of the decline in the quality of information output, but in fact, the decline in the quality of information output is relative. Only the ratio of more authoritative and professional information has decreased under the impact of the information flood. So, fundamentally, it is the result of the information explosion. However, the information explosion has become an irreversible trend of the times, and it is no longer possible to address the cognitive crisis brought about by the information revolution by controlling the amount of information, which is a way to treat the symptoms but not the root cause. Fortunately, information philosophy, which was also developed under the impetus of the information revolution, offers a remedy for this crisis. The theory of information philosophy reveals the unique nature of information as an "indirect being" that coexists with matter. On this basis, the universal mediating character of information can be deduced, thus laying a solid foundation for a new theory of cognition. Instead of considering information as a mere output, the information-mediated cognitive theory emphasizes the interaction between subject and object information. It describes the mechanism and process of cognition through a dynamic and complex multilevel mediating system jointly constructed by subject and object information. This new cognitive theory puts the initiative of resolving the crisis back into the hands of human beings. As Wu emphasizes, the objective thing recognized must not be equated with the objective thing itself because the process of cognition is an objective process by which the subject grasps the object, which must include human subjective activities; it is not a mere mirror mapping process but a process in which the subject selects, processes, and recreates information through the object (Kun, 2005). Although the objective nature of the object must be respected in the process of knowing, the subjective role of the subject must not be ignored as a result (Kun, 2005). Thus, the crisis ultimately becomes an opportunity to enhance the cognitive capacity of the subject (person). In the face of the new situation of information flooding, human beings have to give full play to their enthusiasm and take the initiative to cope with the crisis by optimizing and improving (Kun, 2005). human cognitive structures and enriching and enhancing cognitive means, which in turn will make human cognitive ability step into a new stage.

Regarding paradox three, Xiao believes that it has the greatest impact on persons because the development of information technology directly affects the brain, which is an important organ of human cognition, and its corresponding characteristics, including, specifically, the increase in the amount of information and the decrease in concentration, the enhancement of "extra-brain memory" and "intra-brain memory", the development of figurative thinking, and the decline of abstract thinking, as well as the degradation of the "sensory abilities of the human senses (Feng, 2016)". However, human cognition is a complex and integrated process. The involvement of the mind or psyche, in addition to the brain, cannot be ignored. Many scholars even place them on an equal footing when examining human cognitive activity. Not only that but even in the context of the modern cross-fertilization of disciplines, cognitive psychology has been formed to study the relationship between the mind and cognition. Fortunately, the negative impact of the information revolution on the human brain has not hindered the development of cognitive psychology, and the introduction of information has been instrumental in advancing the issue of representation in contemporary cognitive psychology. For example, Fodor clarifies a series of problems of representation by analyzing symbols and the causal relationship between symbols and entities, but the "synchronic" premise of his theory shields the real cause of the disjunction problem (Fodor, 1987). Dresker is no longer entangled in the causal relationship between symbols and entities but makes an in-depth analysis of information and expounds on the representation problem from the meaning and function of information (Dretske, 1981).

Undeniably, these paradoxes listed by Xiao are sympathetic to us in the information age, which reflects the "information dilemma" of human cognitive ability. However, we should also see the opportunities behind these dilemmas. Although Xiao also believes that these paradoxes provide new opportunities for the development of human cognitive abilities, "the above paradoxes are both new challenges, new dilemmas and even new crises for human cognitive activities in the information technology era, and new opportunities for new development and even breakthroughs in our cognition" (Feng,

2016). Nevertheless, Xiao has yet to give any further advice on how to seize and use these opportunities. He argues that "information in the philosophical sense is an immaterial being, a subject's perception, recognition and construction of objects, and a function of the life control system, especially the nervous system; information is associated with meaning, a phenomenon of human cognition, and there is no such thing as ontological information, but only information in the epistemological sense" (Feng, 2010). However, only by analyzing information at an ontological level can the crisis of human cognition be studied at its source and effective ways of resolving it proposed.

3 Complexity "Information Thinking"

Chinese scholar Kun Wu provides an in-depth analysis of the paradigm shift of human complexity thinking from a more abstract and integrated perspective. He divides the thinking paradigm into three levels: the simplicity thinking paradigm, the local variable complexity thinking paradigm, and the complexity thinking paradigm. He argues that the simplicity and complexity of thinking should be analyzed at the macro and micro levels (Kun, 2002). For example, "entity thinking" has a micro-level invariant feature, which is reflected in the simplicity of the micro, and "ultimate thinking" has a macro-level invariant feature, which is reflected in the simplicity of the macro (Kun, 2002). On the contrary, "energy thinking" has a variable characteristic at the micro level, which is reflected in the complexity of the micro level, and "bifurcation and chaos thinking" has a variable characteristic at the macro level, which is reflected in the complexity of the macro level; however, they are not yet the real complexity (Kun, 2002). True complexity depends not only on the single dimension of "variable" but also on dimensions such as "randomness", "interactivity", and "emergence". The "information thinking" produced in information science and information philosophy has these characteristics (Kun, 2002). It can be seen that "entity thinking and ultimate thinking" is typical of the paradigm of simplicity thinking, while "energy thinking and bifurcation and chaos thinking" reflects the paradigm of local variable complexity thinking and is the most representative of the paradigm of complexity thinking is "information thinking". These basic decisions of Wu are not based on simple subjective division but on specific scientific achievements. To be specific, they are the inevitable result of the rise of new subject groups in the information revolution. According to Wu, the paradigm of the simplicity of thinking is reflected in everything from the ancient natural philosophy of the plain elements to the reductionist thinking of modern science, from the second law of thermodynamics proposed by Clausius to the general system theory of Bertalanffy (Kun, 2002). Although later developments in some disciplines broke the paradigm of simplicity, they embodied the paradigm of local variable complexity thinking. The emergence and development of relativity and quantum mechanics have created new opportunities for the innovation of the human thinking paradigm. However, these theories present only the micro variable complexity, while system self-organization theory and chaos theory reveal the macro variable complexity (Kun, 2002). In Wu's view, only variable complexity is not complexity in the full sense, and the real complexity should be a dialectical synthesis of randomness, interaction, emergence, and other characteristics at the micro–macro level. He believes that the rise of modern complex information systems subject groups provides us with a new picture of the complex world and also provides us with various dimensions of complexity, such as "degrees of freedom", "bifurcation", "dispersion", "nonlinear relations", "chaos", "uncertainty", and "randomness", which are all important standards to measure complexity (Kun, 2002). These are important criteria for measuring complexity, and these indicators are not a single standard but should be considered comprehensively as a "comprehensive integration effect" associated with complexity (Kun, 2002). Finally, Wu turned his attention to "information thinking". What is "information thinking"? Wu clearly defined information thinking in the Theoretical Basis of Complex Information Systems. He believed that the starting point of "information thinking" was the understanding of the "information concept", which was also the understanding of the essence of information, and the revelation of the essence of information would change people's existing way of thinking and form a new way of thinking (Kun, 2010). The way of thinking derived from the nature of information is "information thinking". Specifically, when people take information as a foundational existence, their understanding and awareness of existing things will fundamentally change (Kun, 2010). This change reflected in people will more consciously understand the essence, existence, state, and value of things from the structure and relationship of things and also more consciously grasp the development and change of things from the macroevolutionary relationship and regard this structure and relationship as the carrier of specific information so that they can use this carrier to dynamically analyze the essence, characteristics, historical relationship, development trend of things, etc. (Kun, 2010). Not only that, but people can also artificially symbolize the object and give it specific information content (Kun, 2010). It can be seen that this new mode of thinking embraces the typical complexity features such as randomness, interaction, and emergence (or evolution, creation), which provides a powerful path for the transformation of the complexity thinking paradigm, and it can even be said that "information thinking" constitutes the core of the complexity thinking paradigm. As Wu pointed out, "information thinking" embodies a high degree of "holographic synthesis", which enables people to understand the past, present, and future development of things through evolutionary relationships, thus presenting the essence of things, development trends, etc. (Kun, 2005). "Information thinking" is a reflection of the complex spatiotemporal relations of things in the human brain, an abstract generalization of the complexity of the nature of reality, which provides a new dimension for people to grasp the complexity of the relationship between reality, and the basis of all this is a deep understanding of the nature of information, in-depth analysis of complex information activities, and therefore the new research perspective of "information thinking" is irreplaceable by other thinking paradigms. It is becoming the core paradigm of complex thinking (Kun, 2005). Morin and Wu have affirmed and argued the inevitability of the paradigm shift of human complexity thinking from different perspectives. It can be said that the information revolution provides an excellent opportunity for the innovation of human thinking paradigm, and the complexity thinking paradigm reflects the improvement of human thinking ability, which helps people to examine and grasp the complex world picture from a perspective closer to the real one.

4 Conclusion

The arrival of the information revolution poses challenges and opportunities to the traditional non-complexity thinking paradigm of human beings. On the one hand, the information revolution has led to data paradox, knowledge paradox, and cognitive ability paradox, which has caused a huge impact on human cognitive ability. In this context, human beings should actively meet challenges, seize opportunities, strive to break through the "information dilemma" under cognitive ability, and promote further innovation and development of the thinking paradigm. On the other hand, it provides an excellent opportunity to innovate the human thinking paradigm. The complexity thinking paradigm reflects the improvement of human thinking ability. It helps humans examine and grasp the complex world picture from a closer to real point of view.

All three scholars have paid attention to the changes and challenges of the human thinking paradigm in the modern information age, but each has its own emphasis. Morin emphasizes the possibility and certainty of transforming human thinking from non-complexity to complexity; Xiao pays attention to the impact of the information revolution on cognitive ability and puts forward three paradoxes. He analyzes how to deal with information in an information overload environment. Wu introduced the concept of "information thinking" and re-conceived the complexity thinking paradigm based on the nature of information at the philosophical level.

References

Edgar, M. (1999). *L'Intelligence de la complexité*. Paris: L'Harmattan. pp. 318–325.

Feng, X. (2010). Revisit the philosophical meaning of information. *Social Sciences in China* (4), 32–43.

Feng, X. (2016). On the three epistemological paradoxes of the age of information technology. *Innovat.* 10(1), 47–54.
Fred I. Dretske. (1981). *Knowledge and the Flow of Information.* MIT Press.
Jerry A. Fodor. (1987). *Psychosemantics: The Problem of Meaning in the Philosophy of Mind.* MIT Press.
Kun, Wu. (2002). Change of complexity and science thinking way. *Studies in Dialectics of Nature* 18(10), 46–49.
Kun, Wu. (2005). Complexity and information science study creed. *Chinese J. Syst. Sci.* 13(2), 12–15.
Kun, Wu. (2005). *Philosophy of Information: Theory, System, Method*, Shanghai: Commercial Press. p. 188.
Kun, Wu. (2010). *Theoretical Basis of Complex Information Systems*, Xi'an: Xi'an Jiaotong University Press. p. 76.

Chapter 18

Artificial Intelligence and Creative Education of Future Information: On the Importance of Philosophical Basic Quality Education

Tianqi Wu[*,†,‡] and Ruiyuan Zhang[*,†,§]

[*]Department of Philosophy, School of Humanities and Social Sciences, Xi'an Jiaotong University, Xi'an 710049, China
[†]International Research Center for Philosophy of Information, Xi'an Jiaotong University, Xi'an 710049, China
[‡]tianqi1262016@126.com
[§]Jolyine14@163.com

In pace with the dramatic expansion of artificial intelligence (AI), enhanced human of human–machine fusion may appear in 20 years. So, how to educate the "enhanced human" in the future? The creative education of future information is based on philosophical education. Philosophy education should penetrate primary and secondary education, especially to improve the philosophical thinking ability of primary and secondary school teachers, which is the construction of the philosophy spirit. The dogmatic education model makes it difficult for students to cultivate innovative talents. Traditional education teaches people obedience while philosophical education teaches people the pursuit of freedom, questioning, criticism, and innovation. Therefore, future education should start with the cultivation of philosophical spirit which is the basic quality of future education.

1 Introduction

In the future, with the rapid development of artificial intelligence (AI), the human–machine collaboration might completely get rid of the learning mode of rote memorization. With the development of brain–computer technology, AI devices with powerful information storage, calculation, and search capabilities will be connected with the human brain (Shuyue et al., 2019). In the future era of human brain information explosion, any knowledge with standard answers that need to be recited, calculated, and memorized might be stored by AI, so the related human education and testing methods could also be eliminated, which will lead to an earthshaking educational revolution.

According to the current development direction of AI based on algorithms, it is impossible to achieve real intelligence, and perhaps we can only hope for the future research direction of biological computers and gene (DNA) computers. However, due to the great progress in the collaboration of AI and human beings, the progress far exceeds the upgrading of new AI technology. For example, the relationship between modern people and mobile phones is a primary step of human–machine fusion, the mobile phone as the media will gradually become the extended part of human limbs (Huilin and Baozhang, 2019). The amount of knowledge on the Internet has far exceeded that of any educator, which leads to the fact that imparting knowledge is no longer the greatest role of educators. This is a big trend that can't be stopped in the process. Therefore, when we discuss education issues under the background of the development of AI, we must first start with two important issues.

The first problem is that while AI has greatly improved people's intelligence in the present and near future, we must clearly understand what algorithm-based AI can and cannot do.

The second question, through the evaluation of the limitations of AI, our education should focus on cultivating people's ability to learn and exercise what AI is not good at or lacks so that the next generation has the capability and preparation to face and control the future combination of AI. Only in this way can human beings under the assistance of AI truly exert the greatest potential of man–machine collaboration intelligence in the future.

In view of the two questions raised above, it's necessary to make a bidirectional evaluation of the capabilities and limitations of AI. Philosophy of information may provide a novel evaluation approach. Kun Wu put forward the idea of an information creation system in the epistemology of philosophy of information. As an information processing system, the information creation system includes biological consciousness, including human consciousness and AI (see Fig. 1).

Judging from the functions of the nine subsystems that make up the information creation system, the current AI has completely possessed the functions of four subsystems of the whole and has the

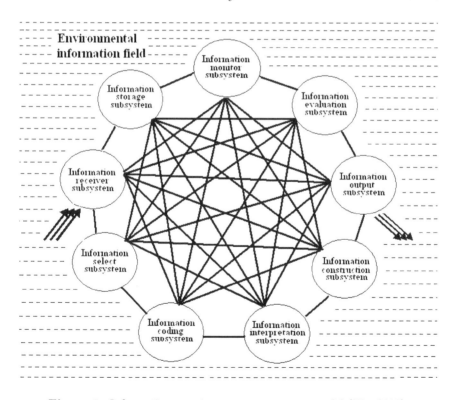

Figure 1. Information creation system structure model (Wu, 2019).

advantage in capability compared with human beings. The weakest part of AI is the following three subsystems: information interpretation subsystem, information evaluation subsystem, and information construction subsystem (Lin and Wu, 2021).

Our education direction should focus on an information-creative education model. We can start with the functions of the three subsystems that AI lacks most and build the focus and core tasks of future education.

First of all, compared with the information interpretation subsystem, AI can't recognize and understand the acquired information because, in fact, the understanding and interpretation are often the designer of the perceptron, not the perceptron itself (Xianfeng, 2016). Therefore, the education of people should focus on the cultivation of the subject's understanding and uniqueness. Avoid letting the understanding and interpretation of knowledge publishers and AI designers excessively influence or even replace the subject cognition after man–machine integration. Train people to understand and explain everything in breadth and depth, that is, the ability to consider problems comprehensively and deeply. Using information technology to increase the education of experience, discussion, and analysis and reduce the education of recitation highlights the relativity, randomness, and uniqueness of cultivating people's thoughts. We must reject the destruction of individual cognition by standard answers and authoritative concepts, which is also the destruction of human innovation ability.

Second, compared with the information evaluation subsystem, with the explosive growth of information obtained by human beings, the wide spread of individual information, extreme information with various purposes and false information, and the rapid iteration of new ideas and theories, the uncertainty of knowledge becomes more and more prominent. It is more and more important for people to evaluate and screen information. These evaluations include authenticity, precision (quality), logic (rationality of processing rules and methods), utility (value), and so on. Education should focus on making the educated learn the ability to filter and judge information, choose one or more items that are more in line with reason and logic among many answers and opinions of the same question, and keep an attitude of pursuing truth, evolution, pluralism, and relativity. This requires an ability to stay out of problems, cross-integrate systematic problems, and consider problems in complexity, and an ability to control many related problems at the same time and think about problems from a broader perspective and a higher concept.

Finally, compared with the information construction subsystem, the work of this subsystem is the most critical and core subsystem of the information creation system. All kinds of creative and novel information about all people is created through this information construction activity. At the same time, this is also the most critical and difficult part of education. Because the construction of this part of the ability depends on everyone's self-organization characteristics, the educational methods that are hetero-organized will only destroy the construction of this system. Therefore, the educator needs to guide the educated themselves into this field, and at the same time, the educator should influence and correct them through speculation, discussion, and ideological collision rather than forcibly instilling the so-called definite concept. The creative information of all intelligent agents comes from the emergence mechanism of self-organizing systems, which itself depends on the evolution and uncertainty of each self-organizing system. This requires focusing on the cultivation of innovative ability in education, reducing the ideological constraints and excessive restrictions on the educated, that is, reducing the external environment only in an open and moderate external environment can self-organization phenomena and behaviors occur. The mechanism of the educated people's innovative ability is similar.

The abilities that this future education focuses on are all philosophical education in essence, so the information creation education we advocate is essentially an educational model induced by philosophical spirit. The education mode of information creation aims at educating people to pursue freedom, dignity, and truth and stick to the spirit of questioning, criticizing, and innovating.

2 The Development of Brain–Computer Interface Technology and the Emergence of "Enhanced Human Beings" Tianqi (2020)

Brain–computer interface (BCI) refers to the direct connection created between the brain of a human or an animal and an external device, in a bid to achieve the exchange of information between the brain and the device. Nowadays, with an Internet-connected mobile phone, almost endless knowledge can be acquired and countless problems can be solved. But this is only the embryonic stage of "human beings enhancement", that is, human–computer interaction stage. Brain–computer interface technology just strives to combine human consciousness with AI.

The earliest attempt of mankind is "microchip implant" technology (that is "human body chip" technology). Its ideological roots can be traced back to the concept of transhumanism put forward by Sir Julian Huxley in 1957, whose connotation was enriched by the development of science and technology. Nick Bostrom, a professor at Oxford University, described the definition of transhumanism as "a possible and desirable intellectual and cultural movement that believes in fundamentally improving the human condition through applied rationality" (Nick, 2003). Transhumanism maintains a positive attitude toward technology and regards technological improvement as the best means to improve the reality of human beings. In 1998, Kevin Warwick (1954–), a British engineer, implanted a chip in his arm and became the first "chip man" in the world. In 2004, John F. Donoghue, a professor at Brown University in the United States, developed a brain microchip technology called "brain control gate". He implanted a biochip the size of a painkiller tablet into the brain of a paralyzed patient, allowing the patient to mentally navigate the television and computer (Lan and Kevin, 2018). Then, in 2008, he enabled two monkeys to feed themselves with candies and fruits using mechanical prosthetic limbs controlled by brain consciousness. Since 2015, Epicenter in Stockholm, Sweden, has implemented the human chip implantation technology for its employees. The implanted chips, such as the size of rice grains, can be used as master keys and identification cards. However, this embedded chip is similar to the randomly generated mobile QR code bound with identity information and is not a real brain–computer interface. Edward Chang, a neurosurgeon at the University of California, San Francisco (UCSF), and his colleagues have developed a decoder that can convert human brain nerve signals into speech and surgically implant electrodes inside a subject's skull to monitor brain activity. On this basis, the researchers used a technology called electrocorticography (ECoG) to directly record and translate the neural activities in the subject's cerebral cortex, which can generate speech and interact with the computer directly through meditation (Anumanchipalli *et al.*, 2019). In July 2019, American entrepreneur Elon Musk (1971−) announced that Neuralink would use a neurosurgical robot to implant dozens of "wires" with a diameter of only 4–6 microns as well as proprietary technology chips and information strips into the human brain, which could then read brain signals directly through the USB-C interface. The new technology does less damage to the brain and transmits more data than the previous one. In April 2021, Neuralink's research made new progress. They published a video on YouTube where monkeys could control the cursor on the screen using only the brain's "mind" and could play games using their "mind". The above-mentioned examples are all invasive brain–computer interfaces.

The prospect of non-invasive brain–computer interface is brighter because it does not need to open the internal environment of the human body. In 2014, researchers at Yale University used functional MRI (fMRI) to read consciousness and reconstruct human faces in the consciousness and tried to reconstruct memories in the consciousness, such as a dreamland Kexun Medical Network (2014). And in July 2019, UCSF's research team of brain–computer interface extracted the deep meaning of a word spoken by humans from brain activity and quickly converted the extracted content into text. Moreover, in August 2019, the World Robot Conference held in Beijing used a non-invasive brain–computer interface for the competition.

In the next step, people might integrate consciousness with machines to create "enhanced human beings".

In the wake of the rapid development of brain–computer interface, it may be very convenient for human beings to obtain information by virtue of brain–computer interface in the next two decades. All the knowledge accumulated by human beings may be instantly presented to anyone who wants to obtain it and may even achieve memory transplantation. The emergence of "augmented human beings" could completely get rid of the era of learning by rote. In the future, any education or test that requires recitation, calculation, and memory might be eliminated, which will lead to an earth-shaking education revolution. If so, where should our education, which lays great emphasis on rote memorization go? Perhaps the answer is philosophy.

3 Misunderstandings in Current Philosophy Education

The public almost knows what science is because the scientific system has been widely incorporated into the basic education of human beings. The public's prejudice and misunderstanding about philosophy are due to the lack of basic education in philosophy. The fundamental reason for this deficiency is the problem of philosophy education itself. What idea can be called philosophy? What kind of philosophy is suitable for education? Especially the enlightenment education. These are all problems that philosophy education should seriously consider before making changes.

Mortimer J. Adler (1921–2001) once mentioned that philosophy is not the exclusive domain of philosophy researchers but something that everyone does. People are born with the ability and tendency of philosophical thinking. Philosophy is formed by the concepts of daily life and thinking activities around us. The concepts contained in the philosophy of each of us are indispensable for us to comprehend ourselves and the world and then make choices and decisions. It decides our thinking activities in tandem with determining our behaviors (Mortimer, 2014).

According to Adler, philosophy permeates our lives and determines our thinking activities, choices, and behaviors. Scientific ability determines what we can do, while who we are and what we should do are determined by our philosophical ability. Nowadays, formal philosophical education generally begins with the history of philosophy and so do the vast majority of philosophical works. Philosophy education has always laid special stress on reciting the history of philosophy and the relevant ideas of scholars in the history of philosophy, and these rote memorization contents have become a subject of a variety of examinations. Even a great number of scholars take it for granted that the history of philosophy is philosophy itself (Genyou, 2019). It is extremely inappropriate to cognize philosophy and interpret it for the general public in this way.

The classic thoughts in the history of philosophy are the ideological achievements created by philosophers using philosophical thinking and carrying forward philosophical spirit. This is just like equating the research results of technology, such as engines and generators, with the technology itself, which is unreasonable. Furthermore, this wrong understanding and method are applied to education, and the text full of technical terms and obscure words is placed in front of beginners of philosophy. Just like putting the formula of advanced mathematics directly in front of the beginners of mathematics, this behavior is extremely ridiculous. This is just like persuading people who have just come into contact with philosophy education. Nowadays, the popularization of mathematics in education is well done. Everyone knows some mathematics knowledge, and the most basic addition, subtraction, multiplication, and division are used in daily life. However, advanced mathematics is much more difficult to understand and use than basic mathematics used in daily life, and only a few mathematicians can control it. Mathematics education is gradual, starting with the simplest numbers, and it is impossible to teach advanced mathematics directly to pupils without mathematical foundation. On the other hand, philosophy education is not divided into elementary, middle, and advanced levels, so the current education mode of philosophy is obviously lacking in the system.

Basic philosophical concepts and thoughts are not too difficult to understand. Basic philosophical thinking has its starting point and logical context. After step-by-step guidance, everyone can understand. Because philosophy is based on the most rigorous logical thinking, it is often easier to understand and accept than those stories that are made up in a muddle and can stand scrutiny. Although it is not easy to achieve a profound understanding, the most important thing is one's knowledge reserve, thinking habits, and life experience. However, these obscure philosophical thoughts are only a part of the history of philosophy, not the core of philosophy, nor philosophy itself. The study of the history of philosophy is a necessary way to study philosophy. It is a treasure house of human knowledge. We can find problems and draw supplies from the study of the history of philosophy. But the real core of philosophy should be a kind of spirit, a kind of life attitude that is good at using its unique way of thinking to deal with problems. It cannot be simply attributed to the philosophical thoughts and achievements of philosophers in history. The real philosophy lies not in the philosophical achievements in the history of philosophy but in the thinking motivation, logical deduction, thinking process, and way of thinking of those philosophers in the history of philosophy.

Many philosophers' thoughts can be extreme, but all classic philosophical thoughts are based on profound philosophical thinking. As long as we can understand their way of thinking and pursuit, we will not feel weird but may be deeply enlightened. For learners, this is a process of understanding the world, accommodating differences and self-growth. It is undeniable that there has always been a so-called "alien" in human groups. However, the inclusiveness of philosophy is just like literature and art, even higher. It gives those "aliens" a proper position, gives them the greatest freedom of thought, leaves their traces in history, and gives spiritual comfort to those who are like-minded today and later generations. We have changed it to a more elegant name, calling this "alien" and "abnormal" as "minority". For this spiritual resonance, we call it "bosom friend". However, some philosophical thoughts or schools can be called "minorities", but the philosophical spirit itself is absolutely something that everyone should have. To some extent, this understanding and tolerance of being different from others is the embodiment of human civilization and philosophical spirit. At the same time, it also embodies the pluralism, openness, innovation, and freedom of a society. The answers to all questions are by no means unique, and the restrictions on the answers to any questions are dogmatic, ignorant, and anti-philosophical. Now, when we think back to philosophy education in China, those behaviors of memorizing the standard answers in the history of philosophy run counter to the philosophy spirit.

Many philosophical thoughts we see, especially some classic philosophical theories, are the ideological feats of the greatest philosophers in history who used philosophical thinking. But philosophical thought is not limited to the history of philosophy, in fact, it has penetrated into every level of our lives. Everyone who is good at philosophical thinking uses a philosophical way of thinking when reflecting on life, thinking about the world, asking questions, and summing up feelings. Philosophy is actually around everyone. In the present and future of knowledge explosion, this ability to think independently is the ability that human beings should really cultivate.

The philosophers in the history of philosophy each have their own unique view and research on philosophy and are all pursuing an independent philosophical system that is different from others. Although their theories are different from each other, they all share the same essence in the study of philosophy, that is, the spirit of philosophy. Even if a bookworm who can only memorize by rote memorizes all the history of philosophy, if he has no habit of philosophical thinking and no philosophical spirit, he is definitely not a philosopher. He is just a recording tool, a knowledge porter, and a computer hard disk. In the future, AI will give people an unprecedented amount of knowledge. Just like AI, it can't handle personal problems, and it can't be a controller, but more like a controlled person (Capurro, 2012). When rote memorization is no longer useful, society will completely eliminate those porters who record as tools. Innovative talents will be highly praised in an unprecedented way, and the taboo of the philosophy spirit is to follow suit, follow the book, and follow the rules and be rigid and dogmatic. The most important point of philosophical spirit is innovation.

4 The Core Foundation of Philosophy: Philosophy Spirit

The core of philosophy is the spirit of philosophy, which is a kind of unique human light that has continued from the ancient Greek era to the present. If the future education needs philosophy, we need to know what the philosophy spirit is at first. What is the philosophical spirit? This should start from the origin of philosophy, that is, the period of ancient Greeks. At the same time, Chinese and Indian philosophy and culture have similar temperaments. This may also be the temperament of the whole Axis era. It is not only a manifestation of ultimate concern but also, more importantly, the respect and pursuit of knowledge, reason, morality, wisdom, and truth. (With the development of philosophy, more elements are added to the spirit of philosophy, such as goodness, beauty, humanity, and freedom.)

Western philosophy comes from the Greek word "Philosophia" (love of wisdom and pursuit of truth), which means the original pursuit of philosophy. In fact, it is consistent with the pursuit of science, but the means are different. Science tends to use empirical methods, while philosophy uses reflection and criticism. But they interact and support each other in the process of pursuing truth. However, philosophy has more freedom than science, with more breakthrough points and deeper and more abstract understanding, while science is more practical and concrete. Science looks for clues to the operating rules of the world in the empirical study, while philosophy focuses on criticizing and reflecting on the existing theories. If mankind wants to seek progress, both are indispensable. Philosophical spirit is a belief, an idea, a worldview value, and an attitude toward life.

Einstein once emphasized that philosophy is the mother of all sciences. The relationship between philosophy and science is so subtle that they should be a symbiont that complements each other and blends with each other. The motive force of philosophy is the most primitive instinct of human beings: intellectual curiosity. Plato emphasized that "Thauma" (surprise) is the beginning of philosophy.

Philosophy happens by surprise which is the nature that philosophers should have. Driven by this nature, philosophers have gained a truly liberating force. Aristotle also emphasized that human nature lies in seeking knowledge. And surprise drives people to engage in philosophical thinking. In fact, the more people study and think, the more they will inspire curiosity and thirst for knowledge, and then modesty and prudence, because only those who study can find their own shortcomings. And people who have only a little knowledge and don't ask for a lot of knowledge often consciously learn.

Ancient philosophers thought about the origin of the world, the meaning of life, and pursued wisdom, truth, knowledge, goodness, and beauty. We are still thinking about what they are thinking, and we are still walking on the road they have taken. At that time, philosophy and science were integrated, not only in the way of thinking, research objects, ways, and means but also in the pursuit of truth. Science is committed to finding the reasons and laws behind natural phenomena, summarizing theories, quantifying and calculating with mathematics, and various disciplines are full of various formulas and mathematical models. Just like Hegel's absolute spiritual intention to unify the whole world, Einstein also spent the rest of his life pursuing the theory of great unity in science. In fact, they pursue the same direction and the same thing, but they work in different fields and use different research methods. To sum up, we can find that the philosophical spirit is a kind of human belief in the pursuit of truth, which is unified with the scientific spirit in this ultimate goal. This belief in pursuing truth is also what every modern person should have.

Regrettably, the philosophical spirit is out of tune with today's social atmosphere. The tradition of the ancient Greek philosophy king has become declining and distorted in the present society, which is an era of lost faith. Just like the separation of science and philosophy, to a great extent, this society attaches importance to science and despises philosophy. Attach importance to economic interests, pragmatism, and hedonism while ignoring thought, ethics, morality, criticism, and innovation.

5 The Importance of Philosophy in Education

The thought foundation of philosophy is philosophical thinking, which is not only the beginning of the operation of philosophy but also the way in which philosophy can be developed. People who can use philosophy still have their natural philosophical accomplishments even if they do not understand the history of philosophy, and they can also sum up their own philosophy of life and world outlook in their lives and carry it out by actual efforts. The focus of philosophy education should not be on memorizing mechanically the standpoints of philosophers in history but on imparting to people how to face problems, life, and research with philosophical thinking. The standpoints of philosophers in history are only rich ideological materials, which enable people with philosophical thinking to taste and think rather than blindly worship and follow. This is the biggest difference between philosophy and theology.

The education of philosophy is different from other subjects. For instance, Mathematics explicitly tells students that "$1+1=2$", while philosophy teaches students to reflect on why "$1+1=2$". How does "$1+1$" equal "2"? Is "$1+1$" equal "2" in all cases or are there any exceptions? To put it simply, dogmatic science or theology in traditional thinking gives you an affirmative answer: "it is this", while philosophy teaches people how to think about a series of problems on their own: "why is it this?", "how does it become this?", and "how to make it not this?" In conclusion, traditional education teaches obedience, while philosophy imparts freedom, questioning, criticism, and innovation. This is not a future problem but actually around us.

There is a case study report from China: after rural students enter famous schools, they will highlight two major defects — lower English level and performance in research study Xiaojuan and Yanjie (2017). The gap between language and academic performance is a minor problem. Rural students are very hard-working, and their academic performance will rise slowly because of their own efforts, narrowing the gap with urban students. The key is that the way of thinking of rural students is different from that of urban students. In the words of news writers, rural students have their "indelible class imprint", and a very important aspect of this brand is the lack of philosophical thinking. There is a child, A, who was admitted to Tsinghua University from a national poverty-stricken country in China. When he entered the school, he discovered many unexpected problems. Because the school-specialized courses are not fully integrated with the high school curriculum, they skip the basic content directly, and the course is very difficult. A can't understand it, but he will only use more diligence to make up for this lack, and he is afraid to ask teachers and classmates for fear that everyone will look down on him. However, he doesn't understand that truly advanced learning is no longer like rote memorization in the college entrance examination but needs to have his own thinking and opinions. The answers to many questions can no longer be obtained from simple books. He began to blame himself for his low grades after the exam. But there are also urban students who are not satisfied with their grades. They will go directly to the professor and check with him whether the score is low. A couldn't believe all this for a while because parents from childhood to adulthood taught them that they should obey the teacher, and they were not allowed to argue with the teacher if they had doubts. "Being afraid to confront the authority" is the "common problem" of all poor students. Their way of learning and thinking is bound by traditional thinking, which obliterates the freedom of thinking, is afraid of questioning and criticizing, and is naturally difficult to innovate. In the long run, when they encounter research-based learning, it will be quite difficult. In contrast, children who grow up in cities have much higher philosophical thinking quality. Perhaps this way of thinking and vision is the most difficult shortcoming for rural children to catch up with. In the future, with the support of man–machine reinforcement technology, this gap will be more obvious.

In fact, this problem is reflected on many levels. The people who live at the bottom of society are more inclined to obey authority and fear to question. The education of philosophy originated from the pursuit of knowledge, which lies in the curiosity of the world and the pursuit of truth. The essence of learning should be self-consciousness, not cramming coercion. Rote memorization brings consumption and fatigue and does not really make people acquire knowledge. Knowledge should be acquired by people's conscious exploration. It happens that the development of AI can finally get rid of the forced memory of a lot of knowledge. This is an opportunity for philosophy education, but it is also a challenge for human education. In the future, knowledge acquisition will be very easy, and powerful databases will be directly integrated with the human brain. And how should people control this power? As a double-edged sword, knowledge can be used for good or evil, and the key lies in the use of people. Without faith, people are incapable to achieve their goals by hook or by crook. Without the pursuit of truth, people will lose their integrity and seek profits. Without reflection, people can't sum up experience and it is difficult to innovate. Without the attention to the spiritual field, people will be materialistic and degenerate into vulgarity. Imagine that if such people gain more knowledge, it will only be a disaster for society! Chinese educational ideas should be reconsidered instead of going backward and walking on the passive old road. China needs to rise, and what it needs is a new person with a philosophical spirit, good at reflection, love of wisdom, and the pursuit of truth and belief. These problems really deserve our deep thinking and review!

6 Philosophical Education of Strengthening Open and Critical Thinking in Primary and Secondary Education

Philosophy is in our life; it determines our thinking activities, choices, and behaviors and determines who we are and what we want to do. "It is almost impossible to think philosophically without learning" (Karl, 1982). The formation and improvement of philosophical thinking capability depends on necessary study and training. Therefore, philosophy education should face all levels of society. Some core ideas in philosophy should be infiltrated from the study of primary and secondary schools, and it has missed the golden age of people's thinking, learning, and growth since the university began to let students understand philosophy.

Take an example that appeared in the first-grade Chinese exam of a primary school in Xi 'an, China, in 2020:

1. The sickle moon looks like — and also like___.
2. The red sun looks like — and also like___.

Judging from the four papers collected, this topic with the most open and innovative thinking actually has a standard answer! The moon must be like a boat and a banana, but the answers such as leaves and moon cakes with a bite are all wrong. The sun must be like fireballs and lanterns, and the answers to balls, persimmons, apples, smiling faces, sunflowers, and oranges are all wrong. And the teacher didn't feel any problems when correcting the papers. This is a test question in Xi'an's leading primary school, not in some schools with backward education level. This is even more shocking.

First of all, the so-called standard answer to this question itself is problematic. It is said that the moon is like a boat and a banana, and boats are not all meniscus-shaped. There are also flat boats such as rubber boats and bamboo rafts, which can also be called boats. Many bananas are straight, not all of them are curved, and millet bananas are thumb-shaped. It is said that the sun is like a fireball and lantern. From the structure of a star, the sun itself is a burning gaseous fireball. It is a fireball. It really doesn't matter whether it is like a fireball or not. There are many kinds of lanterns, and there are all kinds of colors. Kongming Lantern is a white cylinder, not red or round.

Second, there is no standard answer to this question itself, and the so-called imagination is relative. In the test paper, you should let the children speak freely, and you should not set standard answers, otherwise it will stifle the children's cognitive development and innovation ability. From the beginning of primary and secondary education, there has been excessive superstition and worship of the unshakable authority of the so-called standard answer, which has been extended to university education. In addition, in order to facilitate management and maintain the authority of school discipline and classroom order, it is common to criticize and punish students in primary and secondary education, which also leads to the double shackles of students' body and spirit to some extent.

The essence of innovation is to challenge authority (standard answer) because only by thinking what predecessors did not think, doing what predecessors did not do, can we be called innovators. Since innovation is new, it must go beyond the predecessors and beyond the established things. If children's thoughts and courage to question and challenge authority are suppressed or even erased from childhood, and they only know that students are required to obey the established rigid discipline and order, and hope to awaken their innovative ability in adulthood and let them design new systems and new behaviors, then it is simply impossible. This contradictory approach deserves our reflection.

7 Philosophical Education of Cultivating Reflection and Critical Ability in University

In the current education system in China, some innovative talents will also be produced among the students we train. However, such talents can only be a minority, and they spontaneously emerge through rebellion against the existing system. In fact, even such a few "talents" with innovative ability have entered the stage of university and even postgraduate education, but they still retain the deep imprint brought by long-term exam-oriented education in their way of thinking and behavior. These marks often make them have some kind of "schizophrenic" incomplete personality. On the one hand, in order to pass the exam, they memorized the standard answers, tried not to make mistakes, not to mention opinions, not to challenge authority, and to compete with their peers around them, thus achieving proud results in exams at all levels, but there were too many deep depressions of incomprehension, unbelief, and unwillingness in their hearts. Some of them are trying to escape from the system or help their next generation escape from institutional education. These people always yearn for a more relaxed, free, and open education system in their hearts. But they don't know that in the scene of being unable to change the existing education system, this can only be a fantasy, similar to some kind of siege effect.

There are still some students who enter the "rebellious period" when they arrive at the university because the repressed personality is suddenly released after the college entrance examination, and the sense of lack of youth, independence, self-identity, and challenge swarms. Therefore, a large number of rebellious students have been suppressed for a long time in universities. They have been suppressed for too long, but their independent spirit has been released in various deformed ways. The way they challenge authority is even self-indulgence and self-destruction. They are addicted to games, Internet, love, profligacy, comparison, etc., because they understand that learning is rigid and oppressive, and learning has not brought them real happiness and interest. Their starting point of learning is not curiosity, not hope for the future, but ruthless competition for grades, helplessness for all future work, and expectations and pressures from family and school. They didn't learn the habit of truly independent and spontaneous learning through primary and secondary education, but they were always forced to flog their studies, always boring and painful rote learning, and always repetitive and mindless exercises.

Nowadays, many universities even started to use compulsory classroom sign-in, night self-study roll call, and other means to manage students in primary and secondary schools to control these adult students. This will only make these students develop into "giant babies", and their autonomous and spontaneous learning abilities will be even worse. Finally, they will come to the society one day, and they will lead their own study, life, and work. What should they do at that time? Although the problem is manifested in universities, its root cause comes from the system of primary and secondary education. Without the reform of the concept, system, and methods of primary and secondary education, the related problems existing in existing universities cannot be effectively solved.

Philosophy education has been basically popularized in universities now, but it is also problematic in universities. Formal university philosophy education generally begins with the history of philosophy, and most philosophical works also begin with the history of philosophy. Philosophy education has always focused on reciting the history of philosophy and the relevant thoughts of scholars in the history of philosophy, and these rote learning contents have also become the items of various examinations. It is full of standard answers and fixed exam-oriented education content, which is exactly the same as the learning style of primary and secondary schools. People have the original desire for innovation, the idea of being different, independent, and universal. However, the poor education system did not give them the environment for their potential growth and the space for their release. Everyone was produced and shaped according to the standards of a production line. It can be said that this is absolutely anti-human.

The next question is what should philosophy education focus on? Teaching the history of philosophy? We have mentioned that many scholars take it for granted that philosophy is the history of philosophy. But it is extremely inappropriate to understand philosophy and explain philosophy to the public in this way. It is even more impossible to tell the history of philosophy directly to primary and secondary school students. The classic thoughts in the history of philosophy are the ideological achievements created by philosophers using philosophical thinking and carrying forward philosophical spirit. This is just like equating the research results of technology, such as engines and generators, with the technology itself, which is unreasonable. Obtaining the engine and generator or memorizing the installation drawings does not mean mastering the technology. The real technology lies in that people can understand the logic and principle of science and technology and freely implement and deduce it, and then they can have the ability and possibility to transform and develop the existing technology. It is neither a good way to learn technology nor the purpose of mastering technology to memorize and install drawings. Directly teaching the history of philosophy to primary and middle school students who have no philosophical quality foundation, or directly letting them recite the history of philosophy, just like directly teaching the principle of generator machinery to people who have no physical foundation, so the education method without paving the way and gradual process is bound to fail.

The education of philosophy originated from seeking knowledge, which lies in the curiosity of the world and the pursuit of truth. The essence of learning should be self-consciousness, not cramming coercion. Rote memorization brings consumption and fatigue and does not really make people acquire knowledge. Knowledge should be acquired by people's conscious exploration. As a double-edged sword, knowledge can be used for good or evil, and the key lies in how people use it. Without conscience, people will be numb and will do whatever it takes to achieve their goals, and knowledge will lead to destruction and disaster. If you don't pursue the truth, people will lose their integrity, be mercenary, and knowledge will become a profit-making tool. Without reflection, people can't sum up experience and it's difficult to innovate, and knowledge has become people's imprisonment. To sum up, if we don't pay attention to the attitude toward knowledge, knowledge will only bite the human beings themselves. If people who only use knowledge as a tool to seek money and power for themselves had more knowledge, it would be a disaster for society!

The education of philosophy originated from seeking knowledge, which lies in the curiosity of the world and the pursuit of truth. The essence of learning should be self-consciousness, not cramming coercion. Rote memorization brings consumption and fatigue and does not really make people acquire knowledge. Knowledge should be acquired by people's conscious exploration. As a double-edged sword, knowledge can be used for good or evil, and the key lies in how people use it. Without conscience, people will be numb and will do whatever it takes to achieve their goals, and knowledge will lead to destruction and disaster. If you don't pursue the truth, people will lose their integrity and be mercenary, and knowledge will become a profit-making tool. Without reflection, people can't sum up experience and it's difficult to innovate, and knowledge has become people's imprisonment. To sum up, if we don't pay attention to the attitude toward knowledge, knowledge will only bite the human beings themselves. If people who only use knowledge as a tool to seek money and power for themselves had more knowledge, it would be a disaster for society!

References

Anumanchipalli, G. K., Chartier, J., and Chang, E. F. (2019). Speech synthesis from neural decoding of spoken sentences. *Nature* 568(7753), pp. 493–498.

Capurro, R. (2012). Toward a comparative theory of agents. *AI & Soc.* 3/4, pp. 195–201.

Genyou, W. (2019). Philosophy is the history of philosophy: Rethinking the research methods of philosophy and the history of philosophy. *Philosoph. Stud.* 1.

Huilin, G. and Baozhang, Z. (2019). Human-machine fusion analysis based on McLuhan's media ontology. *Stud. Dialect. Nat.* 1.

Kexun Medical Network. (2014). Yale university researchers reconstruct face by reading consciousness with fMRI. *China J. Med. Comput. Imag.* 2.

Karl, T. J. (1982). *Was ist Erziehung?* Beijing: SDX Joint Publishing Company, p. 178.

Lin, L. and Wu, T. (2021). Future information creation education — On the importance of philosophical basic quality education. In *The 3rd International Summit on Information Research & the 4th International Conference on Philosophy of Information*, China Xi'an Branch, September.

Lan, Y. and Kevin, W. (2018). If memory can be transplanted. *Law and Life* 20.

Mortimer, J. (2014). Adler, *Philosophy is Everyone's Business* (translated by Qinghua X., and Sheng, X). Heilongjiang: Northern Literature and Art Publishing House, p. 1.

Nick, B. (2003). *The Transhumanist FAQ*, WTA, p. 4.

Shuyue, Y., Xiang, L., Gongjing, Y., and Jian, S. (2019). Development and prospect of brain-computer interface technology. *Comput. Measure. & Cont.* 10.

Tianqi, W. (2020). Philosophy, philosophical spirit and future education: Reflections on the development of artificial intelligence. *J. Changsha Univer. Sci. Technol. (Social Science Edition)* 3.

Wu, K. (2019). *Philosophy of Information: Theory, System and Method*, Beijing: Commercial Press, p. 125.

Xianfeng, G. (2016). Historical review and development status of artificial intelligence. *Chin. J. Nat.* 36(3). [In this paper, the author described the perceptron as the first neural network defined accurately by algorithm, and the first mathematical model with self-organize learning ability.]

Xiaojuan, Q. and Yanjie, B. (2017). A comparative study on college students' performance between urban and rural areas. *China Youth Study* 3.

Chapter 19

Holographic Thinking of Social Culture from the Perspective of Lost Book Compilation and Calligraphy Inheritance*

Tianqi Wu[†,‡,§], Ruiyuan Zhang[†,‡,¶], and Peiyuan Wang[†,‡]

[†]*Department of Philosophy, School of Humanities and Social Sciences, Xi'an Jiaotong University, Xi'an 710049, China*
[‡]*International Research Center for Philosophy of Information, Xi'an Jiaotong University, Xi'an 710049, China*
[§]*tianqi1262016@126.com*
[¶]*Jolyne14@163.com*

Lost Book Compilation (Ji Yi Theory) depends on information collation and summary. There have been some holographic phenomena in the process of natural evolution, and such holographic phenomena are reflected in the process of cultural evolution in human society. A classic is the embodiment of the spirit of that era, in which thoughts

---

*This chapter was previously published in part in: Tianqi Wu, T. (2021). "Holographic Thinking of Social Culture from the Perspective of Lost Book Compilation", *Journal of Xi'an Jiaotong University (Social Science Edition)*, No. 4, 2015. The latest altered English version was published in: Tianqi Wu, "Holographic Thinking of Social Culture from the Perspective of Lost Book Compilation", The 3rd International Information Research Summit & 4th International Conference on Philosophy of Information, (Japan, Web Conference) Xi'an, China, September 2021. Revised and expanded here.

and discourses bear the brand of that era and have certain correlation and reference with the historical documents of the predecessors of the contemporaries or their successors, which will inevitably reflect the holographic characteristics of a certain social culture. Whether in nature or in social culture, the phenomenon of holographic incompleteness is widespread, and the holographic phenomenon in social culture is reflected in the mutual correspondence, representation, and mapping between books and books, between books and society, and between books and times. The development process of calligraphy style can also be explained as a holographic relationship. It is not only a holography of calligraphy art itself but also a holography of calligraphy aesthetic style, which is also the expression that Chinese culture attaches importance to information thinking in calligraphy art. According to the analysis of the brushwork and aesthetic style in Dunhuang posthumous papers calligraphy in Wei-Jin Dynasties, the significance of calligraphy can be highlighted in its differences and uncertainties. The holographic nature of social culture is reflected in both horizontal and vertical aspects, that is, the holographic nature of contemporary thoughts and the inheritance and development of historical thought.

1 Ji Yi Theory and Chinese Civilization

Chinese Ji Yi Theory is a discipline with Chinese characteristics; there is no similar discipline in the West, probably the most similar discipline is history or archaeology. The Chinese nation is known for highly respecting and cherishing its culture. Chinese civilization is not only one of the earliest civilizations of mankind but also the only human civilization in the world that has continued its cultural development since its inception. As a unique discipline with a long history in China, Ji Yi Theory is proof that the Chinese nation has a strong demand to inherit its own culture, ideology, history, and civilization.

Ji Yi Theory appeared in the Song Dynasty, which is a subdiscipline of philology. The emergence of Ji Yi Theory is closely related to the serious loss of ancient documents. The contribution of Ji Yi Theory is that it can collect and sort out the lost literature materials preserved in other surviving documents in the form of citations so that the lost books and documents can be restored or partially restored. Many ancient books have reappeared in the compilation. For example, according to *Hanshu-Yiwenzhi* records, *Mozi* originally had 71 articles, but only two versions have survived: *Dao Zang* (contains only 53 articles), and the *Si Ku Quan Shu* (contains only 63 articles). The *Si Ku Quan Shu Concise Catalogue* also mentioned that "The original book has 71 articles, but now 8 have been lost". At the end of the Qing Dynasty, Sun Zhirang (1848–1908) gathered the sentences quoted from *Mozi* in the books of previous dynasties rather than the fragments contained in the biography and formed the *Mozi Yi Wen*, attached to the *Mozi Jian Fang*. Therefore, many books and documents that seem lost have actually been preserved or partially preserved in other surviving documents.

From the perspective of information philosophy, documentary materials carry textual information, which can be carried and preserved by various carriers, such as the earliest oracle bones, stone inscriptions, cloth, silk, paper, and so on. The same can be done in different ways directly or indirectly. For example, texts and ideas can be preserved directly in the form of books or indirectly in other materials such as quotations and paraphrases. The multicarrier and multimediated nature of information dissemination and record-keeping allows for the existence of information in multiple dimensions, thus allowing for holographic nature to be reflected in social culture as well.

In fact, a classic work is also the embodiment and carrier of the spirit of its era, and the thoughts and discourses in it all carry the imprint of that era. The holographic nature is reflected in the mutual correspondence and mapping between books and books, books and society, and books and times. It is an obvious fact that books and documents of a certain era will always quote and summarize the ideas and contents of other books and documents however related and will also reflect the cultural ideas and ways of understanding that era from certain specific angles and ways. For example, during the Warring States period, Mencius's Confucianism (372–289 BC) fought against the ideas of his contemporaries Mozi and Yang Zhu (about 395–335 BC) (Chen and Lu, 2013). The relevant ideas of

Mozi and Yang Zhu must be introduced and mapped in the related writings of Mencius. Mencius mentions in *Mencius: Jin Xin Shang*:

> Yang Zhu advocated that everything should be done for himself, and he was unwilling to pluck a single hair of his own even if it would be beneficial to the whole world. Mozi advocated love and harmony, and he would do it even if his head was bald and his heels were broken, as long as it was beneficial to the whole world (Fang, 2003).

Every idea has its origin and inheritance, a lineage of ideas has its development and changes, but it will certainly retain a lot of information about the previous ideas. Therefore, human thought, both vertically and horizontally, profoundly reflects a holographic characteristic. Since ancient times, Chinese writing habits have focused on annotation and quotation so that the textual information of many books can be partially preserved in their original form in other books. With this holographic nature of books, the society, and times, Ji Yi Theory is able to dig out the information of the lost books from the handed down books and restore or partially restore many books by excluding the interference of other information. This also echoes the holographic characteristic of human society and culture laterally.

2 The Characteristics of Information and the Emergence of Ji Yi Theory

It is one of the characteristics of information that information is lost and dissipated at any time when it condenses in the passage of time. Not only have many links in the evolution of the universe ceased to exist, but in the evolution of the biological world alone, too many primitive species have become extinct, and only a very small fraction remains, and the extinction phenomenon continues as the Earth's environment changes. This extinction of biological species can also be seen as the extinction of certain types of biological genetic code information, which is the phenomenon of information dissipation embodied in the biological world. Likewise, this phenomenon is also reflected in social documentation information.

Although geological changes, species extinction, text loss, and information loss are problems studied by different disciplines, respectively, belonging to geology, biology, literature, and communication, which all conform to the properties of the part mapping the whole, intermittent insinuating process, and individual mapping category in holography. The reason is that the research contents related to these disciplines are all related to the change of material existence form and information dissipation, and all depend on the properties of matter and information. In Kun Wu's Information Philosophy, 10 properties of information are mentioned (Wu, 2005). Information as indirect existence has absolute dependence on direct existence, matter is the carrier of information, and matter as direct existence is always in the process of eternal movement and transformation, so what comes from the change of structural mode of information carrier is the change, blurring, and dissipation of information carried by the carrier. Then finally we can deduce that the dissipation of information is eternal and inevitable.

From the point of view of philology, the loss of ancient books is quite common. The Song Dynasty historian Zheng Qiao (1104–1162) once said the following:

> There are too many lost books from the Sui and Tang dynasties, especially the older the era, the more serious the loss (Wang, 2010).

The Yi Wen Zhi was a unique category of book in ancient China, similar to the summary catalog of books currently. The Chinese literati were keen to preserve as much of the valuable cultural heritage of Chinese civilization as possible through the compilation of such historical documents at an early age. Ouyang Xiu, the northern Song dynasty statesman and literary scholar (1007–1072) mentioned in his compilation of *The New Tang Book: Preface to the History of Art and Literature* the following:

> In the history of China, the era with the largest collection of books should be regarded as Kaiyuan period of the Tang Dynasty. At that time, the official collection of books was 53,915 volumes, and the private scholars also collected 28,469 volumes. My goodness, the number of books is really huge! ...... But by the time the books are revised now, the books with

names in the catalog but which are actually no longer found account for 50 to 60 percent of the catalog. Isn't that a great pity (Sheng, 1983)!

Ouyang Xiu lived only three hundred years after the first year of the Tang Dynasty (713), and there were 3,277 books and 53,915 volumes recorded during the Kaiyuan period of the Tang Dynasty. Fifty to sixty percent of the books were already lost when Ouyang Xiu compiled the *The New Tang Book: the History of Art and Literature*, and nearly a thousand years after the Northern Song Dynasty, there were even more books lost in disasters.

Nie Chongqi, a Song historian and cataloguer (1903–1962), also mentions the following in the *Preface to the Twenty Quotations from the Yi Wen Zhi*:

> Summarizing the catalogs of the canonical books included in the 20 kinds of *Yi Wen Zhi* mentioned above, from the pre-Qin period to the end of the Qing Dynasty, there are more than 40,000 books with recorded titles and identifiable previous existence, but I am afraid that even less than half of the books that can still be found today in their original books (Wang, 2002).

Ma Duanlin, a beginner in the late Song and Yuan dynasties (1254–1340), also said the following in his *Wen Xian Tong Kao: Self-Preface*:

> The Chinese Han, Sui, Tang and Song dynasties all had their own *Yi Wen Zhi*. However, if we refer to the bibliographies recorded in the *Han Dynasty Yi Wen Zhi* and then compare the bibliographies in the *Sui Dynasty Yi Wen Zhi*, 60% to 70% of the books can no longer be found; if we then compare and refer to the *Song Dynasty Yi Wen Zhi*, a similar phenomenon will occur (Ma, 1986).

If we examine *Han Shu Yi Wen Zhi* which was published earlier, the percentage of lost books is even greater. The *Han Shu Yi Wen Zhi* was compiled from Liu Xin's *Qi Lue*, which reflected the situation of book collecting in the Western Han period. In his *Han Shu Yi Wen Zhi Jiang Shu*, Gu Shi (1878–1956), a philologist and scholar of various schools, marked each kind of book as "exist", "loss", "remain", and "doubt". Of the 596 books recorded in the *Han Shu Yi Wen Zhi* (there are actually 603 books recorded in *Qi Lue*), only 29 (less than 5%) are clearly marked as "exist" by Gu Shi and 43 "remain", totaling only 72 (12%). Under such circumstances, the various works included in the *Han Shu Yi Wen Zhi* are less than 10% today (Gu, 2009).

There are many reasons for the serious loss of literature. In ancient times, the dissemination level of written information records was limited. Before the appearance of printing, the dissemination of books mainly relied on hand-copying. In the early days, only bamboo slips were carved, and even bamboo slip materials needed to be burned by hand. In short, it was very time-consuming to read at that time. In addition, the document preservation tools at that time were rough and easy to destroy. For these reasons, it was more difficult to preserve books at that time but easy to lose books. There are many reasons for the loss of books, and there are many research results in historical materials and philology. For example, the theory of misfortuned books, the theory of natural elimination, and so on. However, apart from human factors, the dissipation of information is an inevitable process. In fact, the culture in the world experiences the same process. Even many civilizations such as Maya and ancient Egypt have been lost with the passage of time. The external environment is a filter of information retention and dissipation for written books. Only a small part of books can be preserved after being screened by external factors such as war, government, and readers.

However, the loss is relative, as many extinct natural species and natural history periods long ago, which will partially leave traces of existence in the form of fossils and stratigraphic structures. Just as many seemingly extinct natural information has been partially preserved through other ways, ancient civilizations can still leave many traces of their existence and information after they have been lost. For example, in terms of architecture, ancient Egypt left behind magnificent pyramids and sphinxes in the desert, the Maya left behind mysterious pyramids and temples in the primitive jungle, Stonehenge on Easter Island, and so on, as well as a considerable number of ancient written symbols and frescoes and sculptures rich in artistic and cultural heritage, and so on. Similarly, a preserved book may also contain information or part of the information of a book that has been lost. Thus, the holographic nature of information in the natural evolutionary process is also reflected in the social and cultural process.

The reason why certain information of nature and society can be retained is that information has a unique nature different from that of matter, which is the shareability of information. The shareability of information determines the reproducible and transmittable nature of information. Matter is conserved and has no way to share, so the exchange and reaction of mass or energy are one-time, and the sender and receiver cannot have the exchanged or reacted matter at the same time. Information is different, however, the sender and receiver in information reproduction and dissemination have the same information at the same time (although there will be corresponding information loss, distortion, and innovation in the process of information reaction), and the carrier of information is replaceable, and different forms of material existence can carry the same information. These characteristics of information create conditions for multilevel and multichannel preservation and dissemination of information.

There are many differences between properties of information and material so that the objective world has a holographic nature, and Ji Yi Theory relies on these properties of information to become possible, and the previously mentioned disciplines of geology, fossil science, biology, communication, and so on, also take advantage of these special properties of information.

3 Holographic Incompleteness in Ji Yi Theory

Holography is not absolute, and it should not be superstitious. The phenomenon of holographic incompleteness is universal. Holography is based on the interaction and spatio-temporal transformation of things, and historical information is preserved as a kind of "trace" of the passing time, and the spatial structure that disappeared during the evolutionary process is partially stored in the way of integration and reconstruction in the subsequent generation of structures (Wu, 2010). However, this preservation and storage is relative and partial, and the reason why it is partial is that information is being lost every moment. With the passage of time dimension, information is dissipated, and "traces" are blurred, and as subsequent interactions occur and spatio-temporal transformations take place, passing and distortion are inevitable. The second law of thermodynamics shows that the trend of entropy increase in our world determines that the general direction of information loss is irreversible. Therefore, in this world, it is impossible to preserve 100% information in this world.

Obviously, the information mining of lost books by Ji Yi Theory is partial and relative. This is not only because the natural dissipation of information is inevitable, but also because of the limitations of human interpretative capability. It is an obvious fact that the interpretation of information is dependent on human interpretive capability. Although human interpretive capability is advancing, the limited human progress is too small and powerless compared to the infinite information dimension of nature, and the way of interpretation varies from one period of human history to another, which performs especially obvious in social culture. It is difficult for modern people to decipher many ancient scripts, and the meaning of ancient scripts is also very different from the present ones, so it is normal for some distortions to occur in the interpretation, or even the inability to explain them, and the phenomenon of multiple interpretations is also very common. In addition, the language itself has its own limitations, and the expression of ideas is full of incompleteness.

In summary, we can see that Ji Yi Theory is by no means omnipotent; it takes advantage of the holographic nature of the world, but the holographic incompleteness inferred that Ji Yi Theory cannot recover all the lost cultural information. It is also a fact that not only the cultural texts recovered by Ji Yi Theory are limited, but also there exist many errors and omissions. This is why Chen Zhiping (1973–), a professor at the School of Arts of Jinan University, believes that Ji Yi Theory needs the following:

> In order to verify the authenticity of the collected documents through the identification of forgeries, it is necessary to examine the similarities and differences of the collected words and the right and wrong through editions and proofreading (Chen, 2010).

Xu Shaosong's *Shu Hua Shu Lu Jie Ti*, Volume 10, *San Yi: Shu Bu*, collected the "Theory of Bi Xin" and "Theory of Shu Duan" in the brief list of bibliography of ancient lost books, which were noted as "doubt". These doubtful compilation results are either intentionally fabricated by the compilers, or the sources of the documents are unreliable, and many compilation results are found to have problems in later generations. For example, the newfound of ancient books and documents may have conflicts with the narrative of other documents.

Cultural falsification has existed since ancient times. The ancients often relied on the Lost script of relevant books for counterfeiting. If the relevant Lost script can be Ji Yi, it is easy to discover how the counterfeiters could improperly increase and reduce them, for example, this is what our predecessors did to distinguish the authenticity of *Gu Wen Shang Shu* (Sun, 2006). Therefore, in the compilation of lost books, both restoration and falsification can be distinguished. From the perspective of information activities, this kind of forgery can be regarded as man-made information noise. In nature, information noise is caused by the non-stop interaction between natural things, while, in human culture, it is caused by man-made distortion in meaning, or deliberately distortion. This artificial information noise will more easily distort the original information and aggravate the degree of holographic incompleteness. However, some counterfeiting can be identified, which also reflects the holographic characteristics of social culture.

4 The Embodiment of Holography in Wei-Jin Dunhuang Posthumous Papers (Wang and Wu, 2021)

Serial relation holography means that everything maps and defines information on its own history, present, and future. This serial holographic nature is reflected in the historical development of Chinese calligraphy. As the most condensed physical form of traditional Chinese culture, the original form of calligraphy, the oracle bone script, was used by ancient ancestors to perform divination on animal bones and it was considered to be a brand of divine transmission of information, and the origin of calligraphy is related to information. This is consistent with Mr. Yang Weiguo's statement that "Chinese culture values information thinking" (Wang, 2002).

Buddhism originated from India and was introduced into China during the Western Han Dynasty. In the Wei-Jin period, it was widespread and became the "state religion" for a time. Buddhism in the Han Dynasty opened up a key road of cultural and economic exchanges between China and the West, mainly through the Silk Road and economic exchanges. Northwest China became the first point of contact with western Buddhism and its introduction to the Central Plains, forming an early Buddhist gathering and exchange place in the Dunhuang area.

In the Dunhuang posthumous papers, there is a high number of calligraphic works from the Wei-Jin period. From these calligraphic works, we can clearly see that this period of an official script calligraphic style is in a period of transition to a regular script. Bamboo slips from the Han dynasty are shown and compared to the calligraphy style of the period, the similarities lie in the inheritance of history, reflected in the overall layout of chapters and characters, and there are differences between them because of the different styles of times.

The faith worships of Buddhism form the psychological basis of the essence of writing sutra. This mental foundation allows the believer to write with a serious attitude and a devout conviction. Therefore, the aesthetic orientation is plain and forceful. In the early period of copying classics, the brushstroke style of bamboo slips was followed, which gave the Dunhuang handwritten notes of this period their ancient meaning. This led to the preservation of the li script of the Han Dynasty in the process of the evolution of calligraphy, forming a vertical historical holography. The influence of Buddhist thought on calligraphy at that time was a horizontal hologram.

The Dunhuang posthumous papers are mainly regular script volumes. The authors of the Dunhuang posthumous papers typically chose a regular script when copying Buddhist sutras. The existence of the scribe style is the most distinctive feature of Dunhuang posthumous papers, that is, the historical relationship of the pen in the works of Dunhuang posthumous papers in the Wei-Jin period. The process of cursive script and official script gradually evolved into regular script is holographic performance. The fading of the meaning of the official script is the concrete performance of the historical holography. The formation of the regular script is a holographic performance. In the same period, the performance of different styles of calligraphy is the complementary and different performance of horizontal holography.

The plain aesthetic style is mainly inherited from the previous style of writing, which manifested as the holography of historical relations in the series of relation holography. The aesthetic features of this kind of brush are pointed into the brush, gradually increasing from thin to thick, with a brush posture of a thin head and thick tail. These features are inherited from the plain aesthetics of the official script, which is reflected in the holography of historical relations. As a horizontal Dunhuang posthumous paper, it is listed as the aesthetic orientation of the same era as other styles.

Regarding the influence of Buddhist thoughts on the aesthetics of calligraphy, the writing style of Dunhuang posthumous papers in the Wei-Jin period had begun to show the influence of Buddhist thoughts on aesthetics. On the basis of metaphysical ontology, this paper combines "the agreement between god and things" with "no doubt, no change, no change" to form the middle-way consciousness, which is in line with the concept of "dependent origination and the emptiness of nature". This is the holographic expression of calligraphy aesthetics under the influence of Buddhist thought.

5 Socio-Cultural Holography

The term holography originally referred specifically to a technology that allows diffracted light emitted from an object to be reproduced in exactly the same position and size as before. The image displayed changes when the object is viewed from a different location. Thus, the picture taken by this technique is three-dimensional. The meaning of holography in information philosophy is that things map and condense in multiple and complex information relations and contents beyond their own existing nature in their own structure. The holographic phenomenon is the result of the condensation and accumulation of related information that may be achieved by the evolution of complex self-organization. In addition, the true meaning of fractal and chaotic phenomena can also be explained at the level of holographic phenomena, whether it is self-similarity across different levels, or chaotic order in acyclic chaos is some kind of holographic phenomena. The manifestation of holographic phenomena (Wu, 2010).

In the theory of complex information systems, holography is from the perspective of natural evolution, and the evolution of human society and culture is equally holographic. Taking Ji Yi Theory as an entry point, we can see the holographic nature of human cultural products.

Socio-cultural holography is divided into two forms: horizontal holography and vertical holography.

Horizontal holography means that contemporaneous cultural works reflect the spirit and imprint of the era in which they were produced. The social thought of any period is a reflection of the spirit of that era, a reflection of the intermingling, conflict, competition, and absorption of ideas. As mentioned earlier, the writings of Mencius include the ideas of his contemporaries Yangzi and Mozi. At the end of the Warring States period, Xunzi's (BC313–238) theory of human nature as evil was also a direct response to Mencius' theory of human nature as good. Although Xunzi originated from Confucianism, he was a fierce critic of all schools of thought at the time, including the Si-Meng school of Confucianism. Since the ideas of the same period could only dig into the shortcomings of each other's ideas in the struggle, they would only know themselves and each other and could come up with a self-contained academic culture. Therefore, the cultures of the same period were mapping, characterizing, contrasting, competing, and intermingling with each other.

Holography exists not only in the same school of thought but also in very different or opposing ideas because contemporaneous ideas are pulsed and interconnected and inseparable. For example, during the period of the Hundred Schools of Thought, it was the emergence and promotion of worldly ideas such as Confucianism and

Legalism that led to the emergence of worldly ideas such as Taoism. It is the rapid industrial development of the present-day West guided by anthropocentrism that corresponds to the emergence of non-anthropocentrism, ecocentrism, and other ideas. When the cultural thought of an era is taken to an extreme it inevitably gives rise to opposing cultural ideas, the emergence of these opposing cultural ideas may be due to mankind's own cultural reflection, or perhaps due to the negative effects of the reality caused by the previous extreme ideas. The emergence of such opposing ideas seeks to achieve a complementarity, a balance, and the holographic nature is reflected in these complementary competing ideas.

In addition, the holographic nature of social thought is also reflected in the intersection of cultural thought and technology. For example, in slavery and agrarian economic societies, nature was worshiped and revered because of low human productivity. Such socio-economic conditions inevitably gave rise to religious and superstitious ideas. The worship of natural objects devalued the status and value of human beings. However, after human productivity increased, nature began to be transformed by human beings, and the rapid development of science, technology, and economy in the industrial age made the value of human beings begin to manifest, and made human beings' awareness and affirmation of themselves begin to awaken, and anthropocentric ideas emerged and developed in this period. When the negative effects of the anthropocentric ideas cultivated by human beings in the industrial civilization era became more and more obvious, a new kind of non-anthropocentric ideas came into being, which is the ecological civilization trend of taking a path of harmonious development between human beings and nature that has emerged since the second half of the 20th century.

In fact, the anarchism and postmodern thoughts that emerged at the end of the 20th century were also influenced greatly by the contingency and non-deterministic uncertainty of the microcosm revealed by quantum mechanics, the emergent nature of the nature of things, the irreversibility of the time process, the fractal nature of the spatial structure, and the bifurcation and chaos of the direction and outcome of the evolution of things revealed by the theory of complex information systems. In summary, this mutual mapping, correlation, and influence among culture, society, and technology all reflect a socio-cultural holographic nature.

Vertical holography means that all kinds of thoughts in history have rheology and inheritance. It is precisely this diachronic nature that means that human thought can develop. As mentioned above, every era creates ideas that contain the brand of that era, but no matter how an idea is branded by the era, it has its own roots and veins, and the cultural and historical origin of these ideas can be found. For example, in the earliest pursuit of the origin of the world, there were many reveries or speculations about the origin of fire, water, and soil in various ancient civilizations. Democritus first put forward atomism, and modern physics has been pursuing the smallest particle that makes up the world: "the brick of the universe". In fact, this vitality has been reflected since the establishment of various theories and indicates the future development of this kind of thinking. From this, it can also be seen that the vertical holography in human social culture is not only for history but also for the future.

In fact, not only ideas, including the development of the nation, will depend on the national spirit of the nation, at the beginning of the development, it can be predicted whether the nation will stand among the various nationalities or be compatible with the assimilation of other peoples. The vertical holographic nature of cultural thought is such that it is possible to look back at the history of the origin of ideas and at the same time predict the direction of their development: either to survive and prosper or to eliminate and die out.

References

Chen, X. K. and Lu, J. H. (2013). New knowledge of mencius: Confucianism in battle. *Nankai J. (Philos. Soc. Sci.)*. 3.
Chen Z. P. (2010). *Epigraphy Historiography*. People's Fine Arts Publishing House, p. 251.
Fang, Y. (2003). *Mencius*. China Bookstore, p. 271.

Gu, S. (2009). *Lectures on the Art and Literature of the Book of Han*. Shanghai Ancient Books Publishing House.

Ma, D. L. (1986). *Documentary General Examination*. China Bookstore. p. 1.

Sheng, W. (1983). The collation of ancient books in the Northern Song Dynasty, *Month. J. Hist.* 3.

Sun, Q. S. (2006). *Chinese Literature*. Peking University Press, p. 5.

Wu, K. (2005). *Information Philosophy: Theory, System, Method*. The Commercial Press, pp. 65–67.

Wang, X. L. (2010). A brief discussion of Zheng Qiao's theory of documentary cataloging and pseudonymity, *J. Jianghan Univer. (Humanities Edition)*. 3.

Wang, Y. G. (2002). A study of the supplementary bibliography of historical records since the Qing Dynasty, *Library Stud.* 3.

Wang, P. and Wu, T. (2021). A study on the holographic value of calligraphy inheritance-taking dunhuang posthumous (Published in English). Paper of Wei-Jin Period as an Example. *Paper for the 3rd International Conference on Information Studies & 4th International Conference on Philosophy of Information* (Japan, Web Conference). Xi'an, China, September 2021. (Some deletions have been made here.)

Wang, Y. J. (2002). *Calligraphy and Culture of the Six Dynasties*. Shanghai Calligraphy and Painting Press, p. 260.

Wu, K. (2010). *Foundations of Complex Information Systems Theory*. Xi'an Jiaotong University Press, pp. 274–275.

Wu, K. (2010). *Fundamentals of Complex Information Systems Theory*. Xi'an Jiaotong University Press, p. 273.

Chapter 20

Reflecting on the Limitations of Today's Artificial Intelligence from Information Innovation Systems

Tianqi Wu[*,‡], Ruiyuan Zhang[*,†,§], and Ruiqi Jin[†]

[*]Department of Philosophy, School of Humanities and Social Sciences, Xi'an Jiaotong University, Xi'an 710049, China

[†]International Research Center for Philosophy of Information, Xi'an Jiaotong University, Xi'an 710049, China

[‡]tianqui1262016@126.com

[§]Jolyne14@163.com

The realization of artificial intelligence has been one of the goals in the field of machine learning, and the achievements in the field of deep learning have prompted the emergence of a view that the ultimate goal of so-called "intelligence" in artificial intelligence can be achieved by means of deep learning. From the perspective of information philosophy, intelligence should arise from the information activity of the "subject", which is the product of the emergence of dynamic self-organizing systems, while today's artificial intelligence of other-organized nature cannot really emerge from information and intelligence. This chapter compares current approaches to deep neural networks and deep learning with self-organizing information activity systems to point out the limitations of deep learning to achieve artificial "intelligence". Information philosophy proposes the idea of information-generating systems to describe what kind of information-generating capabilities true intelligence should have. Considering the functions of nine of these subsystems, artificial intelligence has complete or simple functions of some of these subsystems, but still seriously lacks the capabilities of three of them.

In the 1990s, artificial intelligence was proposed in the form of multilayer perceptrons but was temporarily shelved due to limitations such as the lack of computing power of computers at that time. In recent years, with the improvement of computing power capabilities, deep neural networks, and deep learning have re-entered the vision as a machine learning approach. Due to the complexity of its algorithm, deep learning has performed well in various fields of data processing, and the achievements in digital image processing, natural language processing, and other fields have enabled machines to "recognize images" and "process languages" like humans. Many of these achievements have been commercialized and civilianized and have largely changed the lives of people today (Wang and Raj, 2017). Nowadays, deep neural networks and deep learning algorithms have become one of the most popular and most advanced directions in artificial intelligence research.

1 The Essence of "Intelligence" as Revealed by Information Philosophy

There are indications that the rapid development of deep learning algorithms seems to be approaching the original and ultimate goal in the field of machine learning — artificial intelligence, i.e., "making machines as intelligent and capable of thinking as humans". From the increasing number of discussions on AI in recent years, it can be seen that a considerable number of people hold the view that the development of algorithms and computing power can make machines generate human-like intelligence or surpass human intelligence. Based on such views, some people have even started to predict and discuss the future challenges and threats of AI to the status of human intelligence. However, a consideration of the nature of intelligence is essential if we are to truly attempt to achieve human-like intelligence. Chinese information philosophy provides a perspective on the nature of human intelligence, from which it can be seen that there is a huge gap between the most advanced deep learning algorithms in current AI technology and true subject intelligence.

According to information philosophy, intelligence is the agential way and method of grasping, processing, creating, developing, utilizing, and fulfilling information by a subject who has the capability to know or practice. In other words, intelligence is the information activity of the agential subject, and if intelligence similar to that of the subject is to be realized, information activity similar to that of the subject is required (Wu, *et al.*, 2021, pp. 160–177).

Since humans are the highest level of subject form in the development of life on Earth, we generally deal with artificial intelligence through the study of the phenomenon of human intelligence. As mentioned earlier, because intelligence is an information activity, it is necessary to explore human information activity first. Human information activity has a bottom-up hierarchical progression: human cognition, as the higher form of information, evolves layer by layer starting from low-level information. The low-level information-in-itself in nature enters the human neural system, generates sensation and perception, and creates human information-for-itself, which is stored in the memory. On the basis of these memories, the agential subject creates information and generates renewed information and then feeds back this renewed information to nature and society through social practice and thus, externalizes it, thereby achieving the unification of information-in-itself, information-for-itself, and renewed social information pattern as a new three-state information (Wu, *et al.*, 2021, pp. 160–177). Human information activity has top-down guidance and inhibition: as the higher information activity starts from its own purpose, it regulates, evaluates, and guides the direction and intensity of the lower information activity, strengthens those parts that are consistent with its own purpose and needs, and inhibits those aspects that are inconsistent. Therefore, intelligence in the true sense must have both "intelligence" and "ability". The purpose of "intelligence" is to receive, store, interpret, process, and create information, while the purpose of "ability" is to achieve and give feedback on new information patterns, and the dual mechanism of orientation and inhibition of information activities. The achievement and feedback of new information patterns are based on correspondence and matching with objects, cognition, and understanding but not simply on mechanical calculation and reaction.

Through bottom-up and top-down synthesis, human information activities at all levels become an organic whole, and the interaction of information at all levels makes human intelligence a complex system with both evolutionary characteristics and internal randomness and the emergent nature of a dynamic self-organizing system so that it can improve itself through the continuous creation of information and continuous self-condensation, and then it can emerge as an intelligence-like phenomenon.

2 The Static Nature of Computer Neural Networks

There are many terms in the field of machine learning that are similar to those used to describe human intelligence, implying the similarity of existing artificial intelligence to human intelligence. For example, it is often assumed that deep neural networks are also "evolutionary" in that they are able to improve their parameters during the process of machine learning. This so-called evolution, however, is fundamentally different from the evolutionary nature of human information activity systems as complex systems. The essence of this is that the "parameters" improved by neural networks are very limited, affecting only a small part of the system.

In machine learning terminology, parameters represent values that will be changed by the system itself during training by the algorithm, as opposed to the concept of hyperparameters, the values of which have been determined before learning begins. The performance of a neural network, for example, often depends on the values of the hyperparameters, which is an important part of the measure of neural network design (Zhou, 2016). One class of hyperparameters determines the shape and size of the network input matrix, the number of layers of the network, the size and shape of the convolutional kernel, and the number of clusters, which basically corresponds to the number of parameters. These hyperparameters determine the size of the allocated memory space when the program is running. In other words, they determine the spatial structure of the neural network statically. Another class of hyperparameters, the most representative ones, include the learning rate, the number of learning rounds, and the threshold of the loss function that determines when the learning process will stop. These hyperparameters determine the performance of the network at different times of training, as well as the time needed for network training.

A dynamically self-organizing system in the process of evolution, the time it experiences is transformed into the structure of space, and the time that passed is stored in the subsequent spatial structure, thus enabling the system to generate new properties (Wu et al., 2021, p. 215). However, in the training process of artificial intelligence neural networks, the hyperparameters that represent the spatial structure are designed to have "other-organized" properties and remain static throughout the process. The length of time experienced by the network training also depends on the temporal hyperparameters that are pre-designed and fixed at the beginning, instead of depending on the initial spatial structure of the network. In neural networks, these parameters which represent time and space cannot be changed subsequently by the so-called machine "learning" and are static. Consequently, the neural network structure lacks the necessary evolutionary and self-organizing properties.

Moreover, as a further consequence of this, neural networks of artificial intelligence lack the emergent quality which is necessary to generate intelligence from the system. For a complex self-organizing system, emergence is the process by which the interaction between the constituent elements of the self-organizing system generates new properties for the whole from the bottom up. Although the training of a neural network yields products that appear to be emergent, unpredictable, and unexplained, this unpredictability is essentially the result of algorithmic deduction rather than emergent new properties.

Modern artificial intelligence systems have some degree of complexity and are incorrectly considered as exhibiting emergence by engineers because the products of their arithmetic have become unpredictable and even unexplainable (untraceable). However, this phenomenon is simply the result of the limitations of people's understanding of the results of the arithmetic they have designed themselves. The truth is that no AI can make a decision in the true sense of the word. A decision is the result of a well-thought-out consideration of reasons; it is not purely derived or arithmetically computed from logic. The judgment of a human being as an intelligent subject is an emergent action, and emergence is characterized by the transcendence of the original thing, the creation of something new with a new nature. This explains why different intelligent subjects or the same intelligent subject at different periods of time have different and unique judgments and decisions on the same problem. The reason lies in the differences and randomness of the emergent new nature.

According to the way AI is developed nowadays, no AI with the so-called autonomous technological systems can generate emergent mechanisms. All potential information processes ultimately involve deterministic mechanisms, each step of the algorithmic process follows a definite rule. Therefore, the decisions claimed by engineers in the AI state can actually be the product of mechanical determinism rather than self-thinking. It can also be deduced that an AI under the same algorithm will only have one judgment and decision for the same problem.

Prof. Rafael Capurro from the Media University in Stuttgart, Germany, and Prof. Wolfgang Hofkirchner also think that AI is not a self-organizing system at the present stage. It is heterogeneously and externally organized. Such an AI cannot have the autonomy of an intelligent body. It cannot deal with the problem of an individual's own decision-making. It cannot serve as a controller as an intelligent subject would do, but would only be a controlled one (Capurro, 2012). According to the contemporary concept of algorithm-based AI design, AI will not have the nature of emergent information no matter whether it is weak or strong (Hofkirchner, 2011).

In summary, the fixed hyperparameters in the neural network, i.e., the static, unchanging overall structure and the essentially deterministic and designed algorithms prevent the individual constituent parts of the neural network from changing the whole to generate new properties. Thus, the neural network itself cannot achieve the entire system of information activities characteristic of an intelligent subject, and intelligence does not emerge from the artificially intelligent neural network that lacks self-organization.

3 What Perceptron Achieves is not Perception

The artificial intelligence terminology has an anthropomorphic nature: the so-called self-perceptron designation is still only a metaphorical reference to human sensations and perceptions. The design of multi-layer perceptual machines originates from neural network theory, which claims that the process it implements is an imitation of human perception, i.e., an imitation of the intuitive recognition process of information by the knowing subject. The object of the intuitive recognition process of information is the information-in-itself. Information-in-itself is the information assimilated and alienated by matter itself at the lower level of interaction, which occurs through the information field generated by matter itself in the form of light, sound, heat, electricity, chemistry, etc. If a process is meant to imitate the process of intuitive recognition of information, the input of this process should also be through information-in-itself, or information fields external to the system. However, the basis of a perceptron is mechanically programmed, and when external information-in-itself is fed into the perceptron system, it is recorded as digital signal information. Afterward, it performs a procedural classification of this data information. However, the basis of this classification is procedural and is only a carrier transformation of the information-in-itself, which is similar to the reflection of light and shadow on the surface of a lake or a mirror. It is completely different from the mechanism by which a subject constructs information-for-itself and renewed information and points to external things.

For example, when a picture of a person is taken, the light emitted and reflected by the person and the surrounding environment passes through the camera lens and reaches the camera's sensor; the sensor obtains discrete samples of this light according to its own resolution; if this is a 24-bit photochrome, the natural light reaching each sampling point will be divided into red, green, and blue light, and the brightness will be recorded separately. Then, these brightness values are discretized as integers from 0 to 255. Eventually, the brightness information about the whole photographic environment is sampled and encoded into some integers from 0 to 255. If these integers are arranged according to a pre-agreed number of channels and resolution and displayed with a specific light intensity, then the information input system of the intelligent subject can receive the information carried by the light and use it, for example, to identify the gender of the person whose photo was taken by sampling and recording the reflected light in the environment (Szeliski, 2005, 2012). However, if these integers are fed into a perceptron for training it and then for utilizing these data to determine the gender of the person as a material entity, all that the perceptron gets is the recoded data information. From the point of view of information philosophy, the photographic reconstruction of this scenario is essentially the reproduction, transmission, and recoding of information-in-itself. According to the conventions of photographic techniques, sampling, resolution, and color channels, the data on the photograph or the variability of the reflected light wavelengths presented constitute a virtual reconstruction of the original scenario information.

Although the perceptron is called by this name, it only imitates the process of information input, recording, and storage as carried out during the process of perception of intelligent subjects. For example, the information carried by external light enters the lens of a camera and is captured, sampled, and encoded by the photoreceptor, which is similar to the "sensing" process in subject perception, and this process is not self-organized by the perceptron. This is more similar to the natural information recording phenomena such as the reflection of the moon on a lake, the fossil lines in a rock formation, and the annual rings of a tree. This process is also apparently different from the human perception process in that no one believes that there is or needs to be intelligence involved in this process.

This process is neither "sensing" (sensation) nor "knowing" (perception). For the intelligent subject, sensation is accompanied by the subjective presentation and representation of individual information, while perception is about the structural integration, understanding, and interpretation of the subjective presentation and representation of individual information; and the whole process of perception is accompanied by the subjective presentation and representation of information. The most important features of perception are objectivity, wholeness, and structure. Perception cannot be regarded as a mechanical sum of individual sensations, it reflects the great dynamism of the subject. The process of perception can be considered as the process of regulation and integration of individual sensory information by sensory centers, as well as the process of synthesis and abstraction of different sensory information aspects by each sensory center during interconnection and action, which is performed not only within individual sensory systems but also in the primary contact and integrated information network of each sensory system (Wu, 2005). Perception can also be understood as an emergent phenomenon, and the corresponding information structures resulting from this emergence are accompanied by a series of internal processes of presentation, representation, cognition, and interpretation.

In summary, the perceptron simply captures, samples, encodes, and stores information, and in the process does not perform cognitive and interpretative behaviors accompanying the internal presentation and representation of any information and has no object with integral, structural, or emergent characteristics. Therefore, it is inappropriate to say that AI machines have human-like perceptual behaviors. In fact, it is the designer of the perceptron, not the perceptron itself, that understands and interprets. Therefore, the perceptual machine does not actually have perceptual behavior similar to that of an intelligent subject through "anthropomorphic" information manipulation.

4 Logical Reasoning, Grounding, and Subject Cognition of Symbolic Information

Given that perceptrons mimic the functions of capturing, sampling, encoding, and storing information, is it possible to claim that the process of artifactual complexity of computing and producing final results after obtaining information from a perceptron is equivalent to some kind of symbolic information logic reasoning performed in the human cognitive system? Although symbols are often used as synonyms of patterns in fields such as pattern recognition, in the subject's information activity, the concept of symbols needs to include the process of referring one information pattern to another. This act of referring must be based on the subject's cognition, while at the same time, it also relies on some subjectively agreed meaning, the information about which must be internalized in the information base stored in the cognitive structure of objectified perception already established by the subject. Only then can the subject understand the meaning of the symbols and proceed to the next logical derivation of the symbolic information on this basis. Associating symbols with the things they refer to requires the subject's perception to objectify and interpret these symbols. For example, the subjective agreement on the symbolic information of "sun" in Chinese requires the subject to establish a cognitive basis for the object of the sun before the agreement and also requires the cognitive basis of the Chinese language system. If a person has never seen the sun before, the meaning of the word needs to be explained in order to achieve the effect of subjective agreement on symbolic information.

The perceptron does not include the process of constructing this part of subjectively agreed information nor can it use this agreed information to correct the learning process. Therefore, from the perceptron's point of view, due to the lack of perception in the information activity process, the input and output information lacks the connotation of symbolic information, and this information cannot be associated with the object to which the information refers, thus degenerating into a simple data-information model. Consequently, the perceptron does nothing more than the computation of complex data.

Stevan Harnad, a cognitive scientist at McGill University in Canada, raised the problem of symbol grounding in the development of artificial intelligence, i.e., the problem of how symbols acquire meaning in a symbolic system. If the logical deduction of symbolic information is to be achieved, symbolic information must be available first. The system needs to contain information related to the agreed reference of the symbol, i.e., the connotation of the symbol, so that the symbol can become a symbol (Harnad, 1990). From the perspective of information philosophy, it is a question of how the information patterns used by artificial intelligence become symbolic information. The construction and creation of symbolic information, as the structure of human higher-level information activities, must be based on various information activities at lower levels, and the construction and creation of symbolic information depend on the condensation of information in the evolution of the information activity system. The subjective conventions related to the referent symbols are also closely related to the lower level of information activity, i.e., "perception". When an artificial perceptron acquires the meaning of symbolic information from its significance, the qualitative lack of perceptual information leads to the fact that this referential relation cannot be reduced from the information pattern itself.

Harnad also claims that top-down, symbolic research paths (e.g., natural language processing that starts from analyzing syntactic structures and word associations) are unable to solve the symbol grounding problem. He likens this approach to the process of having someone who does not know any language learn a language with only a dictionary written in that language (which contains the meanings of all the symbols of that language, but which are also written in that language), without any other material and real-life perceptual experience, which is almost impossible. Because the learner does not understand the real meanings of the words through the process of perception and usage of these words to interpret the phenomena that occur in the lower-level information activity system, the information patterns referred to by these word symbols do not become part of the learner's cognitive system. Thus, even if the learner remembers the associations between words, he or she cannot really grasp the connotations of these symbols.

And the only feasible path to address symbolic grounding is to study the bottom-up, constructing path of perception. Only when a realistic information system with the capacity of information-creating activities for direct information interaction with the environment by sensory, perceptual, and memory storage processes, is taken for granted, an abstract schema can rely on the conventions of the system itself to produce its referent. In other words, symbolic information creation comes from the emergence of the information subject's activity. A person is able to create new symbolic information with the help of symbolic information reasoning, and the process of symbolic information reasoning cannot be separated from other parts of the information system, or to put it more clearly, this process cannot be separated from the self-organized system of information activity.

5 Lack of Self-Organization and Emergent Properties in Machine Information Flow

Reviewing the information flow associated with AI neural networks and comparing it to the human information activity system, it becomes more evident that the information flow associated with machine learning is not systematic. From the low-level aspect of information activity, the "sensory" source of the neural network is the data sampled from various external sensors, which are often chosen arbitrarily because only the data are important to the neural network. As a result, the mutual assimilation and dissimilation of information from the sensors and the surrounding environment, and the condensation of information in these processes are not reflected in the "learning" of the neural network, and the neural network acquires only part of the static information of a particular sensor at the moment when a particular set of data is recorded. On the other hand, these sensors are often controlled by humans or placed in locations that are not relevant to the neural network, i.e., these sensors and the associated information are other-organized, and the so-called "learning results" of the neural network often do not translate into feedback and control of these sensors. Furthermore, this information is not interpreted or directly symbolically related through the "perception" process of these sensors. In this case, real "artificial intelligence" cannot emerge from the lower level of "perception" activity. Thus, the "low-level information activity" of this information flow is not systematic and emergent.

At a high level, the renewed information created by humans can directly become the purpose and plan of the subject, creating a real physical entity through the process of social practice. On the other hand, AI neural networks are not social, and their "symbolic reasoning" often stops there and cannot be fed back into the environment through social practices. The purposes of neural networks cannot be modified in the process of "learning". They often depend on the design of scientists and engineers and are expressed in the form of pre-mandated deterministic algorithms, hyperparameters, and data set labels, which are not part of the parameters that can be modified by neural networks.

In essence, the "decisions" made by these so-called "artificial intelligence" systems cannot be considered as a real decision because a real decision is far from conclusive with only preconditions but requires a kind of information emergence progress based on the purpose of the decision. In the process of "decision-making" by algorithms such as neural networks, the relevant factors and processes are other-organized and deterministic, and essentially depend on the pre-design rather than on the actual "judgment" process of the neural network, and thus, cannot produce a high-level information activity similar to that of the subject. Therefore, these "high-level information activities" also lack self-organization and emergence properties. The IEEE Global Initiative on Ethics of Autonomous and Intelligent Systems (A/IS) has released a comprehensive document on ethically consistent design, and the following statement:

> When understanding the relationship between humans and autonomous intelligent systems, it is of particular concern that the uncritical anthropomorphic approach to autonomous intelligent systems, which is now being used by many industries and policy makers. This approach wrongly blurs the distinction between ethical controllers and ethical controllees, or understands the distinction between 'natural' self-organizing systems and artificial non-self-organizing devices. As noted above, autonomous intelligent systems cannot, by definition, be as autonomous as humans or creatures. That is, the autonomy of a machine, when strictly defined, refers to how a machine can act and operate independently in a given environment by taking into account the order of execution that arises from laws and rules. In this sense, autonomous intelligent systems can be defined as autonomous, especially in the case of genetic algorithms and evolutionary strategies. However, attempts to implant true morality and emotion in autonomous intelligent systems and thus take responsibility, i.e., autonomy, would blur the distinction between the controller and the controllee and could encourage anthropomorphic expectations

of humans in designing and interacting with autonomous intelligent systems for machines (The IEEE Global Initiative on Ethics of Autonomous and Intelligent Systems, 2019).

It is clear that the deep learning approach we currently use and the information structures embodied in this approach are short of the complexity, integrality, dynamism, self-organization, and evolution that characterize information activity systems that produce human intelligence. Mechanically determined deep learning and other related design methods will not be sufficient if the original goal of AI is really to be achieved to build information systems that exhibit human-like intelligence.

6 Constructing an AI Architecture Based on Dynamic Self-Organization

Deep learning outperforms many traditional design approaches to artificial intelligence when it comes to solving specific problems. Deep learning is also useful in some areas where currently no other technology can offer any help. However, the design of intelligence is qualitatively different from simply solving a specific problem. In this section, we make certain reflections on future design directions for and approaches to AI from an information philosophy perspective.

In current design approaches to deep learning and other AI methods, the "intelligence" part of the design is often built on top of an existing computer architecture, whether it is the input information, the "decision" process (algorithms), or the overall architecture, which is built on top of the other-organizational system. For a system to be able to emerge as intelligent, it should be dynamically self-organizing. In order to go beyond the pre-determined other-organizational nature of the input information stored in the computer system, AI should abandon the traditional mechanical programming based on the passive reception of artificial information input and be able to create appropriate information patterns and program itself during the process of directly interacting with the environment.

A similar point is made in the inclusive architecture for AI design introduced by Australian roboticist Rodney A. Brooks, who states that AI should be designed in reference to intelligence or robots with independent behavioral capabilities, combining the upper algorithmic part with the lower input–output part. In his view, AI can thus solve the symbol grounding problem from the bottom up, i.e., by generating meaningful symbols from sensory information from sensors and other information in the environment (Brooks, 1999).

Of course, his design still does not allow for true artificial intelligence, because he has not abandoned the use of what is essentially a pre-designed algorithm, and an other-organized machine body with current technology. For true intelligence to emerge from the system, the AI must abandon the pre-processed data and use its "senses" to interact directly with the world's information such that symbolic information (in the true sense of "symbolic grounding"), decision-making processes, subject purposefulness, and the higher-level information activities associated with them emerge from this sensory information. In neural networks and other current approaches to AI design, these structures are static and other-organized and do not enable evolutionary and emergent processes. Intelligence cannot come from an immutable designer's predetermination, from pre-determined algorithms, predetermined hyperparameters, and pre-determined symbols; such contemporary design thinking is only a disguised condensation of the designer's intelligence in artificial intelligence, not the intelligence of the machine itself. Intelligence itself should emerge from the interaction of a self-organizing system with its environment and the process of symbol creation.

Perhaps the future of biological forms of computers, and genetic (DNA) computer research direction can allow for the ultimate breakthrough in artificial intelligence.

7 Information Creation System Model Proposed by Information Philosophy

The idea of information creation system is proposed in Kun Wu's epistemology of information philosophy. At the same time, this creation system can also be understood as an intelligent creation system. From the functions of the subsystems of this creation system, we can find the subsystems that are lacking in the existing artificial intelligence and thus need to make targeted research in the design. An information creation system is a system that generates new information by processing existing information. The work of such systems is different from the simple encoding and decoding activities in the process of information transfer and reception that are typically exhibited by algorithm-based AI. The focus of information creation is not to simply maintain, transform, reconstruct, or copy the patterns and contents of existing information, but to make innovative changes to it. According to the general characteristics of information activities, information creation is achieved through information integration and reorganization, and this process is consistent with the creative activities of recombining, selecting, matching, constructing, and virtualizing existing information (Wu and Da, 2021) (as shown in Fig. 1).

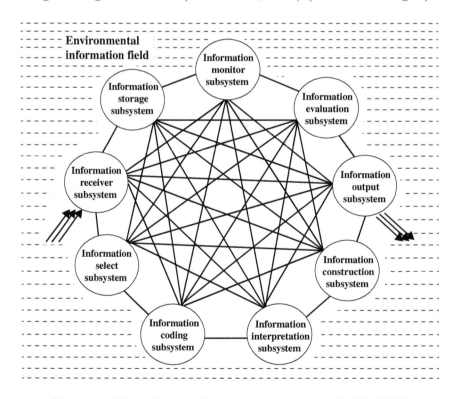

Figure 1. Information creation system structure model (Wu, 2005).

An information creation system is a complex system with multiple information processing functions. Figure 1 represents an information creation system with internal randomness of human subject's thought processing, which is formed by nine functional subsystems connected in a network (Wu, 1998).

The activities of the nine subsystems in the structural model of the Information Creation System are synergistic, supportive, and mutually contextual and conditional. No matter what level of information processing activity is at hand, and no matter which subsystem's activity is unfolding, all subsystems need to function as a whole.

8 Enlightenment of the Information Creation System Model for Current Artificial Intelligence Design

According to Fig. 1, among the functions of the nine subsystems that make up the information creation system, the current AI has four relatively complete functions and has advantages over human beings in terms of capabilities: (1) *Information receiver subsystem*: It is the input side of external information responsible for obtaining information from the environment and reconstructing it for recognition and cognition. (2) *Information storage subsystem*: Its function is to store the information received from the outside and the information generated internally for subsequent information processing activities. (3) *Information coding subsystem*: It can encode all kinds of information in the system's own way, and transform the object information into a form that can be easily transmitted, stored, and manipulated by the system through encoding. (4) Information output subsystem: It is the information output to the external environment and is responsible for transferring the corresponding information from the system to the outside of the system.

The primary artificial intelligence system simply has the functions of two subsystems: (1) *Information select subsystem*: It provides information screening and extraction for various levels of information processing activities so that the system's information processing activities have the corresponding selectivity. (2) *Information monitor subsystem*: It can provide universal awakening and attention background for various information processing activities and can provide purposeful supervision and control over the direction, manner, speed, quality, process, and results of various information processing.

However, the current artificial intelligence system lacks the functions of the three most critical subsystems: (1) *Information interpretation subsystem*: It is responsible for deciphering, interpreting, and expressing the information content represented by the patterns or symbols in various codes, so as to accurately and necessarily grasp and understand the information processed. (2) *Information evaluation subsystem*: It can evaluate the content of various types of information, the way of encoding and interpretation, the effect of monitoring, and the process and results of various information processing activities. (3) *Information construction subsystem*: It has the functions of specifically matching, integrating, and reorganizing various types of selected information, and constructing and virtualizing the corresponding innovative information patterns.

First, due to the lack of an information interpretation subsystem, the current AI acquires information through the information input from the perceptron and processes the corresponding information according to a fixed procedure that has been defined by the designer. This leads to a situation in which it is neither necessary nor possible for the AI to recognize, understand, and interpret the acquired information. In fact, the only person who can understand and interpret is the designer of the perceptron, not the perceptron itself. This results in the current lack of truly autonomous intelligent behavior of the AI machine; instead, only the processing of information by means of other-organization is possible. Compared to human beings, different perceiving individuals as intelligent subjects, the same perceiving individual in different perception fields and perception stages, perceive, understand, and interpret the same thing often with relativity, randomness, and uniqueness, while the current AI is unable to do so.

Secondly, the information evaluation subsystem is a key element of the intelligent subject's ability to implement "decision-making". The basis of information evaluation is the evaluation of facts (truth) and values (utility), while the measurement of positive or negative values comes from the intelligent choice and judgment of the subject, as well as the evaluation of the relevant stakes and the purposeful design to be pursued. All the activities of artificial intelligence lack autonomous choice of judgment and the pursuit of spontaneous purpose; its so-called purpose is only imposed by the designer's own purpose. At the same time, the autonomy of this subsystem also profoundly affects the cooperative operation of the "information select subsystem" and the "information monitor subsystem".

Finally, the information construction subsystem is the most critical and core subsystem for the information creation system to finally achieve its innovative function. It is through this information construction activity that all kinds of creative and novel information can be created. As mentioned earlier, in deep learning and other "artificial intelligence" design approaches, the "intelligence" part of the design often builds on existing computer structures, whether it is the input information, the "decision" process (algorithms), or the overall architecture, it is built on top of other-organized systems. Such an AI system cannot accomplish a true information emergent construction. In order to be capable of intelligent emergence, a system should have dynamic self-organizing properties. All information created by intelligent subjects originates from the emergent mechanism of self-organized systems.

In summary, the current AI design concept should focus on the three subsystems mentioned above that the present-day AI systems lack the most. Perhaps, this is the key breakthrough in whether AI can approach true intelligence under the current technical limitations.

References

Brooks, R. A. (1999). *Cambrian Intelligence: The Early History of the New AI*. Cambridge: MIT Press, pp. 63–68.

Capurro, R. (2012). Toward a comparative theory of agents. *AI & Soc.*, 27(4), pp. 479–488.

Hofkirchner, W. (2011). Does computing embrace self-organisation? In Dodig-Crnkovic, G. and Burgin, M. (eds.), *Information and Computation*. Singapore: World Scientific.

Harnad, S. (June 1990). The symbol grounding problem. *Physica D: Nonlinear Phenomena*. 42 (1–3), 335–346.

Szeliski, R. (2005, 2012). *Computer Vision — Algorithms and Applications*. The Commercial Press, pp. 57–63.

The IEEE Global Initiative on Ethics of Autonomous and Intelligent Systems. (2019), *Ethically Aligned Design: A Vision for Prioritizing Human Well-Being with Autonomous and Intelligent Systems* (1st edn.). IEEE, p. 41. https//standards.ieee.org/content/ieee-standards/en/industry-connections/ec/autonomous-systems.html.

Wang, H. and Raj, B. (2017). On the origin of deep learning, arXiv preprint arXiv:1702.07800. pp. 12–18.

Wu, K., Wang, J. and Wu, T. Q. (2021). *Introduction to Information Philosophy*. Xi'an Jiaotong University Press, pp. 160–177.

Wu, K. (2005). *Information Philosophy: Theory, System, Method*. The Commercial Press, pp. 115–116.

Wu, T. and Da, K. (June 2021). The Chinese philosophy of information by Kun Wu. *J. Document*. 77(4), 871–886.

Wu, K. (2005). *Philosophy of Information: Theory, System, Method*. Beijing: The Commercial Press, p. 125.

Wu, K. (1998). A general model of information systems. *J. Sys. Dialect*. 2.

Zhou, C. H. (2016). *Machine Learning*. Tsinghua University Press, p. 28.

Chapter 21

The Nature of Differentiation of New Human Forms and Intelligent Ecosystems

Haisha Zhang

Department of Philosophy, Xi'an Jiaotong University
Xi'an 710049, China
zhanghaisha@stu.xjtu.edu.cn

Nowadays, the development, application, and popularity of new human forms have made a qualitative leap compared to the past, and new human forms are gradually appearing in the social life of human beings. The new human form is a dual existence of matter and information, the essence of its differentiation is the replaceability of information carriers and the variability of information of matter, and its differentiation dynamics and its mode depend on the continuous overcoming of various crises by the development of science and technology. Therefore, the new human form's ability to perceive, acquire, collect, and process information and the level of its intelligent behavior is consistent with the development of human science and technology. In this sense, only by realizing highly adaptable connections between "new human forms" and "human intelligence", i.e., creating an intelligent ecosystem of the new human form, information, and ecological synergy, can we promote the harmonious development of new human forms and human beings to the greatest extent.

1 Introduction

Research on artificial intelligence is in full swing, accompanied by the emergence of new human forms. However, people have been arguing about the rationality of the existence of this new thing. Will human intelligence benefit or destroy the natural person? Can all human forms coexist harmoniously in the future? Should we prohibit its further development so as not to jeopardize the survival of humanity or continue to promote it, and in what way? By analyzing these questions in depth, this chapter tries to give a more reasonable answer and provide solutions.

2 The Nature of Differentiation of New Human Forms

The universality of material interactions is inevitably accompanied by the generation, transmission, exchange, and transformation of information. Information is the matter showing its properties. There is neither isolated information nor matter without information. All existence is both a material and an informational body (WU, 2005). The same kind of information can be carried by different material carriers. For example, the same piece of music can be presented by many carriers such as musical notation, CD, instrument playing, and vocal singing. In other words, the material carriers of information are characterized by replaceability, and the material information is constantly updated and changed in the interaction of assimilation and dissimilation, that is to say, the material carriers of information have the characteristic of variability. The new human form is a dual existence of matter and information. In short, the differentiation of the new human form essentially develops based on the replaceability of the material carriers of information or the variability of the information in the material carriers.

The differentiation of new human forms goes through three paths: the first is the self-enhancement of natural persons, which is achieved by changing human physiological information, such as gene editing. The second is the human–machine combination. This way is to change the physiological structure and information characteristics of human beings by integrating inorganic matter into the human body and to act the purpose information of human beings on inorganic matter to make it play functions. The third way is to apply human information to inorganic matter so that inorganic matter can produce human-like or superhuman intelligent behavior.

The specific process of morphological differentiation from natural persons to new humans can be represented by the following models:

Stage 1: Generation of human information and intelligent behavior. Through the interaction with the physical environment, the natural person receives, stores, accumulates, and creates information from the physical environment to form their information groups (in-itself information, for-itself information, regenerative information, and social information (Wu and Wang, 2019)) and produces intelligent behavior.

Stage 2: Human intelligent behavior creates two kinds of existence: new information (e.g., new information generated by gene editing) and new carriers (whose nature is the reconstruction of material structures, e.g., the material structures of robot carriers and the material structures of cyborgs carriers).

Stage 3: A part of the new information acts on the natural person and combines with the carrier of them to produce a superhuman. The original human information groups are combined with the new information, and part of them is transformed into primitive information compatible with that new carrier and then combined with the new carrier in a certain way to produce a new human form (robots, cyborgs, etc.) and generate intelligent behavior and new information (a change of structure of matter is inevitably accompanied by the generation of new information).

Stage 4: Superhumans continue to interact with the environment to produce new information; the primitive information of new human forms is continuously updated based on the original foundation, and its new information is continuously updated and upgraded in assimilation and alienation with environmental information and generates intelligent behavior.

Stage 5: Each human form influences and restricts each other, and their intelligent behavior is constantly changing the physical environment so that information and matter are always in dynamic change.

3 The Realistic Differentiation Dynamics and Differentiation Patterns of New Human Forms: Arising and Developing in the Scientific, Technological, and Cognitive Revolution of the Natural Person

In terms of historical development, the purpose of humans is to improve the efficiency of labor and liberate their hands throughout the history of human development. Initially, it was maintained by using simple tools, enslaving others, purchasing human labor, etc. With the advent of the Industrial Revolution, stimulated by capital, the ability of machines and their efficiency have far surpassed that of human beings in some aspects, so human beings have tried to create humanoid intelligent creatures to fulfill their own needs to be able to liberate their hands. Each stage of the development of an intelligent body is also a stage of differentiation, which needs to constantly break various crises to achieve its qualitative leap. The way to overcome the crisis is through scientific progress and technological development. The development of an intelligent agent roughly goes through these stages as follows:

The first stage is pure natural human intelligence. Humans have learned to make simple wooden and stone tools as an aid for their hands. What is faced at this stage is a practice crisis, which is mainly overcome by a series of practical activities carried out by humans and interaction with nature.

The second stage is the combination of the intelligence of natural persons with simple mechanical tools. The first crisis to be overcome in this stage is the complexity crisis and then is the energy crisis. The lifting of this crisis started from the invention of the steam engine in the First Industrial Revolution and continued with the construction and development of electric power science and electromagnetism in the Second Industrial Revolution. Thus, intelligence is upgraded to the third stage.

The third stage is that of complex mechanical. The overcoming of the information crisis in this stage is based on the revolution and development of computer technology, quantum mechanics, relativity, etc. Since then, intelligence has been upgraded to the fourth stage.

The fourth stage is the machine–electric integration. At this stage, the problem that AI is not intelligent enough is what human beings need to do their best to overcome. Therefore, intelligence is the main crisis to be overcome at this stage.

The fifth stage is the development from special-purpose robots to all-purpose robots, biochemical people, etc. After that, the crisis is survival.

From the developmental level of functioning, the new human form may have the possibility of eventually reaching or even surpassing the level of human intelligence, i.e., they can design, repair, and improve themselves through self-design or even design themselves to be more advanced. As a result, the survival crisis arises.

However, from the perspective of the development process of the new human form and technology, the ability to perceive, acquire, collect, and process information as well as the level of its intelligent behavior is synergistic with the development of science and technology of natural man. The development of the new human form, which is constantly improved in the cognitive revolution of the natural person, is inseparable from the development of science and technology of the natural person. The need for a natural person to liberate his/her hands and the cognitive revolution will inevitably give birth to the revolution of science and technology, which will inevitably be accompanied by the change of intelligent environment to promote the differentiation of intelligence that will further deepen the understanding of human intelligence to nature and the grasp and application of the laws of nature. The development of science and technology is the main driving force for the formation and development of the new human form, and it is also the main axis for the synergistic development of the natural person and the new human form, while science, technology, capital laws, etc. all belong to the information groups of the natural person, which makes the social information of the natural person has long played a central role in the process of the formation and development of the new human form.

No matter from the nature of the differentiation of the new human form, or its development model, the natural person and the new human form are in a synergistic relationship. The new human form, as a derivative of human intelligence, its development and progress also represent the development and progress of human intelligence, therefore, the nature of the differentiation of the new human form is still the unfolding of human intelligence at a higher level. It can be said that whether or not a survival crisis will come depends entirely on human beings themselves because the steering wheel controlling the direction technology will develop is in the hands of humans. In the future, the way to overcome the survival crisis is to link the new human form with human intelligence to create an intelligent ecosystem.

4 Sustainable Development for New Human Forms: Intelligent Ecosystems

The reorganization of material structure and flow of information between human forms and between human forms and their intelligent information environment and the natural environment are constantly undergoing changes and are in a dynamic equilibrium of interaction and mutual influence, thus constituting an intelligent ecosystem. In the whole ecological chain, they play different roles and occupy different ecological positions due to their different advantages, thus fulfilling their respective functions. The relationships in the intelligent ecosystem are mainly manifested as follows:

(a) The most obvious relationship between human forms such as natural humans and new human forms (e.g., robots and biochemicals) is competition, competing for a larger share of limited resources and living space. However, the difference from the cut-throat competition is that, due to the need for each other in their respective development, this competition mode is more inclined to a relationship of both dependence and constraint. To avoid competition in which the strong replace the weak, there is bound to be an arms race among new human forms, and the result will still be that each side is evenly matched.

(b) The interaction between human forms, the intelligent information environment, and the natural environment is another constraint on the development of human forms. Human forms utilize and change the natural environment and the intelligent information environment, which in turn constrains or influences the development of new human forms, thus forming a negative feedback loop that can inhibit the collapse of the system caused by the overdevelopment of certain elements.

(c) Each human form always exists in an intelligent ecosystem, and its respective development is achieved in the context of the interactions of many elements of that system. There is no isolated development of a single element. A similar view has been expressed by Haken (2018) that individuals need to rely on collective behavior to determine their destiny, either using direct competition or indirectly by taking a synergistic approach.

A human form is not only influenced by other human forms but also constrained by the environment. The development of a human form is manifested by the interaction and coordination between the human form, other related human forms, the intelligent information environment, and the natural environment to maintain a relatively balanced and stable synergistic development. The result of synergistic development within the intelligent ecosystem is coexistence, i.e., to become the fittest in their respective specialized fields through specialization, thus avoiding competition to achieve symbiosis.

(d) The "choice" under the dual role of intelligent ecological environment and each human form is the driving force for the development of all human forms. While the development of all human forms takes their purposes as the development trend, they need to adapt not only to the current physical environment but also to the information environment created by themselves. Only by achieving dual adaptation can they achieve their development.

(e) Human and intelligent environments are always interdependent. First of all, people design intelligent agents according to their own needs or designers design them to make people have needs. It is the initial purpose of human beings to create and design intelligent agents according to their own needs, but most of the time, it is human beings who do not have needs, behaviors, and experiences, while market and capital, etc. create and design these needs, which greatly stimulate the development of the intelligent agent. Second, an intelligent environment selects people. An intelligent environment can screen people and users who can adapt to the intelligent environment. Therefore, natural persons are in a constant state of differentiation. The natural person and the new human form are competitive in space, but they can coexist at the information level due to their specialization, mutual needs, and mutual influence. Third, human development also remains inseparable from the influence of the natural environment. The nature of human beings, including new human forms that are born and grow up in nature, cannot be eliminated. Once the collapse of the natural ecosystem is caused, the human intelligent information ecosystem will also be gradually disintegrated. The way to maintain its stability is to synergize human development with an intelligent information ecosystem and natural ecosystem. The relationship between human forms, their intelligent information environment, and the natural environment will always be a coexistence of synergistic development, and only in this way can self-development be maximized.

(f) The intelligent ecosystem needs to maintain its balance at all times, i.e., to maintain the dual balance between the intelligent information ecosystem and the natural ecosystem. Taking natural persons as an example, they are developing themselves in the process of creating an intelligent agent while the natural environment is changing, i.e., the self-information of the natural environment is changing, which makes the natural human constantly create new human information groups. In this sense, the natural person is not always unchanged. In reality, they will have multiple differentiation options: Maintaining the status quo (natural person), self-enhancement, becoming a cyborg, etc., to build an intelligent environment that includes new human forms, which in turn will select, influence, and reshape the natural person.

However, although we believe that an intelligent ecosystem needs to maintain its balance and stability, "need" is not equal to "will happen", and in the course of the operation of the system, there will inevitably be a series of social problems as well as concerns about the future fate of natural persons. This is not a threat of the new human form to human intelligence and the natural person, but the progress of society. According to Wu and Yuan (2019), in the process of achieving intelligentization, the social environment also prompts human beings to change themselves and create their new nature. Human beings constantly overcome the single nature and develop toward richness, diversity, and complexity, making it possible for each person to develop comprehensively and freely. The process of human social development is always accompanied by the desire for liberation from nature and society, and the liberation of human beings has been realized in the intelligentization of society.

References

Haken, H. (2018). *The Secret of Nature's Success: Synergetics.* China: Shanghai Translation Publishing House.

Wu, K. (2005). *Philosophy of Information.* China: The Commercial Press.

Wu, K. and Wang, J. (2019). *Introduction to the Philosophy of Information.* China: Xi'an Jiaotong University Press.

Wu, K. and Yuan, Y. (2019). The demands of the intelligent social system and the new evolution of human nature. *Studies in Dialectics of Nature* **35**(1), 123–128, ISSN 1000-8934.

Chapter 22

The Charm of the Complexity of Innovation of Zhao Zhiqian's Official Script

Li Ziyao

Philosophy Department, Xi'an Jiaotong University,
School of Humanities and Social Sciences,
No.28 Xianning West Road, Xi'an 710049, China
xiao_luobo298@stu.xjtu.edu.cn

Calligraphy has the characteristics of complexity, and the essence of complexity lies in embodying emergence. The essence of emergence is the creation realized in the multidimensional interactive synthesis. Zhao Zhiqian, as a representative of the official script innovation in the Qing Dynasty, his calligraphy not only selectively inherited the style characteristics of the ancients but also made a new construction in combination with the calligraphic style of the times, thus realizing emergent creation. By studying Zhao Zhiqian's official script calligraphy, we can reveal the complex characteristics of its brushwork, structure, and composition, which have a multifaceted meaning and value for the development of contemporary Chinese calligraphy.

1 Introduction

The continuous discovery of ancient epigraphy literature made the literati and scholar-officials in the Qing Dynasty gradually turn their attention to epigraphy academically, and this shift in academic research objectively promoted the prosperity of epigraphy calligraphy in the Qing Dynasty. As far as the calligraphic characteristics of the unearthed cultural relics are concerned, most epigraphy that appeared in this period were Han tablet inscriptions and epitaphs inscribed on tablets during the Wei, Jin, and Southern and Northern Dynasties. Among them, the inscriptions in the Han Dynasty were mostly in official script, while the inscriptions in the Wei, Jin, and Southern and Northern Dynasties were mostly in regular script, but they still had strong characteristics of seal script, which greatly inspired calligraphers in the Qing Dynasty. Therefore, the models used by calligraphers and scholars are mostly Lishu and Wei tablet inscriptions. In addition, scholars such as Gu Yanwu and Wanjing also collected many Han tablet inscriptions, which theoretically provided insights into the development of calligraphy, thus creating certain conditions for the development of official scripts in the Qing Dynasty. Among them, Zhao Zhiqian was a representative figure who followed the Han Dynasty official script and achieved breakthroughs in the calligraphy circle at this time.

Zhao Zhiqian (1829–1884), formerly known as Tiesan, was later changed to Yifu, and only after his middle age was named Zhiqian, also known as Huishu. He is a knowledgeable seal carver and calligrapher. Zhao Zhiqian is very good at regular script, cursive script, official script, and seal script, and he especially likes Wei tablet inscriptions. His Wei tablet inscriptions are smooth and gorgeous when writing, which is delicate and attractive. Therefore, his other calligraphy styles are more or less affected, and the official script is no exception. Zhao Zhiqian's official script had imitated Deng Shiru, but he was ingenious. His calligraphy widely imitates other classics, such as "Liu Xiong Tablet", "Wu Rong Tablet", and "Yuanping Tablet", and further incorporates the inscription techniques of the Southern and Northern Dynasties. Eventually, a book style with his own unique style was formed-the official script with the "wei tablet inscriptions style", which became a model of the calligraphy of the times.

Ancient Chinese philosophy emphasizes the generation and process of natural things, emphasizes the change, contingency, and indecision of things, and emphasizes the inseparable interconnection between parts, between parts and the whole, and between the whole and the environment. At the same time, it also emphasizes the overall structure, relevance, coordination, etc. and emphasizes the mutual penetration, reflection, implication, and unity among different things. Ancient Chinese documents, such as the four books and five classics, elucidate the existence and nihility, gossip, yin and yang, the five elements, the number of things, inaction, and the correspondence between heaven and man, and other concepts all embody rich and profound thoughts of information, systems, and complexity.

As a form of human cultural creation, calligraphy is a cultural information phenomenon created by the main body existing in the way of social information. During the Ming and Qing Dynasties when Zhao Zhiqian lived, his academic background was deeply influenced by Chinese classical philosophical thoughts. In addition, his calligraphy works not only integrate the strengths of many people but also have unique innovations. Therefore, his calligraphy works naturally have the characteristics of information complexity. In Santa Fe's research philosophy, the emergence concept and the complexity concept are often equally important concepts. "The essence of emergence is to generate complexity, and the essence of complexity is to embody emergence". The essence of emergence is to create new things through the synthesis of multiple factors.

From the perspective of contemporary complexity theory, this article reveals the complexity characteristics of his calligraphy works through the analysis of the innovative value of Zhao Zhiqian's official script calligraphy. In this way, contemporary creators are guided to rationally consider the complex and unified relationship between inheritance and development, tradition and innovation, time and style, work and spirit. As a complex system, calligraphy must be composed of a large number of agents, and these agents are all different in form and performance. At the same time, as the three elements of calligraphy, brushwork, structure, and composition have naturally become the most representative agents in this system. Therefore, this research starts from the brushwork, structure, composition, and other techniques to conduct a comprehensive analysis.

2 The Complexity of the Brushwork

In calligraphy, horizontal stroke, vertical stroke, left-falling stroke, and left-falling stroke, as the basic strokes of writing, are the most representative active agents in the complex system of calligraphy strokes. Therefore, the complexity of brushwork is analyzed from horizontal stroke, vertical stroke, left-falling stroke, and right-falling stroke.

2.1 *Enrich the form of starting and ending the brush*

Comparing Zhao Zhiqian's official script with the round head and tail of the classic official script of the Han Dynasty, it can be seen that his official script is a circle in the square, combining square and circle. For example, his regular script and seal script are calm and sophisticated, simple and rich. Let's take "The Four Calligraphy of Zhang Heng Lingxian" as an example. Among them, "Tian", "Ke", "Yuan", and other characters, the horizontal stroke is mostly cut sideways to form a square stroke, and then spread the brush to start from the middle, which is broad and natural, giving people a capable and simple feeling. At the end of the stroke, the horizontal stroke is usually lightly turned at the end of the stroke to form an oblique section and then slightly is emerged along the most prominent section of the oblique section, such as "Tian", "Yi", "Bu", and other horizontal stroke endings. The vertical stroke is similar to the end of the horizontal stroke, and it is also a vertical stroke with words such as "You", "Wai", and "Hua" after forming a chamfered surface. The left-falling stroke often pauses at the end of the stroke and then exerts force and strikes to the upper left, which is full of rhythm. For example, the left-falling strokes of the characters "Bu", "Yong", and "Tian" all stop at the end and move toward the upper left. The right-falling stroke ending pen is the most distinctive. At the end of the right-falling stroke, I didn't hesitate to pause at the end of the right-falling stroke and went straight forward. The right-falling stroke of words such as "Zhi", "Tian", and "Jiu" is full of "sharpness".

2.2 Increase the transition between "paving" technique (when the brush is in motion, the lower end of the brush is laid out as a flat line. The force of "paving" is mainly "pushing") and "wrapping front" technique (when the brush is in motion, the lower end of the brush head gathers into a conical tip)

Zhiqian Zhao's mature period of the clerical script with the brush starts side fronts with paving hair, wrapping front and turn, and writing with the central brush technique. Compared with Shiru Deng's round starting and round ending, Zhiqian Zhao's clerical script was more square strokes, using the "pen" to copy the "knife". For example, in the clerical part of Moon-shaped Fan Made for Henian with Seal Script and Clerical Script, the starting part used paving technique, which is soft outside and strong inside. The wrapping front turns with many square strokes. With the natural transformation of the wrapping and paving techniques, the "silkworm's head and swallow's tail" posture of the ancient and simple Chinese script was slowly formed. The early part of the clerical script was full of regular script influence: the starting strokes are hidden, the closing strokes are subtle and introverted, the strokes are calm and steady, and there is little difference in the thickness of a paving and a wrapping technique. When the wrapping front is turned, it is usually round and smooth. The influence of Shiru Deng's teachings still remains, such as Clerical Script of Eight-word Couplet for Hequan.

2.3 Enhancement of the variation of lifting and pressing

In Zhiqian Zhao's early works, since he was still in the stage of learning to copy, the use of lifting and pressing was separate. For example, his copy, Liu Xiong Tablet Inscription, has more than four versions, and each version has a different emphasis on the use of strokes: some focus on lifting strokes and some focus on pressing strokes, in a variety of forms and in the late works of clerical script, Zhiqian Zhao's strokes were in a wave of twists and turns, flexible and charming. In his later works, Zhiqian Zhao's brush strokes were heavy, with many times of lifting and pressing. In the closing strokes, he often pressed heavily to emphasize the sharp edges, blending the "silkworm's head and swallow's tail" posture of Chinese calligraphy. The richness of lifting and pressing in a single stroke was not mechanically in a certain position, making the strokes flexible, such as the Four-column Clerical Script of Zhang Heng's Ling Xian (astronomical works written by Zhang Heng, an astronomer of the Eastern Han Dynasty).

Agents' behavior is proactive, so they can feel environmental information and accumulate self-learning through experience, choose, adjust, and change their own behavioral rules in order to actively adapt to the environment. On the basis of repeated copying and practice, Zhao Zhiqian selected the extraordinary points in the Han inscriptions to form a distinct personal style. At the same time, after accepting the idea of steleology and Bao Shichen's idea of "writing with force" and other concepts, he incorporated the vigor and uprightness of the stele from the Southern and Northern Dynasties into the official script From posts to monuments, writing pens have undergone major changes. The formation of the new style of calligraphy is not about the calligrapher's blind inheritance of a certain calligraphy work, but the collection of the strengths of calligraphy masters. If it is only a simple inheritance, it will not be able to jump out of the original style of the work (Liu, 2009).

3 Structural Complexity

The complexity of the structure of Zhao Zhiqian's official script is mainly reflected in the identification of the structure. Identification is a signal, a program, and a banner provided for gathering and boundary generation. Identification can promote selective interaction, allowing the agent to choose among some difficult to distinguish between the agent and the target.

3.1 Main part

Most of Zhao Zhiqian's early official script works have the style of Deng Shiru, that is, the composition of the characters is relatively flat, and the main part of the characters is stable in the middle. For example, the characters "Bu", "Wu", and "Er" in "Eight-character Couplets of Lishu for Hequan" have square fonts with small changes and the main part is in the middle. And Zhao Zhiqian's official script in his later years has a mature calligraphy style. Compared with Zhao Zhiqian's early official scripts and the Han Dynasty official scripts, the horizontal paintings of his late official scripts are more stretched, and the fonts are flat and thin, and the center of gravity is slightly higher than that of the early official scripts and the Han Dynasty official scripts, such as the words "Bu", "Wu", and "Er" in "The Four Calligraphy of Zhang Heng Lingxian". As a master of seal cutting in the Qing Dynasty, Zhao Zhiqian's innovation in the structure of official script not only was reflected in the paper calligraphy but also has existed in the design of the seal. He specially marked "scribe" in the name of Qian Songzi. Zhao Zhiqian directly explained in the printing style of "Ling Shou Hua Museum" that the printing method was taken from "ChuJun kai Bao Slope Road Stele". Compared with the font structure of the original stele, it can be seen that the font structure center of Zhao Zhiqian's "Ling Shou Hua Museum" seal is on the side, which further proves the innovative path of Zhao Zhiqian's official script structure center of gravity shift (Wang, 2013).

3.2 Enhancement of the top tightness and bottom looseness

In the early period, Zhiqian Zhao's clerical script was tightened in the middle place, and the main stroke of the right-descending stroke was nearly the same as Shiru Deng's, spreading outward and retaining a strong shadow of Shiru Deng's. In the later period, he fused the Han Dynasty Inscription and tablet of the Wei Dynasty with the regular script, making the horizontal stroke stretch out horizontally and the direction of the right-descending stroke change. However, the overall structure featured top tightness and bottom looseness, for example, in the *Seven-word Couplet in Clerical Script for My Children*, which distinguished his clerical script from Shiru Deng's.

3.3 Exaggerating the misshapes of the characters themselves

In Zhiqian Zhao's early clerical works, the characters are mostly flat and square, without much misalignment, so this paper mainly analyzed the misalignment of the characters in Zhiqian Zhao's later clerical works.

In Zhao Zhiqian's later years, the characters of the left and right structures in the official script were compared with the Han Li, the left is lower and the right is higher, and the scattered areas are obvious, such as the words "wai", "gang", and "yin" in "The Four Calligraphy of Zhang Heng Lingxian". The glyphs are staggered and interspersed with avoidance, making the glyphs more dangerous, such as "yin", the radical is lower than the right side, and the right-side skewers and crosses are inserted into the radical, making the structure of the word more compact. The characters in the upper and lower structures are compared with Han Li, the upper part of the stroke is to the right, the lower part is to the left, and the upper and lower strokes are avoided. For example, "Se", "Ding", "Shi", and other characters, the left and right swings add a different kind of fun to the font.

In addition to drawing from numerous gold and stone inscriptions, Zhiqian Zhao often incorporated his own interests found in painting and seal carving into his writing, such as exaggerating the structure and character momentum to make the clerical script present a novel character of elegance and flappy (Liu, 2009). Xizeng Wei of the Qing Dynasty wrote in his book *A Collection of the Treatises of the Performance Language Classroom* that "if a book is written from a seal, a seal is also written from a book", which is to illustrate the interdependence between seal engraving and calligraphy. Not only does seal engraving require a foundation in calligraphy, but calligraphy also draws its essence from seal engraving, especially when the pale and simple flavor of seal engraving is incorporated into the clerical script, which is rich in "the majestic and ancient features of ancient bronzes and stone tablets". Zhiqian Zhao wrote in his poem on seal carving in the margin of his seal *Julu County Wei Clan*, "The ancients had a pen and ink, but today we have a knife and a stone". The method of seal engraving on this seal was mostly the brushwork of the Han and Wei dynasties' copies, with full features of ancient bronzes and stone tablets (Wang, 2003). As a great seal carver of the Qing Dynasty, Zhiqian Zhao not only incorporated his clerical script into the seal but also incorporated the interest he had inadvertently found in seal carving into the writing of his clerical script, which gave his clerical script a novel character. The analysis of the formation of Zhiqian Zhao's clerical style shows that the formation of a new style of calligraphy is not an all-round blend of various styles, but rather one of them is the main one, otherwise the style formed will become featureless.

4 The Complexity of Constitution

Through the interaction of Zhao Zhiqian's brushwork and nonlinear structure, the recycling of resources triggered by the gathering of all kinds of agents has produced the diversity of the constitutional level.

4.1 *Blank area relationship*

In the early years of Zhao Zhiqian's official script, the constitution of the official script was similar to that of the official script of the Han Dynasty. For example, the "Liu Xiong Stele", which he copied in his early years, has short left and right kerning and large upper and lower kerning. In his later years, Zhao Zhiqian incorporated the majestic style of many Wei inscriptions in the official script. Coupled with the continuous improvement of writing proficiency, he no longer succumbed to some inherent restrictions, making the whole work look more relaxed. For example, in "The Four Calligraphy of Zhang Heng Lingxian", there is no obvious boundary between the rows and the characters are interspersed and avoided, which makes the whole work more flexible and natural.

4.2 Writing rhythm

Zhao Zhiqian incorporated the pen on the inscriptions of the Northern, Southern, and Northern Dynasties into the official script, reflecting the sense of movement of the strokes through his decisiveness and determination. The strokes are twisted, flexible, and colorful, making it present the novel features of ever-changing, elegant, and moving shapes. For example, in the character "Xian" in "Zhang Heng Lingxian's Four Calligraphy", the last stroke is drawn with heavy pressure, and after a quick step, the heavy press is especially emphasized at the tail of the falcon and then swept out by the side to make it look charming and smart.

4.3 Increase the use of variant Chinese characters

Zhiqian Zhao received a good education from his childhood. He was fond of research on copperware and stone inscriptions and had close contacts with scholars of copperware and stone inscriptions and collectors of the time, such as Xizeng Wei and Zuyin Pan, with whom he often examined and reviewed newly discovered stone inscription inscriptions, and wrote books such as *Completing the Huanyu Visiting Tablet Record and The Records of the Six Dynasties with Different Characters* (Zou, 2003),which shows that Zhiqian Zhao had general opinions about different characters. In addition, he strived for "uniqueness", not only using the brushwork of the northern monuments in his clerical style but also the use of variant Chinese characters, which gave his clerical style new and unique characters, such as "Xian", "You", "Bu", "Pi", "Rui", "He", and "Yang", in the Four-column Clerical Script of Zhang Heng's Ling Xian and The Copy of Liu Xiong Tablet Inscription.

The Qin, Han, and Northern Dynasties inscription plates and stone inscriptions from which were taken in Zhiqian Zhao's clerical scripts were far from the Qing dynasty in which Zhiqian Zhao lived for a long time. The social and cultural background had changed greatly, so there were great differences in the aesthetic approach and standards. Most of the works of the Han Dynasty's clerical script show the full-bodied and simple, elegant and neat characteristics with strict law (Zhong, 2017), while in the Qing dynasty, due to "literary inquisition" and other socio-political turmoil, calligraphy embarked on the gradual decline, so Zhiqian Zhao with the civilian class origin needed to consider socio-political factors and their own stylistic characteristics in the creation, who couldn't completely copy the characteristics of the style of the Han Dynasty's clerical script. Instead, he combined the style of the times and especially reflected the innovation in the stroke organization. Zhiqian Zhao's clerical script is more obvious than the neat stroke organization of Han Dynasty's clerical script in terms of individuality in the handling of the stroke organization, which is not yielding to the rigid frame and boundaries but interspersing and avoiding between words and characters, randomly evolving, resonating with the times and thus having distinctive characteristics.

5 Zhao Zhiqian's Rules of the Complexity of Official Scripts and Contemporary Enlightenment

Zhao Zhiqian's official script is a complex system, and his rules are selected and generated in random interactions. The agents have diversity in the rules of environmental information, so there will be advantages and disadvantages in the response to a certain situation among the many existing rules. It is impossible for agents to prepare an established rule in advance to adapt it to every situation encountered. Rules can only be selected through trial and error in the process of interaction with the environment. If there is no corresponding adaptive rule in the existing rule base, then it is necessary for the agent to generate a new rule in order to adapt to the new situation.

Zhao Zhiqian imitated the calligraphy of the ancients, instead of accepting it entirely but choosing according to his own style and needs. In addition to the evolution of Zhao Zhiqian's calligraphy from the seal script to the official script, the most important thing is to incorporate the style of the inscriptions of the Southern and Northern Dynasties into the official script. When the ancients generally imitated the calligraphy works of their predecessors, they mostly traced back from the late calligraphy to the early calligraphy, such as "from seal script to regular script". However, many calligraphers in the Qing Dynasty chose to incorporate late calligraphy into the early calligraphy. Zhao Zhiqian, who "incorporates the style of the inscriptions of the Southern and Northern Dynasties into the official script", is one of them. Therefore, today's calligraphers can be more diversified when choosing objects to imitate.

In recent years, as the Chinese Calligraphers Association and local calligraphy associations in various provinces and cities have held more and more exhibitions, calligraphy has flourished and a large number of calligraphic talents have emerged. It can be seen from the works of the exhibition awards and exhibits, today's calligraphers take a variety of objects of calligraphy: some of them take bamboo and silk manuscripts for writing in ancient China, Han Dynasty's inscriptions, Ming and Qing Dynasties' styles of calligraphy. However, the most popular is still the Han Dynasty's inscriptions. It is true that the best way to inherit the Han Dynasty's clerical script is to copy the Han Dynasty's inscriptions, which have standardized techniques, mature chapters, and rich aesthetic paradigms and which can encourage today's calligraphers to inherit better. But at the same time, it also brings limitations such as taking a single approach, copying the zeitgeist of the Han Dynasty's inscription, and having small but not large characters (Zheng, 2017). The same is true of the direct learning from the Qing Dynasty's clerical script, which is more innovative than the Han Dynasty's clerical script. It is very difficult to come up with a "new" script on the basis of the Qing Dynasty's clerical script, and it is easy to fall into the dilemma of imitation. The value of Qing Dynasty's clerical script for contemporary calligraphers is reflected in its creative spirit of following the old and bringing forth the new. The innovative path of Zhiqian Zhao's clerical script is worthy of learning.

The book style of an era is closely related to the background of this period. We are the creators of calligraphy works but also people living in this society. We breathe the breath of this era all the time, so the spirit of the times will be unconsciously expressed in our works.[8] The environment and background of calligraphers are different, which means that in addition to studying the works of the ancients, if they do not add the characteristics of the calligraphy of their time, it will be difficult to create a masterpiece. Nowadays, when imitating Han inscriptions, many calligraphers stick to the font and brushwork of the official script of the Han Dynasty. They did not conduct in-depth research on the background of the times and aesthetic orientation behind Han inscriptions, so their works are stagnant and stable, lacking "potential" movement. Zhao Zhiqian's innovation in the official script inspired us to learn and create realistically in accordance with the current background in our continuous learning and practice. Only in this way can we create calligraphy works that are more suitable for us.

References

Liu, H. (2009). *History of Chinese Calligraphy, Qing Dynasty Volume.* Nanjing: Jiangsu Phoenix Education Publishing House.
Shen, S. (2005). *Looking at Zhao Zhiqian's Creative Personality and Artistic Characteristics from the Fusion of Rubbings.* Henan: Henan University.
Wang, D. (2003). *On the Official Script of the Qing Dynasty.* Shanghai: Shanghai Painting and Calligraphy Publishing House.
Wang, X. (2013). *Qian Song's Epigraphy and Seal Cutting Art Research.* Zhejiang: Hangzhou Normal University.
Wu, K. (2010). *The Theoretical Basis of Complex Information Systems.* Xi'an: Xi'an Jiaotong University Press.
Zheng, P. (2017). *Collection of the 11th National Exhibition of Calligraphy and Seal Cutting Works, Part One.* Beijing: Calligraphy Press.
Zhong, M. (2017). *History of Chinese Calligraphy.* Shaanxi: Shaanxi People's Fine Arts Publishing House, Shaanxi Xinhua Publishing and Media Group.
Zou, T. (2003). *Chronicle of Zhao Zhiqian.* Beijing: Rongbaozhai Publishing House.

Index

A

abduction, 113, 126
abductive cognition, 113
ability to make choices, 144
abstraction, 293–296
abstraction level, 292–296
academic integrity, 425, 427
adaptability, 144
additional knowledge, 279
advanced AI, 243, 271, 279
advanced AI theory, 263
aesthetic, 561
agency, 139, 144
AI systems, 136–138, 178
Alexander the Great, 142
algorithmic complexity, 170
alphabet, 446, 448, 451–452, 454–461, 463
alphabet, -evolution of, 447
alphabet, -Greek, 447–448, 451, 456–457, 461
alphabet, -Latin, 448
alphabet, -Phoenician, 446, 454–456
alphabet, Carian, 448, 451, 458
alphabet, Etruscan, 448, 457
alphabet, Latin, 452, 457–458, 461–462
alphabet, Phoenician, 447, 459
alphabetic machine, 124
analog and digital information, 169, 178

analog type, 176
analogue information, 93
analogy of software and hardware, 138
analytic philosophers, 217
ancient Chinese philosophy, 550
Andrey Kolmogorov, 170
anthropo-relational humanism, 372–373
anthropocentrism, 188
anthropology, 261
anti-Cartesian metaphysics, 221
aristotelian formal logic, 189
aristotelian philosophy, 189
aristotelian principles, 194
Aristotelianism, 204
aristotle, 164
artificial agent, 138, 255
artificial consciousness, 139
artificial intelligence (AI), 58, 63, 69, 134, 136, 138–139, 146, 240, 412, 432, 480–482, 484, 520, 523–524, 527–528, 530–537, 540, 543

B

Bachelard, Gaston, 133
Bachelardian epistemological rupture, 164
Bachelardian rupture, 133
Bacillius subtitilis, 359
background, 551, 560–561

background of calligraphers, 560
basal cognition, 57, 70, 78
Bateson's information theory, 34
Bateson, Gregory, 151, 354, 358, 361, 363
behavioral approach, 256
Bell, John, 170
Bertalanffy, 196
bi-partitions of the brain, 148
big bang, 14–15
binary approach, 177
binary formal logical system, 211
biological entities, 201
biological neural science, 261
biosemiotic information, 159
biosemiotics, 21, 158
bit from it, 355
blank area relationship, 557
Boltzmann–Planck equation, 90
Boolean algebra, 171, 176
bottom looseness, 555–556
Brenner, 186
Brentano, Franz, 157
brushwork, 549, 551–552, 557, 561
brute force learning, 268–269
Burgin, Mark, 87, 89, 97–99, 101, 299, 300, 310–311, 338, 363

C

calligraphy, 512–514, 549, 551
capable, 552
Carnap, Rudolf, 155, 171
carrier medium, 96
cartesian dualism, 187
cartesian duality, 157
cartesian epistemology, 207
cartesianism, 187
catastrophe theory, 109, 127
category, 173
Chalmers, A.F., 110, 195
Chinese classical philosophical, 551
Chiribella, G., 112
circuits, 36
circular information processes, 34
Clark, A., 116

closure space, 174
coevolution of matter and information, 220
cognition, 2, 4, 13, 57–60, 63–79, 108, 113–116, 118, 123, 125–126, 139, 233, 259, 263, 282, 411, 422, 482, 521–522, 527–528, 535
cognition process, 267
cognition-oriented nature, 218
cognitive behavior, 197
cognitive logic, 57
cognitive paleanthropology, 122
cognitive science, 261
collective intelligence, 144
common-sense criteria, 141
common sense knowledge, 276
common-sense understanding, 133
comparisons of the theories of information, 156
complex cognitive phenomena, 201
complex information, 145
complex system, 197, 559
complexity, 133–134, 140, 145–146, 150, 172, 177, 192, 549, 554
complexity of information, 140
complexity science, 262, 283
complexity theory, 210
composition, 549, 552
comprehensive bases, 259
comprehensive cognition, 269
comprehensive information, 259, 271, 282
comprehensive knowledge, 268, 270–271
comprehensive knowledge base, 265
comprehensive knowledge graph, 273
computation, 5–7, 16, 57–58, 63–71, 74–75, 77–79, 108, 114, 116–117, 121, 124–126, 134, 177, 312–313, 413, 528
computational domestication of ignorant entities, 116
computational revolution, 125
computing, 140
conation, 139

concept, 288, 291–293
concept of information, 152, 173–174
concept representative, 288
concept, blurry, 290
concept, class, 290
concept, ensemble, 290
concept, extended Peircean model, 289
concept, fuzzy, 290
concept, general, 289–290
concept, impossible, 289–290
concept, individual, 289
concept, representational model, 288–289
concept, set, 290
concept, strict, 290
concept, symbolic, 289
concept, transdisciplinary, 295
concepts, 287
conceptual framework, 296
conceptual integration, 293
conceptual network, 295–296
conceptual space, 287–288, 290–294
conceptual space, attributive, 288
conceptual space, mental, 290–291
conceptual space, structural, 288
conceptual space, substantial, 291
conceptual space, symbolic, 291
conceptual structure, 291
conceptual system, 293–294
conceptualization of information, 156
conditional probabilities, 154
connotation, 163–164, 288
conscious artificial intelligence, 139
consciousness, 116, 133–134, 138–139, 146, 150, 157, 172, 177, 190
constitution, 557
contemporary sciences, 283
content knowledge, 268
context, 290
conventionality of the assessment, 142
convivialism, 375
Copenhagen interpretation of QM, 99
copperware, 558
copy, 560

correlation, 265
cosmos, 13–14
creative knowledge, 431
creativity, 4, 11
cultural bias, 142
cybernetics, 34, 295

D

Daoist, 199
Darwin, 199
Dasein, 216
data, 212, 267
decision-making mechanism, 196
decisiveness, 558
Dedekind cuts, 176
dedonema, 208
defining intelligence, 140
definition, 152, 287
definition of information, 150–151, 173–174
definitions of information, 156
degrowth, 432
delayed choice experiment, 99
Deng Shiru, 550
denotat, 288–289, 293
denotation, 157, 160, 163–164
Descartes, René, 141, 389–393, 403, 409
described object, 161
description, 287
determination, 558
dichotomy, 217
dictionary, 287
difference that makes a difference, 151
different norms and values, 142
differentia, 164
digital and analog computing, 169
digital competencies, 434–440
digital humanism, 368, 372–373, 378, 380
digital imperative, 373
digital literacy, 437, 439
digital machines, 125
digital skills, 436, 438
digital type, 176

direct existence, 219
dis-matching, 248
discipline, 123
discrete state machine, 124
discretization of continuous magnitudes, 169
disembodiment of the mind, 111, 115–116
distance, 168
diversified, 560
divide and conquer, 249, 282
Dretske, 191
Dreyfus, 202
dual-existence, 189
dual-existence theory of matter and information, 200
dualism, 5, 11
Dummett, M., 113
Dunning–Kruger, 140
dynamics of information, 138

E

eco-cognitive computationalism, 116–117
ecological chain, 243, 246, 281
ecological framework, 159
ecological psychology, 109
ecosystems, 214
education, 121–122, 426, 480, 482–483, 486–487, 489–490, 492–499
effectiveness of action, 143
Einstein, 199
elucidation, 295
emergence, 15, 57, 71, 79, 192, 397, 400, 416, 467, 473–475, 483, 486, 504, 515–516, 523–524, 530–531, 537
emergentist panpsychism, 222
emotional intelligence, 142, 276
empty names, 273
encoding, 155, 162
encoding information, 165
encoding of information, 154, 164
encyclopedia, 288, 291, 295

energy, 10, 14
energy science, 262
energy–matter equivalence principle, 199
engineering, 2
engineering problem, 153
entropy, 4, 6, 8, 10, 15–16, 152–153
entropy-noise, 12–13
entropy-signal, 2–5, 7, 9, 11–16
entropy-thermodynamic, 3–4, 7, 10, 14
environment, 159, 560
epigraphy, 549
epistemological foundation, 204
epistemological information, 244
epistemological obstacles, 132–133, 177
epistemology, 15
epistemons, 216
equal probabilities, 154
erroneous methodology, 149
error, 278
evolution, 3–5, 11, 13, 198, 282
evolution, alphabet *see also* alphabet, 447
evolutionary processes, 210
evolutionary sciences, 190
exceptionalism, 187
exhibition, 560
experienced knowledge, 276
experiencing, 211
expert system, 250, 278
extended, 309, 337
extended conditional probability, 308
extended evolutionary synthesis, 67, 72
extended probabilities, 299–304, 306–310, 312, 314, 317–320, 322–323, 325, 327–328, 330–338, 343, 346–349
extended probability, 299–300, 302, 304–307, 311, 318–319, 322–325, 328–335, 337, 343–344, 347–348
externalization of the mind, 111, 115

F

false belief, 167
feedback, 279
filters, 175
flexible, 554, 557
Floridi, 186
Fodor, J., 110
force, 10
form, 244
form-utility-meaning trinity, 253
formal knowledge, 267, 271
formal matching, 253
formal similarity, 90
forms of life, 150
foundational theories, 262
framework, 261
framework, conceptual, 296
Fresco, N., 126
from many to one, 174
from seal script to regular script, 560
function, 3, 6–7, 11–14
function-degrees of freedom, 6–10, 15–16
functional approach, 256
functional method, 143
fundamental concepts, 264
fundamental law, 261
fuzzy set theory, 262

G

Gadamer, Hans Georg, 398–401
Galilei, Galileo, 166
Galois closure, 176
general closure space, 175
general concept, 290
general evolutionary theory, 191
general study of information, 150
general theory, 281
general theory of information (GTI), 89
generality, 294
generalized cognition, 202
generative AI, 137, 178
generative AI systems, 153
genetic codes, 19

genus, 164
genus-species (classical) concept of definition, 164
Gibson, J.J., 109
global challenges, 369
global model, 254, 282
global optimization, 282
GlossaLAB, 294
goal, 241, 254, 266, 274
goal implementation, 258
goal-reaching, 241
Goyal, P., 112
Greek philosophy, 204
Grice, 216
guiding force, 281

H

Hameroff, Stuart, 147, 224
Hartley, Ralph, 154
Herbert Spencer, 160
hermeneutics, 389–390, 394–398, 401, 416
hidden variables, 175
high quality intelligence, 281
higher education, 422, 426, 430, 439, 441
higher stage, 254
Hofkirchner, 219
holistic, 282
holistic approach, 246, 267
holistic perspective, 207
holographic, 505–506, 510–512, 514, 516–517
holographic analogy, 147
homeorhetic, 357
homeostasis, 22
homunculus fallacy, 148
homunculus' watch, 148
horizontal stroke, 552
horizontal stroke, vertical stroke, left-falling stroke, and left-falling stroke, 552
Huishu, 550
human ability, 241
human brain, 255

human cognition, 188, 470, 472–473
human cognitive functions, 143
human cognitive structures, 471
human competencies, 441
human control, 137, 139, 177
human goals and values, 139
human intelligence, 138, 140,
    142–144, 178, 241, 255, 422, 426,
    431, 522, 532, 539–540, 542–544,
    546
human judgment, 143
human purpose, 241
human thinking, 232, 468, 474, 476
human wisdom, 241, 254
human–machine collaboration,
    480–481
humanities, 261
Husserlian phenomenology, 203

I

idealism, 185
identification of a structure, 174
identification of a variety, 174
identity information, 485
Igamberdiev, 203
incorporate the style, 559
inductive cognition, 269
infant brain, 122
inference, 277
inference rule, 276–277
info-computational framework, 64
infoautopoiesis, 354–355, 357–358,
    360–363
informatio, 205
information, 1–2, 4, 6–7, 11, 13–16,
    57–73, 75–79, 87, 93–94, 97,
    108–110, 114–116, 121–127, 140,
    146, 149–150, 152, 159–160,
    162–164, 170, 174, 230–233, 287,
    291, 315, 317–318, 321, 346–347,
    354–355, 357, 361–363, 403, 412,
    414, 422–424, 426, 430–432, 434,
    480, 482–484, 486, 506–507,
    509–513, 515, 517, 520, 522,
    524–528, 530–531, 533–537

information civilization, 234, 236–237
information complexity, 145–146
information conversion, 257, 260,
    278–279, 282
information discipline, 240, 280–281
information dynamics, 138
information ecology, 246, 257, 260,
    282
information entropy, 90
information function, 242
information has mass, 88
information is physical, 88
information literacy, 434
information paradigm, 234
information philosophy, 185, 283
information revolution, 467–472,
    474–476
information science, 247, 261
information science and technology,
    186
information society, 422
information storage, 98
information structures, 532
information system, 174
information technology, 242
information theory, 158, 170, 246
information thinking, 196
information world, 230
information, definition of, 423–424
information, meaningful, 358
information, philosophy of, 423
information, process, 362
information, self-production, 360
information, semantic, 359–360,
    362–363
information, structural organization
    of, 291
information structural realism (ISR),
    207
information, syntactic, 359–360, 362
information, useful, 358
information–mass–energy
    equivalence, 88
information-abstraction, 6–8, 10, 15
information-content, 5, 7, 10, 14

information-context, 5, 10, 14
information-data, 6
information-meaning of, 2–10, 12, 14–16
information-metadata, 4–7, 10–11, 13, 16
information-noise, 14
information-philosophy of, 7
information-semantics, 2
information-symbol, 2, 4–5, 8, 10
information-types, 3–6
information-types, information-semantics, 2
informational content, 164
informational content of the universe, 95
informational entities, 161–162
informational medium epistemology (IME), 200
informational phenomena, 220
informational relationship, 163
informational structures, 162
informational structures of entities, 162
informational system, 191, 197
informosome, 197
inherit, 560
initiative, 123
innate knowledge, 276
innenwelt, 22
instances of the information, 175
integrate knowledge, 292
integrated information, 147
integration of information theory of consciousness, 147
intellectual environment, 173
intellectual intelligence, 276
intelligence, 1, 3, 16, 58–59, 64, 67, 69–71, 75, 133–135, 140, 145–146, 150–151, 172, 177, 231–233, 236, 248, 266, 413, 424, 426, 433, 481, 521, 525, 533
intelligence creation, 257, 260, 278–279, 282
intelligence of cells, 144

intelligence, artificial, 389, 409, 411–412
intelligence-artificial, 2, 11
intelligence-general, 13
intelligence-types, 6, 12–13
intelligent actions, 150, 255–258, 278
intelligent behavior, 144
intelligent decisions, 273
intelligent strategy, 257, 259, 275
intelligent strategy creation, 274
intensional performance, 294–295
intention, 163
intentionality, 289
interaction, 280
interdisciplinarity, 288, 294, 296
interdisciplinary glossaries, 288, 294–295
interdisciplinary studies, 262
intermediate relation, 290
international society for the study of information (IS4SI), 218
interpretant, 157
interrelationships, 273
interspersing, 559
invariance of information, 162, 165
isomorphic, 220
isomorphic relationship, 198
It from Bit, 99, 354

J

James, William, 146
Ji Yi, 506, 510–512, 515

K

key concepts, 264
keyword, 265
knowledge, 231–232, 241, 254, 257, 267, 287, 424–425, 430–433
knowledge domain, 296
knowledge expression, 273
knowledge graph, 273
knowledge integration, 293–294, 296
knowledge integration, assessement, 296
knowledge representation, 263, 271

knowledge structure, 288
knowledge system, 287, 291–292, 429
Kripke semantics, 171–172
Krzanowski, R., 97

L

lack of competence, 141
Landauer, Rolf, 98, 170
Landauer's principle, 91
large language models (LLMs), 178
large world models, 178
Laszlo, 196
lattices of closed subsets, 176
law, 260
law without law, 99
left is lower, 556
left-falling stroke, 552
level of abstraction, 161
Levin, Michael, 144
life, 177, 363
life forms, 151
lifting and pressing, 553–554
linguistic semantics, 162
literary inquisition, 559
Liu Xiong Tablet, 550
Lloyd, S., 112
logic, 5, 11, 14, 175
logic in action, 58–59
logic in reality, 203
logic of information, 176
logic theory, 262
logic-adaption, 11–13
logic-dialectic, 5
logical, 5, 10
logical foundation, 209
logical monism, 223
logical thinking, 488
Longo, G., 111–112, 124
loss of generality, 146
lost book compilation, 504
lost control, 139
lower part, 556
lower stage, 253

M

machine, 57, 64–68, 74–76, 390, 392, 402, 408, 426
machine intelligence, 143
machine intelligent, 300
machine learning, 269, 520, 522–523, 530
machine, intelligent, 409, 426
machines, 241
Magnani, L., 113, 116, 118–119, 121
magnitudes, 168
major principles, 264
man–machine integration, 482
man–machine reinforcement, 493
Marijuán, 192
matching element, 265
material phenomena, 220
material science, 262
materialism, 185
materialistic perspective, 202
materiality, 213
mathematical formalism, 174
mathematical logic, 262
mathematics, 2–4, 7, 9–10, 165, 167, 262
matter, 14
matter–mind problem, 359
meaning, 157, 159, 163–164, 203, 244, 288, 291, 362–363
meaning of information, 155–156, 158
meaning of meaning, 156
meaningful information, 123
measure, 168
measure Φ of consciousness, 147
measure of information, 153–154
measure theory, 177
mechanism, 258, 277
mental phenomena, 158
metaphysics, 206, 211
methodological poverty, 148
methodology, 240, 246–247, 280
microelectronics, 262
Mikkilineni, R., 89, 98
Mill, John Stuart, 163

mind, 288
mind and brain, 138
mind–body problem, 354–355, 362–363
mind–matter problem, 360
mismatch, 240
misshapes of the characters themselves, 556
Mithen, S., 115
model, 288
model of reality, 178
models, large language, 389
modern science, 261
modes of existence, 132
Monad, 216
monadology, 221
morphological computing, 57–59, 65, 76–77, 79
Morris, Charles, 159
multiple levels of complexity, 162
multiple-component partitions, 148
multiple-regional mutual information, 149

N

Nagel, 201
Nalimov, Vasily, 423
natural, 557
natural entity, 200
natural sciences, 247
naturalism, 191
naturalistic framework, 186
naturalistic information epistemology (NIE), 202
naturalistic information philosophy (NIP), 187
naturalization, 203
negative consciousness, 149
negentropy, 34
network theory, 293
network, conceptual, 296
neural network, 21, 250, 277
neutral monism, 223
Newtonian mechanics, 214

noise, environmental, 357
non-Aristotelian logic, 195
non-semantic form, 212
normative characteristic, 144
normative idea, 142
normativity, 139
notion, 287
number theory, 165–166, 177
numbers, 168

O

object, 253, 266, 291, 293
object information, 244, 255, 257–258, 264, 278–279
object, mental, 289
object, physical, 289
objective factors, 280
objective information, 219
objectivity, 2–3, 5, 7, 10–11
oblique section, 552
observable, 169–170
official script, 549–550, 552, 554, 559, 561
Ogden, Charles Kay, 158
one-many opposition, 173
ontological information, 244
ontological status, 148, 160–161
ontological status of AI, 136
ontology, 14–15
ontolons, 215
open science, 427–431, 440–441
open science infrastructure, 428
open science, definition of, 427
Orch-OR, 224
organism-in-its-environment, 354, 359, 362
over-generalization, 150

P

pair set, 245, 265
pan-humanism, 371–373
pan-informationalism, 222
panprotopsychism, 206
panpsychism, 147, 195
paper machines, 120

paradigm, 239, 280
paradigm change, 251, 264, 279–280, 283
paradigm of philosophy, 230
paradoxic, 4
participatory universe, 99
patterns of reality, 178
paving, 553
Pavlov, I.P., 114
Peirce, Charles Sanders, 111, 113, 157, 288, 360
Penrose, Roger, 147, 224
perceived information, 243–244, 257, 259, 264, 267, 275, 277
perception, 110–111, 263–264, 267, 282
perceptions of the information, 161
periodic table, 3–5
perspective, disciplinary, 294
pervasive mechanism, 279
phenomenology, 203
Philosophia Prima, 195
philosophical category, 231
philosophical spirit, 487–491
philosophical thinking, 486–487, 489, 492
philosophy, 262
philosophy of information, 101, 469, 471, 481, 505, 514, 521, 529
phylogenetic mechanisms, 123
physical determinism, 219
physical discipline, 240, 280
physical nature of information, 99
physical nature to information, 88
physical sciences, 247
physical states, 169
physicalism, 198
physics and information, 99
plagiarism, 425, 427
plastic cognition, 110
platonic ontology, 208
platonic world, 193
Platonism, 204
poly-crisis, 367–368, 380–381, 383

possibility of escape, 138
possible worlds, 171
Post-Cartesian era, 221
post-processing unit, 268
postmodernists, 213
potential, 561
practice, 232–233
pragmatic explanations of meaning, 159
pragmatic information, 265
pragmatic theory of reasoning, 59
pragmatic views of intelligence, 143
pragmatics, 159, 246, 289
pre-processing unit, 268
predictability of the crowd's actions, 145
pregnancies, 109–111, 114
Prigogine, 214
primary consciousness, 276
prime filters, 175
principal filters, 175
principle, 260
principle of dynamical opposites (PDO), 210
priori knowledge, 193
probability, 152
probability theory, 170, 177, 262
problem, 254
problem-solving, 141–142, 241, 256, 274
problem-solving process, 275
process, infoautopoietic, 360
processes, 213
program, 123
pseudoscience, 149
pseudothoughts, 113
psychological considerations, 155
punched cards, 88
punishment, 123
pure formalism, 249, 253, 282

Q

Qing Dynasty, 549–551, 559
qualified type of data, 160
qualitative characteristics, 132

qualitative methodologies, 168
quantitative methodologies, 165, 168
quantitative methodology, 132
quantitative vs. qualitative methodologies, 177
quantities, 168
quantum, 12, 14, 224
quantum computing, 171
quantum information theory, 170
quantum logics, 175–176
quantum relativistic information, 93
quantum theory, 170
qubits, 171, 176
qudits, 171, 176
Quine, 193

R

Raftopoulos, A., 110
Raja, V., 109
randomly evolving, 559
rational cognition, 198
rationality, 209, 431
raw information, 244
raw resources, 243
real line, 167–168
real numbers, 168
recursiveness, 214
referent, 158
regular knowledge, 276
relation, conceptual, 293
relation, inner, 291–292
relation, intermediate, 290–292, 295
relation, pragmatic, 293
relation, semantic, 293
relation, syntactic, 293
relational structure, 287
representation of numbers, 169
representation, conceptual, 291
retrieval, 265
reverse semantics, 162
revolutionary change, 256
reward, 123
Richards, Ivor Armstrong, 158
right is higher, 556
right-falling stroke, 552

Rota, Gian-Carlo, 170
Rothschild, Friedrich Salomon, 158
rough set theory, 262
rule applications, 277–278
rules, 559
Russell, Bertrand, 288

S

salience, 38, 109, 112
Schrödinger, Erwin, 158
science, 1, 3–6, 11
scientific discipline, 240, 281
scientific ethos, 431
scientific inquiry, 165
scientific knowledge systems, 223
scientific realism, 212
scientific structure, 261
scientific worldview, 240, 246–247, 280
script, 446, 448–455, 458–461, 463–464
script, evolution of, 454, 459
script, Cretan Hieroglyphs, 449
script, Linear A, 448–450, 461
script, mirror symmetry, 446–447, 453–458
second law of infodynamics, 95
selection, 174
self-consciousness, 139, 193
self-organising systems, 374, 382
self-produced information, 355
self-similarity, 215
Sellars, 193
semantic information, 208, 265
semantic information theory, 163
semantic theories of information, 177
semantic web, 273
semantical theory of information, 160
semantics, 154–155, 159, 245, 289
semantics and syntactic of information, 164
semantics of information, 156, 159, 165
semiosis, 108, 360
semiotic niche, 40

semiotic process, 157
semiotics, 245
sensation-information-action, 356–357, 362
sense, 289
sensing, 266–267
sensing system, 265
sensor-motor system, 250, 278
set theory, 4
SFI, Santa Fe Institute, 551
shallow knowledge, 271
Shannon theory, 20, 251
Shannon, C., 90, 126–127, 152
sharpness, 553
sign, 111, 157, 287–288
social background, 123
social consciousness, 248
social existence, 248
social information, 368, 374, 377, 382
social intelligence, 142
social relations, 370–371, 373–374
social science, 261
soul, 205
space, conceptual, 287
space-time, 7–8, 14, 16
species, 164
speech, 123
square, 552–553
Stéphane Lupasco, 209
standard model, 4–5, 14
standard model of particle physics, 3
starting and ending the brush, 552
statistical approach, 278
statistical learning, 269
Stiegler, Bernard, 425
stone inscriptions, 558
Stonier, 222
storage and retrieval, 271
strategy, 254, 274
strategy creation, 263
strategy execution, 260, 263, 278
strategy-making, 282
structural, 554
structural analysis, 168–169

structural analysis of information, 153–154, 156, 173
structural approach, 256
structural characteristics, 175
structural information, 154
structural organization, 288
structure, 8, 58–59, 61, 64–67, 75–76, 78–79, 132, 392, 402, 406–407, 414, 475, 495, 510, 515–516, 522–524, 527–528, 540–541, 544, 549, 551, 554, 557
structure of information, 164
structure of real numbers, 168
structure, conceptual, 291
structure, information-types, 3
structure, knowledge, 288
structure, relational, 287
structures, 168
structures governing the behavior, 145
structures of these encodings, 164
studies of information, 177
study of information, 139, 150, 165, 170, 177
study of signal transmission, 155
subconsciousness, 276
subgoal, 277
subject, 253, 258, 266
subject–object interaction, 246, 253, 280, 282
subjective factors, 243, 280
subjectivity, 2–3, 5, 10–11
supernatural qualities, 205
superorganism, 144
symbiotic nature, 206
symbol, 158, 160
symbolic entities, 164
symmetry, 162
syntactic, 159
syntactic analysis of information, 155
syntactic information, 265
syntactics, 245
syntax, 289
syntax, conventional, 289
syntax, generalized, 289

system, 521–530, 532–537
systems capable of action, 140
systems philosophy, 196
systems science, 210, 295

T

Tarski's theorem, 160
technical intelligence, 115
techno-eco-social transformation, 373, 377, 382
teleonomy, 144
tertiary relation, 157
testing, 266
text, artificial, 389–390, 393, 401, 405, 407–408, 413, 416
the center of gravity, 555
theologians, 213
theories, 152
theories of information, 156, 177
theory of communication, 152
theory of information, 155
thermodynamic entropy, 90
thinking, 391, 394, 396, 520
thinking outside of the box, 142
Thom, René, 109, 111–113, 127, 173
Thomson, William, 166
thought, 158
threshold of perception, 34
Tiesan, 550
Tononi, Giulio, 147
top tightness, 555–556
total entropy, 91
traditional philosophy, 211
traditional thinking, 492
transcendental epistemological standpoint, 190
transdisciplinarity, 296
transdisciplinary knowledge, 288, 294
transfinite, 215
transformation of information, 146
triangular relation, 177
triangular scheme, 158
trinity, 259
tripartite models of semiosis, 157

true data, 160
Turing, A., 117–125, 127, 143
type theory, 4–5

U

ultrafilters, 175
Umwelt, 22
unavoidable challenge, 248
uncertainty, 152
undefinability of the truth, 160
undefined concepts, 151
understanding, 241, 250, 253, 266–267, 294
understanding ability, 283
unified AI, 256
unified approach, 256, 260
unified field theory, 10, 15
unified information science (UIS), 185, 195, 218
unified theory, 250, 256, 282
uniqueness, 558
universal, 260
universal logical computing machines (LCMs), 120
universal mechanism, 260, 264, 282
universal practical computing machines (PCMs), 120
unorganized brains, 122
unpredictable individual behavior, 145
untying the Gordian knot, 142
upper part, 556
utility, 244, 258, 267

V

value, 551, 560
value knowledge, 267, 277
variant Chinese characters, 558
vertical stroke, 552
via sensing, 266
Vienna Circle, 188
volition, 139
Vopson, 87, 91–92, 100

W

well-defined concept, 151
Wheeler, 99, 194
Wheeler, John Archibald, 171
Wheeler, William Morton, 144
Whitehead, 213
Wiener, Norbert, 194, 361, 423
wisdom, 433
Wittgenstein, Ludwig, 159
Wolfram, S., 112
wrapping front, 553
writing rhythm, 558
writing with force, 554

Wu, 186
Wu Rong Tablet, 550

X

Xizeng Wei, 556

Y

Yehoshua Bar-Hillel, 155
Yifu, 550
Yuanping Tablet, 550

Z

zeitgeist, 560
Zhao Zhiqian, 550–554, 556–559, 561

www.ingramcontent.com/pod-product-compliance
Lightning Source LLC
Chambersburg PA
CBHW050353090625
27790CB00004B/27